Undergraduate Texts in Mathematics

Springer

New York
Berlin
Heidelberg
Barcelona
Budapest
Hong Kong
London
Milan
Paris
Santa Clara
Singapore
Tokyo

Undergraduate Texts in Mathematics

Anglin: Mathematics: A Concise History and Philosophy.
Readings in Mathematics.
Anglin/Lambek: The Heritage of Thales.
Readings in Mathematics.
Apostol: Introduction to Analytic Number Theory. Second edition.
Armstrong: Basic Topology.
Armstrong: Groups and Symmetry.
Axler: Linear Algebra Done Right.
Bak/Newman: Complex Analysis. Second edition.
Banchoff/Wermer: Linear Algebra Through Geometry. Second edition.
Berberian: A First Course in Real Analysis.
Brémaud: An Introduction to Probabilistic Modeling.
Bressoud: Factorization and Primality Testing.
Bressoud: Second Year Calculus.
Readings in Mathematics.
Brickman: Mathematical Introduction to Linear Programming and Game Theory.
Browder: Mathematical Analysis: An Introduction.
Cederberg: A Course in Modern Geometries.
Childs: A Concrete Introduction to Higher Algebra. Second edition.
Chung: Elementary Probability Theory with Stochastic Processes. Third edition.
Cox/Little/O'Shea: Ideals, Varieties, and Algorithms. Second edition.
Croom: Basic Concepts of Algebraic Topology.
Curtis: Linear Algebra: An Introductory Approach. Fourth edition.
Devlin: The Joy of Sets: Fundamentals of Contemporary Set Theory. Second edition.
Dixmier: General Topology.
Driver: Why Math?
Ebbinghaus/Flum/Thomas: Mathematical Logic. Second edition.
Edgar: Measure, Topology, and Fractal Geometry.Elaydi: Introduction to Difference Equations.
Exner: An Accompaniment to Higher Mathematics.
Fischer: Intermediate Real Analysis.
Flanigan/Kazdan: Calculus Two: Linear and Nonlinear Functions. Second edition.
Fleming: Functions of Several Variables. Second edition.
Foulds: Combinatorial Optimization for Undergraduates.
Foulds: Optimization Techniques: An Introduction.
Franklin: Methods of Mathematical Economics.
Hairer/Wanner: Analysis by Its History.
Readings in Mathematics.
Halmos: Finite-Dimensional Vector Spaces. Second edition.
Halmos: Naive Set Theory.
Hämmerlin/Hoffmann: Numerical Mathematics.
Readings in Mathematics.
Hilton/Holton/Pedersen: Mathematical Reflections: In a Room with Many Mirrors.
Iooss/Joseph: Elementary Stability and Bifurcation Theory. Second edition.
Isaac: The Pleasures of Probability.
Readings in Mathematics.
James: Topological and Uniform Spaces.
Jänich: Linear Algebra.
Jänich: Topology.
Kemeny/Snell: Finite Markov Chains.
Kinsey: Topology of Surfaces.
Klambauer: Aspects of Calculus.
Lang: A First Course in Calculus. Fifth edition.
Lang: Calculus of Several Variables. Third edition.
Lang: Introduction to Linear Algebra. Second edition.
Lang: Linear Algebra. Third edition.

(continued after index)

Peter Hilton
Derek Holton
Jean Pedersen

Mathematical Reflections

In a Room with Many Mirrors

With 134 illustrations

Springer

Peter Hilton
Mathematical Sciences Department
SUNY at Binghamton
Binghamton, NY 13902
USA

Derek Holton
Department of Mathematics and Statistics
University of Otago
Dunedin
New Zealand

Jean Pedersen
Department of Mathematics
Santa Clara University
Santa Clara, CA 95053
USA

Mathematics Subject Classification (1991): 00A06, 00A01

Library of Congress Cataloging-in-Publication Data
Hilton, Peter John.
Mathematical reflections : in a room with many mirrors / Peter Hilton, Derek Holton, Jean Pedersen.
p. cm—(Undergraduate texts in mathematics)
Includes bibliographical references and index.
ISBN 0-387-94770-1 (hard : alk. paper)
1. Mathematics—Popular works. I. Holton, Derek Allan, 1941– . II. Pedersen, Jean. III. Title. IV. Series.
QA93.H53 1996
510—dc20 96-14274

Printed on acid-free paper.

Production managed by Steven Pisano; manufacturing supervised by Joe Quatela.
Photocomposed copy prepared from the authors' LaTeX files.
Printed and bound by R. R. Donnelley and Sons, Harrisonburg, VA.
Printed in the United States of America.

9 8 7 6 5 4 3 2 1

ISBN 0-387-94770-1 Springer-Verlag New York Berlin Heidelberg SPIN 10524250

To those who showed us
how enjoyable mathematics can be.

Preface

Focusing Your Attention

The purpose of this book is (at least) twofold. First, we want to show you what mathematics *is*, what it is *about*, and how it is *done*—by those who do it successfully. We are, in fact, trying to give effect to what we call, in Section 9.3, our ***basic principle of mathematical instruction***, asserting that "mathematics must be taught so that students comprehend how and why mathematics is done by those who do it successfully."

However, our second purpose is quite as important. We want to attract you—and, through you, future readers—to mathematics. There is general agreement in the (so-called) civilized world that mathematics is important, but only a very small minority of those who make contact with mathematics in their early education would describe it as delightful. We want to correct the false impression of mathematics as a combination of skill and drudgery, and to reinforce for our readers a picture of mathematics as an exciting, stimulating and engrossing activity; as a world of accessible ideas rather than a world of incomprehensible techniques; as an area of continued interest and investigation and not a set of procedures set in stone.

To achieve these two purposes, and to make available to you some good mathematics in the process, we have chosen to present to you eight topics, organized into the first eight chapters. These topics are drawn both from what is traditionally described as applied mathematics and from what is traditionally described as pure mathematics. On the other hand, these are not topics that are often presented to secondary students of mathematics, undergraduate students of mathematics, or adults wishing to update and upgrade their mathematical competence—and these are the three constituencies we are most anxious to serve. Naturally, we hope that the teachers of the students referred to will enjoy the content of our text and adopt its pedagogical strategy.

Thus, we have chapters on spirals in nature and mathematics, on the designing of quilts, on the modern topic of fractals and the ancient topic of Fibonacci numbers—topics which can be given either an applied or a pure flavor—on Pascal's Triangle and on paper-folding—where geometry, combinatorics, algebra, and number-theory meet—on modular arithmetic, which is a fascinating arithmetic of finite systems, and on infinity itself, that is, on the arithmetic of infinite sets. We have tried to cater to all mathematical tastes; but, of course, we do not claim to be able, through this or any other text, to reach those unfortunate people for whom all mathematical reasoning is utterly distasteful. Pythagoras inscribed on the entrance to his academy, "Let nobody who is ignorant of geometry enter." We might say, at this point, "Let nobody who abhors all mathematics read any further." We see this book as a positive encouragement to those who have already derived some satisfaction from the contact they have had with mathematics. We do not see it as performing a therapeutic function on the "mathophobic"—unless, as is often the case, their mathophobia springs purely from a mistaken view of what mathematics *is*.

You will see that the eight chapters described above are largely independent of one another. We are not at all insisting that you read them in the order in which we have written them. On the other hand, we do also want to stress the *unity* of mathematics, so that there is bound to be a considerable measure of interdependence in the material we present. Cross-referencing will help you to find material from another chapter relevant to the chapter you are currently studying. We should add that Chapter 2 is particularly rich in ideas which play a part in the other chapters of the book.

In our final chapter, Chapter 9, we set out our views of what mathematics *should be* in action, that is, how it should best be done. Naturally, this chapter is quite different in nature from those that precede it. It is not, however, different in tone; we continue the informal, friendly approach which, we very much hope, shows up clearly throughout our text. But we do believe that our readers may find it helpful to have available to them, in easily digestible form, some suggestions as to how to raise their expectations of success in tackling a mathematical exercise. That is our main purpose in including Chapter 9.

We would like to say a few words here about our notation and terminology and our expository conventions. First, we number the items in each chapter separately but as a unity; that is, we start again with each chapter, but, within a chapter, we do not take account of the various sections in our numbering system. Moreover, we have two numbering systems within a chapter; one is the numbering system for theorems, corollaries, examples, and so on, and the other, appearing on the right side of the page, indicates the sequence of displayed formulae of special importance.

Second, we adopt certain conventions and practices in this text to help you to appreciate the significance of the material. From time to time within a chapter we introduce a **BREAK** which gives you the opportunity to test your understanding of the material just presented. At the end of some chapters there is a **FINAL BREAK**, testing your understanding of the entire chapter, which is followed, where appropriate, by a list of **REFERENCES** and the **ANSWERS** to the problems in the final break.[1] As for the references, they are numbered 1, 2, 3, . . . , and referred to in the text as [1], [2], [3], ... (You are warned that, in Chapter 2, [3] may be the residue class of the integer 3! The context will make this quite clear and, as we say at the end of the Preface, no notation can be reserved for eternity for one single idea.) From time to time we introduce harder material, which you may prefer to ignore, or to save for a second reading; the beginning and end of such material are marked by a star (⋆) in the left-hand margin and the extent of the difficult material is indicated by a wavy line, also in the left-hand margin,

[1] Our readers are, of course, to be trusted only to consult the answers after attempting the problems.

connecting the two stars. Certain key statements and questions appear displayed in boxes; the purpose of this display is to draw your attention to ideas whose pursuit is going to determine the direction of the subsequent development.

Third, despite our informal approach, we always introduce you to the *correct mathematical term*. In particular, *theorems* are important assertions (not necessarily of a geometrical nature) which are going to be *proved*. *Corollaries* are less important assertions which follow fairly quickly from the previous theorem. *Conjectures*, on the other hand, are hypotheses which may or may not be true, but which we have some rational grounds for believing. It is the standard means of making progress in mathematics to survey what one knows, to make conjectures based on such a survey, to prove these (by logical argument) or disprove them (very often by finding *counterexamples*), to find consequences of the theorems thereby established, and thus to formulate new conjectures (compare Principle 4 in Chapter 9, Section 1). We use LHS and RHS as abbreviations for *left-hand side, right-hand side*. We employ the phrase "if and only if" when we are claiming, or proving, that two propositions are *equivalent*. Thus "proposition A if proposition B" means that "B implies A," usually written "$B \Rightarrow A$"; while "proposition A only if proposition B" means[2] that "A implies B" or "$A \Rightarrow B$." The obvious, simple notation for "proposition A if and only if proposition B" is then "$A \Leftrightarrow B$." As a final example, we may use the term *lemma* to refer to an assertion that is to be established for the explicit purpose of providing a crucial step in proving a theorem.[3] Where a proof of an assertion is given rather formally (probably introduced by the word ***Proof***), the end of the proof will be marked by a *tombstone* (□).

It may also be helpful to say a word about the use of the letters of our alphabet, or the Greek alphabet, to represent mathematical

[2]"Proposition A if proposition B" appears sometimes as "B is a *sufficient* condition for A"; while "proposition A only if proposition B" may appear as "B is a *necessary* condition for A." These terms, however, often create difficulties for students.

[3]In fact, the literature of mathematics contains many examples of lemmas that have become more famous than the theorems they were originally designed to prove (e.g., Zorn's Lemma in set theory, Dehn's Lemma in topology, and the Snake Lemma in homological algebra).

concepts like numbers, functions, points, angles, and so on.[4] There are a few—very few—cases in which, throughout mathematical writing, a certain fixed letter is always used for a particular concept; thus, the ratio of the circumference of a circle to its diameter is always π, the base of natural logarithms is always e, the square root of -1 (regarded as a complex number) is always i, a triangle is always Δ, a sum is represented by $\sum$, and so on. However, this does *not* mean that every time these letters appear they refer to the concept indicated above; for example, i may be the subscript for the ith term a_i of a (finite or infinite) sequence of numbers. By the same token we cannot reserve, in the strict sense, any letter throughout an article, much less a book, so that it always refers to the same concept. Of course, different usages of the same letter should be kept far apart, so that confusion is not created; but the reader should remember that any particular usage has a *local* nature—local in place and time. There are far too many ideas in mathematics for each of them to be associated, for all time, with a particular letter of one of four alphabets (small or capital, standard or Greek)—or 40 alphabets, for that matter.

As a final remark in this Preface, it is a pleasure to acknowledge the essential assistance given to us by Kent Pedersen in assembling the Index.

Now you're ready to start, and we hope you have an interesting and enjoyable journey through our text.

[4]You will find one use of a letter of the Hebrew alphabet. This is the standard use of the first letter ℵ, pronounced "aleph," to represent the cardinality (size) of an infinite set in Chapter 7.

Contents

CHAPTER 1 Going Down the Drain

What have *Helianthus annuus* and *Helix pomatia* got in common? First of all, you probably need to know what these things are. *Helianthus annuus* is generally known as the (common) sunflower, while *Helix pomatia* is the common or garden French snail that finds its way onto dinner plates in fancy restaurants all around the world.

We suppose there's a sense in which both the sunflower and the escargot are edible. The one provides seeds to go in snacks and salads and edible oil which is used in margarine and for cooking, while the other provides what some people believe is a delectable source of protein. But the gastronomic connection is not what we had in mind.

1.1 CONSTRUCTIONS

While you're working on that conundrum, try doing something more practical. In Figure 1 we have a spider web grid for you. You might like to photocopy or trace it, because we want you to start drawing all over it. While we're not against defacing books if it's in

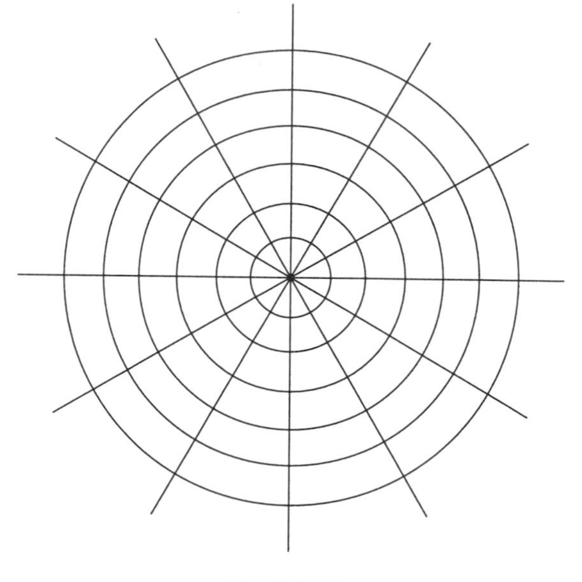

FIGURE 1

a good (mathematical) cause, you may want to use Figure 1 several more times. It's best to start with a clean version each time.

What do you see in Figure 1? There is a series of concentric circles whose radii are increasing at a constant rate. In fact, the radii are 1, 2, 3, 4, 5, and 6 units, respectively. Then there is a series of straight lines all of which pass through the central point. The angle between neighboring pairs of these straight lines is 30°. Actually, you'll notice that these lines go off to infinity in only one direction. We call such half-rays ***rays***.

Some of you may recognize Figure 1 as ***polar graph paper*** but we won't worry about that for a moment or two. What we are interested in is that you go off and find a rectangular piece of cardboard. You'll need a pencil too. We'll wait here while you go and get them.

Now look at Figure 2. Choose a point P_1, anywhere on one of the rays of Figure 1. Now put the cardboard on your polar graph paper so that one side touches P_1. Then slide the cardboard so that the adjacent side of the card touches the next ray (see Figure 2(a)). When you've done that, mark the point on this next ray which is at the corner of the right angle in your card. Call this new point P_2.

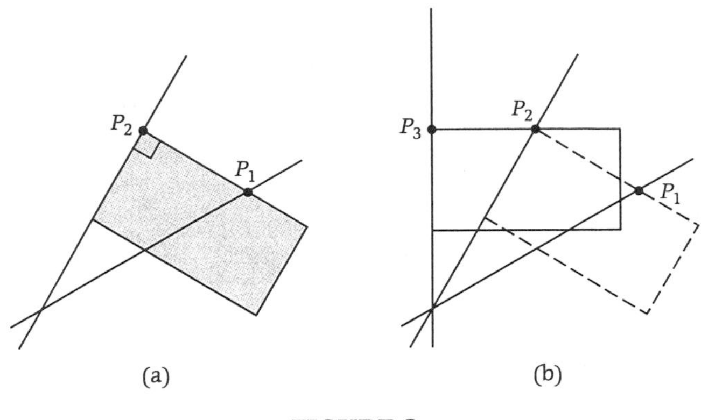

FIGURE 2

When you've got that organized, do the same thing again but this time start at the point P_2. So now one side of the card touches P_2 and the adjacent side of the card runs along the next ray around (see Figure 2(b)). Mark the point where the right angle touches this next ray and call it P_3.

Once you've got the idea, continue until it's no longer physically possible to add any more points. Suppose that P_n is the last point that you were able to mark on your copy of Figure 1. Now join the points P_1, P_2, P_3, up to P_n in as smooth a curve as you can manage. You should produce a spiral similar to the one in Figure 3.

It's worth reflecting for a moment on what you have just done. You have just been involved in an *iterative* geometrical procedure

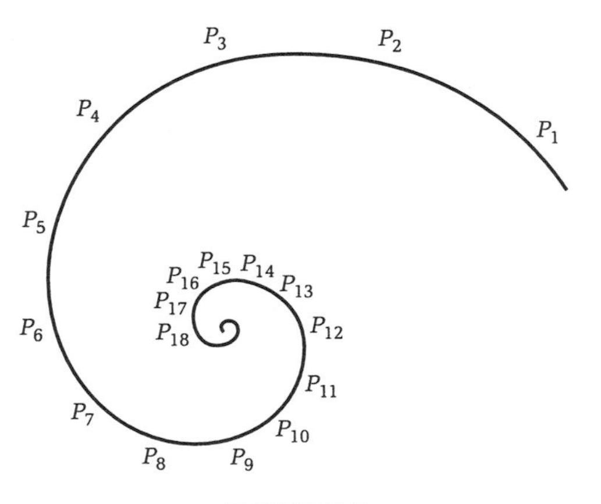

FIGURE 3

which generates a sequence of points. This means that we perform an operation on one point (P_1 here) to get another (P_2). We then perform the same operation again but this time on the new point (P_2), to get the next point (P_3). We keep doing this over and over again.

In Chapter 3, you'll see us playing around with Fibonacci and Lucas numbers. There we will be iterating *numbers*. Here we are iterating *points*. Later on in this chapter, we'll tie up these two ideas.

In the meantime, we just want to stop for a minute because some of you may have got a different spiral from the one we've drawn in Figure 3. Our curve is spiraling inward in a *counterclockwise* (anti-clockwise if you don't have a North American education) direction. The different spiral that we've just mentioned would be spiraling in toward the center in a *clockwise* fashion!

• • • **BREAK**

You might like to think for a minute how that could possibly have happened, given the exquisitely accurate directions that we described above.

Well, while you were thinking, we have looked back at our iterative instructions and have discovered that, although we pointed you to Figure 2, we didn't *actually* say that the ray that the right angle touched had to be the one in a counterclockwise direction from the ray the point P_1 was on. The misinterpretation that we noticed clearly put P_2 on the ray that was the next *clockwise* around from P_1. Obviously, this was the work of a left-handed person!

OK, so things can be done that way. For those of you who followed the implied counterclockwise direction of Figure 2, have another go, but this time do it clockwise. And for the people who did it clockwise the first time, would you mind having a try in the other direction now, please?

• • • **BREAK**

Can you manage to make your spiral go the other way?

FIGURE 4

Fine! So now everybody should have *two* spirals, one with a clockwise decline and the other with a counterclockwise decline. This left-handed version we've shown in Figure 4.

But what we would dearly like to know is: Why is the spiral heading for the *center*? What are the alternatives? The points P_1, P_2, etc., could spiral *in* to the center, they could keep the *same distance* from the center, they could spiral *away* from the center, or they could exhibit erratic, exotic behavior not yet described in the pages of this *magnum opus*.

• • • **BREAK**

Why do the points spiral *in*?

Before we start our erudicious explanation, you must write down a quick reason of your own. Nothing too elaborate, mind. Something like "the hypotenuse of a right-angled triangle is longer than either of the other sides" will do. In fact, if that's what you wrote, then you're on top of the game. That's exactly what's going on. Look at the counterclockwise iteration shown in Figure 5(a) and let C be the center of the polar graph paper. Then you'll see that ΔCP_1P_2 is right-angled at P_2. The hypotenuse of this triangle is CP_1. So clearly $CP_2 < CP_1$. This means that the point P_2 is closer to the

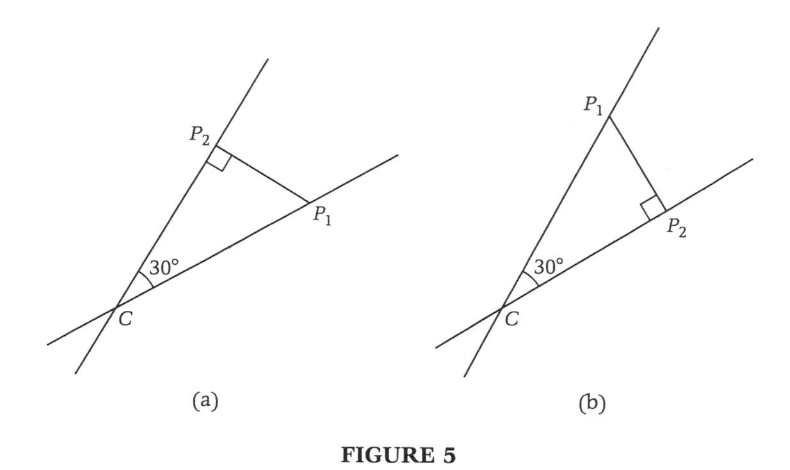

FIGURE 5

center C than P_1. Hence the points go spiraling in as we move in a counterclockwise direction.

For the left-handed among us, the clockwise situation is dealt with in Figure 5(b). Of course, we haven't yet used all of the information available from the precise rules of the construction.

Now when you're on a good thing, stick to it. We'll just vary the iteration slightly. Take your card and a pair of scissors and cut off a right angle as shown in Figure 6. Make the angle α any size you want. Keep one part of the card to use straightaway. Call this part A, and the other part B, and put B aside somewhere. We won't need it for the moment but we will use it later on.

Now get hold of another copy of Figure 1 and use the A part of your card to go through the iterative process described above, all

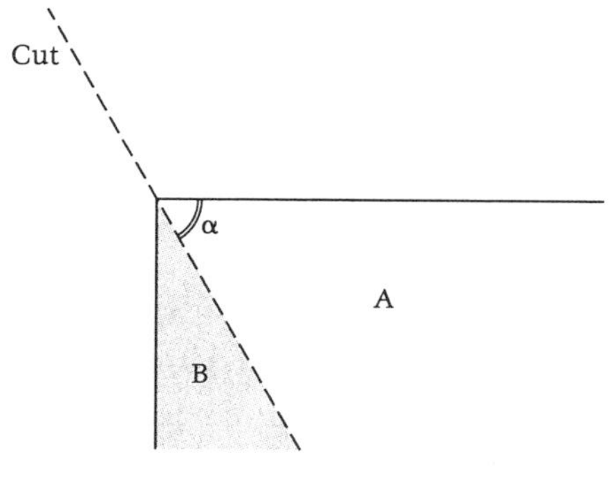

FIGURE 6

over again. The only difference now is that this time the angle α goes where the right angle went before. If you take an arbitrary point P_1 in any ray, and have one side of the card touching P_1 and the neighboring side of the card along the next ray, then P_2 is at the vertex of the angle α. You should be able to see how to continue from here. It's the same old routine.

• • • **BREAK**

The big question now is: "What sort of a curve did you get when you put a smooth curve through the points $P_1, P_2, \ldots$?" Did you get another spiral? Did it spiral in or not? Did it stay a constant distance from the center? Did it exhibit some exotic, erotic behavior? If so, what sort of behavior?

So what happened? First of all, we'll assume that you all adopted the Figure 2 approach so that P_2 was counterclockwise from P_1 and so on. (Perhaps there is still the odd person who went the other way!) We've listed some possible outcomes in Figure 7. Which, if any, did you get?

The thing that interests us is that we can get *any* of the shapes in Figure 7! Those with some other sort of erratic behavior should go back to the drawing board. The answer is definitely one of the curves in Figure 7, as we will now show.

Perhaps a diagram like Figure 5 will be of some help. We may be able to sort it all out with a simple right-angled triangle. Except

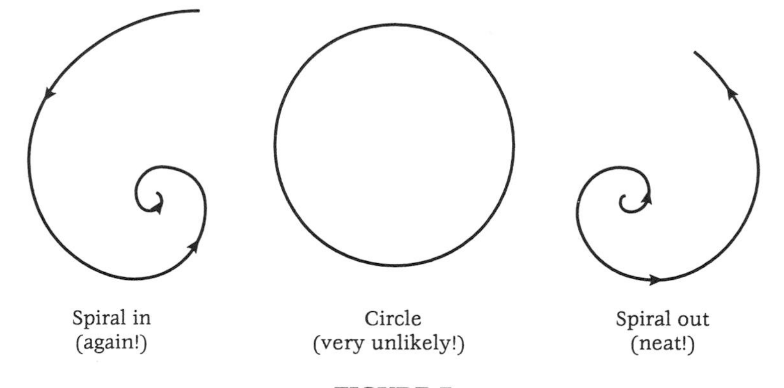

FIGURE 7

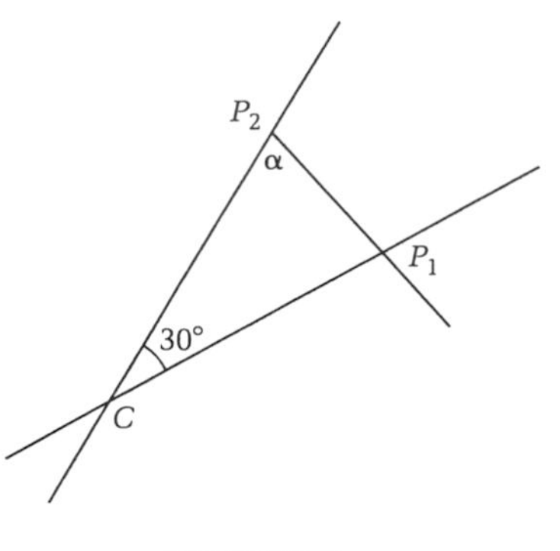

FIGURE 8

when we look at Figure 8 there don't appear to be *any* right-angled triangles!

How can we compare CP_1 and CP_2? Is it possible that CP_1 could be *bigger* than CP_2 for some value of α? Could CP_1 actually *equal* CP_2? We know already that if $\alpha = 90°$, then CP_1 is *smaller* than CP_2 so that ought to be a possibility too.

Ah! Is that the clue? What do we know about the relative sizes of sides and their opposite angles? Surely the bigger side is opposite the bigger angle. So if the angle at P_2 is *bigger* than the angle at P_1, then CP_1 is *bigger* than CP_2.

Now if the angle at P_2 is α (as we know it is), then $180° - 30° - \alpha$ is the angle at P_1. So whether the iterative curve spirals in or out, depends on whether α is bigger or smaller than $150° - \alpha$! So when is α bigger than $150° - \alpha$?

$$\text{Now} \quad \alpha > 150° - \alpha,$$
$$\text{is equivalent to} \quad 2\alpha > 150°,$$
$$\text{or} \quad \alpha > 75°.$$

Those of you who had cut your card so that α was *bigger* than 75° found that your curve spiraled *in* because $CP_2 < CP_1$. Those of you who had α *smaller* than 75° has a spiral going *out* ($CP_2 > CP_1$). And one of you may have fluked a circle by taking α exactly equal to 75°.

α	curve
bigger than 75°	spiral in
equal to 75°	circle
smaller than 75°	spiral out

It's actually interesting to play around with α very close to 75° and see how long it takes for your spirals to move away from the circle.

• • • **BREAK**

Instead of using the A part of the card with the angle α, try using the angle $90° - \alpha$ from the B part (see B in Figure 6). Is there any connection between the A and B curves? What about a right-handed A curve and a left-handed B curve? It's worth looking at Figure 1 again too. There we had rays that were 30° apart. What happens if you repeat the card construction with rays that are only 10° apart? What is the critical value of α for this case?

If you use a 10° gap between rays you'll find it much easier to get a smooth curve than in the 30° case. However, it all takes a bit longer and you will have to be more careful with your construction because small errors mount up.

1.2 COBWEBS

We've drawn the cobweb of Figure 1 for you again in Figure 9(a). Compare it to the rectangular grid of Figure 9(b). In Figure 9(b) we've put in the x- and y-axes. You're probably used to this. It's easy to locate a point in the plane using the x- and y-coordinates. Anything that's x units horizontally away from the origin O and y units vertically away from O, is given the coordinates (x, y). The streets of many North American cities are laid out on such a rectangular grid, perhaps with the x-axis called Main Street and the y-axis called State Street. It makes it very easy to find your way around.

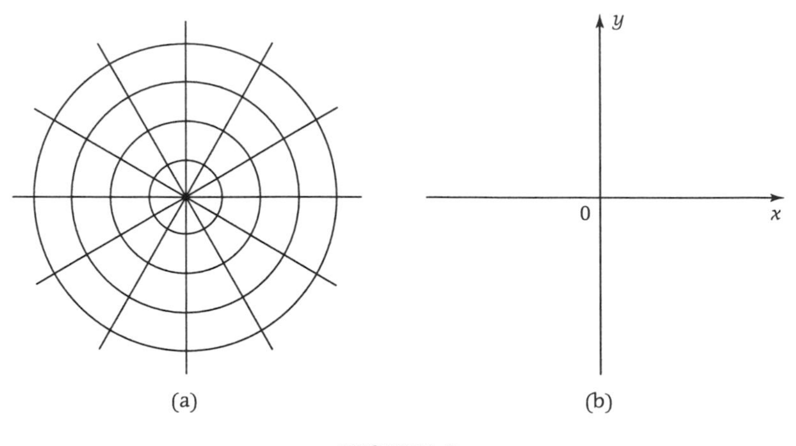

FIGURE 9

On the other hand, if you are a spider and you have just captured a particularly delicious *Musca domestica*, what you'll probably do is park it for a while to let it mature. Of course, you would like to remember where the *Musca domestica* is for future gastronomic purposes. It doesn't make any sense to superimpose a rectangular grid on your cobweb. Why not use what you've got directly? You've got a polar graph situation, why not use ***polar coordinates***? A fairly simple approach, using the web of Figure 9(a), would be to say, well, the *Musca domestica* is 15 units (probably centimeters but we won't bother to specify them precisely) from the center C and $60°$ around from the window ledge. (We're assuming here that the ledge has a ray that you, as the spider, are particularly fond of and that you have decided to use this as your reference point.) All you now have to do is to store the polar coordinates of the point M as $(15, 60°)$ in your brain next to the *Musca domestica* and you'll know exactly where your next meal is coming from.

In Figure 10 we've shown the position of the *Musca domestica* as M. We also notice that you've gathered a few other interesting specimens in your web. For instance, there is a *Diptera culicidae* at D (reference $(20, 150°)$) and a poor *Bombus bombus* at $B = (25, 300°)$.

But there is one thing that you need to know straightaway. The more educated spiders amongst you use radians for angle measurement rather than degrees. This came about because you realized that when you walked once around your web one unit out from

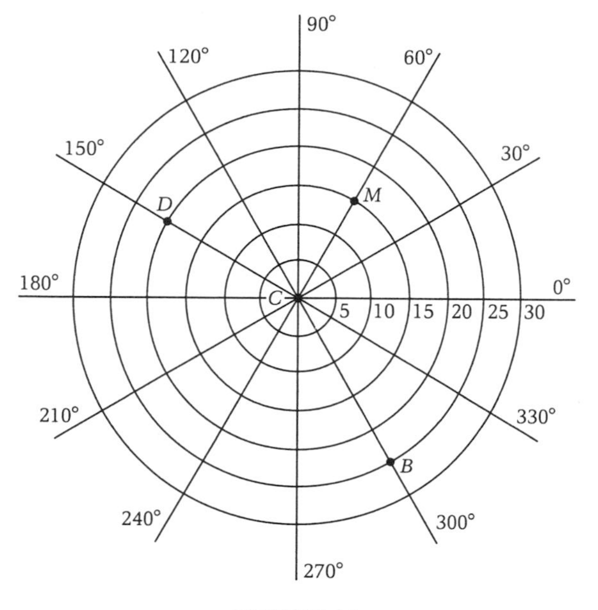

FIGURE 10

C, you actually traveled 2π units. So you thought of this as having turned through an angle of 2π radians. So, for spiders, 2π radians equals 360°. This means that $180° = \pi$ radians, that $30° = \frac{\pi}{6}$ radians, and so on.

• • • BREAK

Locate the positions of the *Diptera culicidae* and the *Bombus bombus* using polar coordinates (r, θ), where r is the distance from C and θ in radians is the angle turned through, starting from the ledge already mentioned.

Actually when you think about it, the place where the *Musca domestica* is stored cannot only be described as $\left(15, \frac{\pi}{3}\right)$, but also as $\left(15, \frac{7\pi}{3}\right)$, and $\left(15, \frac{13\pi}{3}\right)$, and indeed $\left(15, \frac{\pi}{3} + 2n\pi\right)$, for any value of n, positive or negative! So, unlike Cartesian coordinates, polar coordinates are not uniquely defined. It's always going to be possible to monkey around with the angle part to the tune of multiples of 2π. Now it's possibly a minor complication that there is more than one way to locate every point, but it does seem to be an easier way to locate objects on your web than using Cartesian coordinates. You never know, there may be some other advantages. Who knows?

Now if you were a particularly intelligent spider, you might be interested in the card construction of the last section. What's more, you might even start to draw spirals on your web. If you could manage a colored thread, then, no doubt, insects would be attracted from miles around you, and you and your descendants would therefore have an evolutionary advantage over the rest of your species. You might even take over the world eventually. We can just imagine huge webs, with colored spirals, attracting members of the species *homo sapiens* to their doom in droves.

But, as you know, being a spider, it's a little hard to carry a card and pencils around with you to mark out the position of the next point in the spiral. It would be much easier to know the location of the next point so that you could lay out your colored spiral thread in that direction.

The big question then is, given the first point P_1, what is the location of the point P_2? Let's make life easier for you and put P_1 at $(5, 0)$ and use rays that are $\frac{\pi}{6}$ radians (or 30°) apart.

• • • BREAK

We'll also use the first card construction, where the card has a right angle at the corner as shown in Figure 11. If P_1 is at $(5, 0)$, where is P_2?

The coordinates of P_2 have to be found, right? Now we know that P_2 is on the $\frac{\pi}{6}$ ray. So $P_2 = \left(r, \frac{\pi}{6}\right)$. All we have to do is to find r. But ΔCP_1P_2 is a right-angled triangle. We know all the angles in this triangle (after all, $P_1CP_2 = \frac{\pi}{6}$, so $CP_1P_2 = \frac{\pi}{3}$). So we only

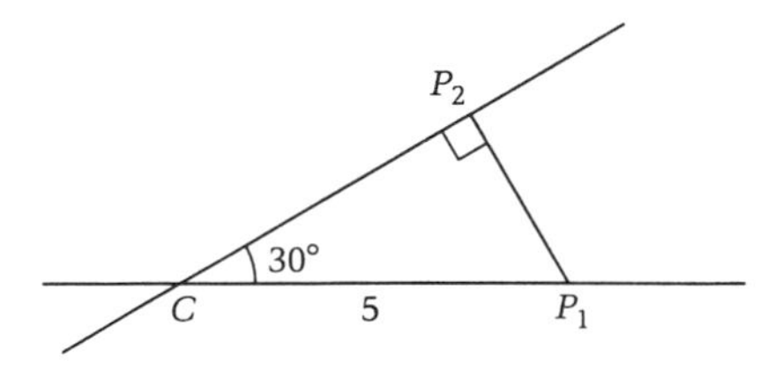

FIGURE 11

need use a bit of trigonometry to see that $\frac{CP_2}{CP_1} = \cos \frac{\pi}{6}$. Therefore, $CP_2 = 5 \cos \frac{\pi}{6} = \frac{5\sqrt{3}}{2} \simeq 4.3$. So P_2 is approximately $\left(4.3, \frac{\pi}{6}\right)$.

Using the same method, your spiderness could calculate the positions of a whole collection of points P_3, P_4, and so on. You could do this forever if you liked, though this might delay the colored spiral thread manufacture and your inheritance of the Earth.

But maybe you could find a formula which would give you all these points in one fell swoop. What a savings that would be! What a colossal evolutionary advantage. Soon spiders of all genera would be at your door for the rule that would provide the key to everlasting lashings of fast food, fully self-delivered to the table.

Before you get too many dreams of arachnidic grandeur you'd better find an equation for the spiral. What you need to be able to do is to find a relation between r and θ so that any point with coordinates (r, θ) lies on the spiral and no other points do. First, of course, we must find the relation satisfied by all the points P_n.

Let's have a look at the situation in Figure 12. This supposes that we know $P_n = (r_n, \theta_n)$ and we want to find $P_{n+1} = (r_{n+1}, \theta_{n+1})$. Once again, of course, $\theta_{n+1} = \theta_n + \frac{\pi}{6}$. So it's easy enough to find the angle part of the coordinate. But we have another right-angled triangle here. So $CP_{n+1} = CP_n \cos \frac{\pi}{6}$. This means that $r_{n+1} = r_n \cos \frac{\pi}{6}$. In other words, $P_{n+1} = \left(r_n \cos \frac{\pi}{6}, \theta_n + \frac{\pi}{6}\right)$.

Now that's all very well, and we know that you are only a spider, but if you want to get on in this world you are probably going to have to find an equation linking r and θ for the general point (r, θ).

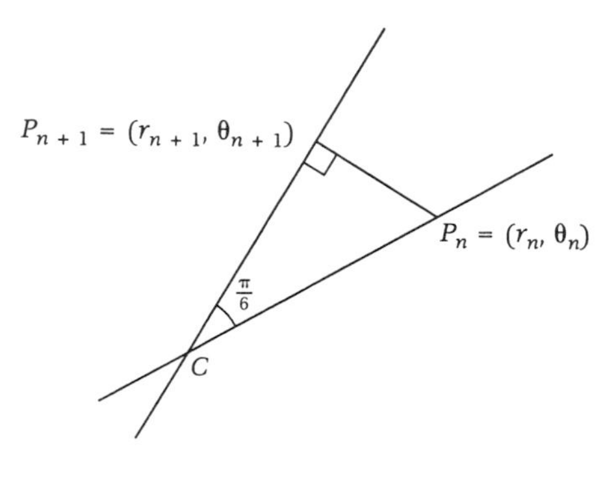

FIGURE 12

All you've been able to do is to give us an iterative relation between the coordinates of P_n and P_{n+1}.

Forget about that for a minute and let's see what we *can* work out. If we add $\frac{\pi}{6}$ to the angle every time we move on, then $P_1 = (r_1, 0)$, $P_2 = \left(r_2, \frac{\pi}{6}\right)$, $P_3 = \left(r_3, \frac{\pi}{3}\right)$, and so on. So the angle part of P_{n+1} should be just a multiple of $\frac{\pi}{6}$. Probably $P_{n+1} = \left(r_{n+1}, \frac{n\pi}{6}\right)$. Check that out to make sure it's OK. It is, so $\theta_{n+1} = \frac{n\pi}{6}$.

So can we calculate the distance from C in the same way? Let's tackle it the same way. We know that $P_1 = (5, \theta_1)$ and $P_2 = \left(5\frac{\sqrt{3}}{2}, \theta_2\right)$. From what we know about the way P_n and P_{n+1} are related

$$P_3 = \left(\left(5\frac{\sqrt{3}}{2}\right)\left(\frac{\sqrt{3}}{2}\right), \theta_3\right), \quad P_4 = \left(5\left(\frac{\sqrt{3}}{2}\right)^3, \theta_4\right),$$

and so on. In general then, $P_{n+1} = \left(5\left(\frac{\sqrt{3}}{2}\right)^n, \theta_{n+1}\right)$. This means that we can at last give the complete polar coordinates for P_{n+1}. They are $\left(5\left(\frac{\sqrt{3}}{2}\right)^n, \frac{n\pi}{6}\right)$.

But how do we get a formula linking r and θ for the general point (r, θ)? Let's think what's going on for a minute. Suppose we let $r = 5\left(\frac{\sqrt{3}}{2}\right)^n$ and $\theta = \frac{n\pi}{6}$. Now both of these last equations have an n in them. What if we eliminate n? Won't we then have a relation between r and θ, satisfied by the coordinates of all points P_n?

Well, $\theta = \frac{n\pi}{6}$, so $n = \frac{6\theta}{\pi}$. Substituting for n in the r equation gives $r = 5\left(\frac{\sqrt{3}}{2}\right)^{\frac{6\theta}{\pi}}$. What a mess! Let's write it out large to see if it looks any better

$$r = 5\left(\frac{\sqrt{3}}{2}\right)^{\frac{6\theta}{\pi}}. \tag{1}$$

It certainly is a mess but it does seem correct. After all, the points P_n all satisfy it. The other points on the spiral are just what we get by smoothing between the P_n points. As well as that, we can easily see that as θ gets larger r gets smaller. This is because $\frac{\sqrt{3}}{2}$ is less than 1. As t approaches infinity $\left(\frac{\sqrt{3}}{2}\right)^t$ approaches zero. This means that r will approach zero as θ gets larger and larger. So this curve will definitely spiral *in* as we've already seen.

Young arachnid, we think you're on a winner here. You'll get so many insects in your new colored spiral web that you'll be able to sell them to all the spiders in the neighborhood. Just think of it. Arachdonalds! Selling fries and juicy Big Arachs!

1.3 CONSOLIDATION

The work of the last section has, of course, only opened up a can of worms. What equation would we get if we had used rays which were only $\frac{\pi}{18}$ apart? What equation would we find for cards which have corner angles α equal to $\frac{4\pi}{9}$, $\frac{7\pi}{18}$, $\frac{\pi}{3}$, and especially $\frac{5\pi}{12}$? How could we explain the right-handed and left-handed versions of all the spirals? There are clearly a lot of mathematical questions that are still unresolved here, not to mention the sunflower–escargot conundrum.

Now we have been looking for a relation between r and θ, starting from curves that we knew something about. We certainly knew how to construct them. Why don't we turn the questions around and look at (r, θ) relations to see what curves *they* produce? It's probably a good idea to start with something simple. We'll then say goodbye to you and let you explore to your heart's content.

So what could be simpler than $r = k$, a constant? In such a curve, the points are always a constant distance from the origin. Hence they must lie on a circle, center C.

Another simple equation that needs to be dealt with is $\theta = k$. Any point on the graph of this relation makes with C the same fixed angle from the initial direction. Hence we get a straight line which starts at C and heads off to infinity. Notice that we get a ray, not a complete line through C.

Another simple equation is $r = \theta$. What does the spider web graph of this relation look like? Well, there are at least three ways to go about answering that question. We could plot lots of points and join them all up, or we could use a graphing calculator, or we could think about what could happen.

• • • BREAK

See if you can make any progress with the graph whose polar equation is $r = \theta$?

If you've plotted points you may have found that they disappeared off your web pretty quickly. We hope that you changed your scale so that you were able to get points whose values of θ were larger than 2π.

Is there very much to say about $r = \theta$? As θ increases so does r. So the curve formed by the points (r, θ), where $r = \theta$, must spiral out from the center. It'll have to look like the curve in Figure 13.

Normally in polar coordinates, we only allow r to be positive or zero. After all, it is the *distance* of the point from the pole. However, in some books you will see r being allowed to be negative. We won't though, because we have an aversion to negative distances. Of course, θ is *generally* allowed to be *any* real number but, for any particular relation, we only allow those values of θ which make $r \geq 0$. Naturally, the point with coordinates (r, θ) is the same as the point with coordinates $(r, \theta + 2\pi)$.

Getting back to relations between r and θ, the next obvious things to try are the *linear* relations—things like $r = m\theta + c$, where m and c are fixed real numbers.

• • • **BREAK**

Why not see what curves have equation $r = m\theta + c$? You may need to use a combination of point-plotting and thinking. But thinking is always preferable if there's a choice. You might like to try the special cases $m = 0$ and $c = 0$.

You've probably realized by now that all the relations with m positive give spirals. Since we allow θ to be negative, the spirals

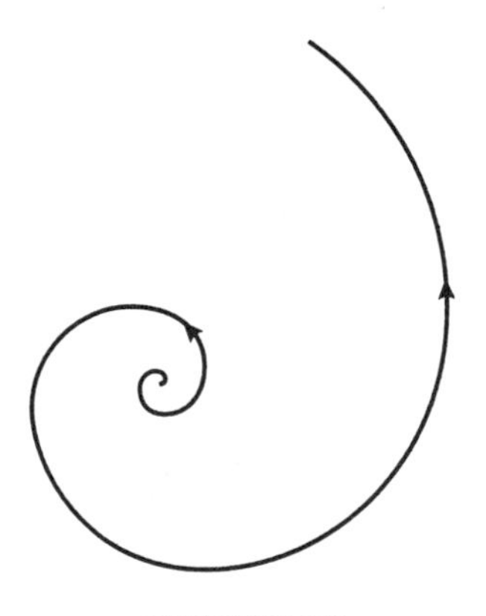

FIGURE 13

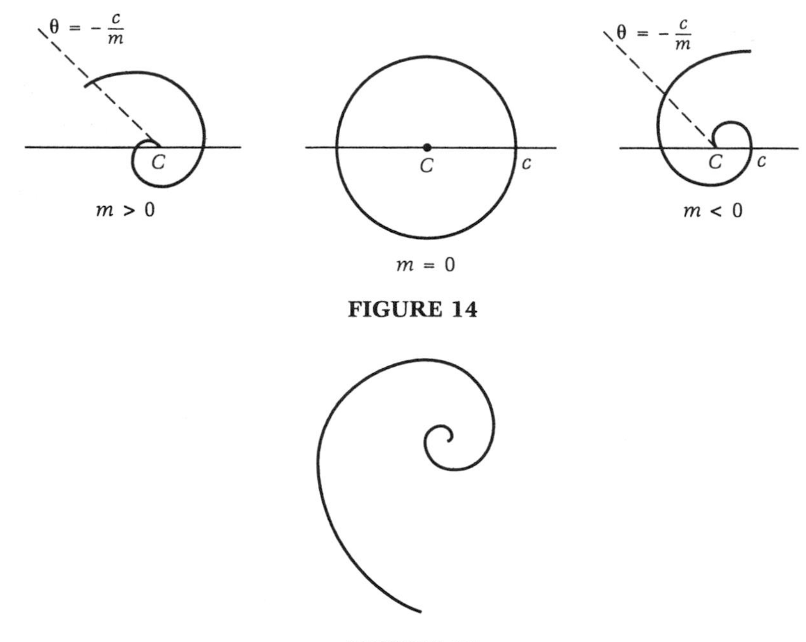

FIGURE 14

FIGURE 15

start at $\theta = -\frac{c}{m}$, because r can't be negative, that is, we require $\theta \geq -\frac{c}{m}$. We already know that if $m = 0$, we get a circle of radius c. For m negative, we require $\theta \leq -\frac{c}{m}$. All of these situations are shown in Figure 14.

Let's have a little deeper look into a special case of these ***linear*** spirals. So let $r = 2\theta$. The polar curve with this equation is given in Figure 15.

The interesting thing that we want to point out here is the constant nature of this curve. Look what happens every time it crosses the initial line.

From the table you can see that the value of r increases each time by 4π. But the same thing happens no matter what ray we look at. As the curve spirals out, every time it crosses a fixed ray, it is 4π further out than the last time. To see this constant increase for

TABLE 1 $\quad r = 2\theta$

θ	0	2π	4π	6π	8π
r	0	4π	8π	12π	16π

r, take any ray, $\theta = \theta_1$, say. When the curve crosses that ray again, θ has increased by 2π to $\theta_1 + 2\pi$. At the first crossing, $r_1 = 2\theta_1$ and at the second $r_2 = 2(\theta_1 + 2\pi) = 2\theta_1 + 4\pi$. Clearly, the difference between r_1 and r_2 is 4π. And that constant difference applies no matter which ray the curve crosses.

You probably also managed to show that the same thing happened for any polar curve of the form

$$r = m\theta + c. \tag{2}$$

The argument is the same. At the ray $\theta = \theta_1$, we get $r_1 = m\theta_1 + c$ at first. The next time past this ray

$$\theta = \theta_1 + 2\pi \qquad \text{and} \qquad r_2 = m(\theta_1 + 2\pi) + c.$$

So the difference between the two values of r is

$$r_2 - r_1 = (m\theta_1 + m2\pi + c) - (m\theta_1 + c) = 2\pi m.$$

Again, a constant increase. Again, the same increase occurs for *every* ray. Such curves are known as ***Archimedean spirals*** (see [2] and [3], for example).

• • • **BREAK**

Can you think where you might have seen Archimedean spirals?

If you have a non-zero constant c, in your Archimedean spiral, see (2), the curve looks as if it might follow the surface of some sort of material on a roll—dress material, for instance. But this isn't quite right. Material certainly winds around the roll adding a constant width once every time round. However, the start isn't quite right. On the other hand, if the constant is zero, the spiral above is just the kind of curve you get when you roll a length of something tightly up onto itself. Tape measures are sometimes rolled this way.

Now when we're dealing with Cartesian coordinates, polynomial relations give some interesting curves. But if we allow, for example, $r = \theta^2 + 2\theta + 2$, then we find we don't get anything very exciting—just more spirals. So we'll try something different.

• • • **BREAK**

What do you think the polar curve with equation $r = \sin\theta$ looks like? Have a guess and then try to sketch it.

Before we do $r = \sin\theta$, let's have a look at $r = \cos\theta$. In fact, we'll show it in Figure 16. It's a circle, with center at $\left(\frac{1}{2}, 0\right)$ and radius $\frac{1}{2}$. How does that come about?

If you can't see how we got this, draw up a table of values. You should find that as θ goes from 0 to $\frac{\pi}{2}$, r goes from 1 to 0. There are no values for r with θ between $\frac{\pi}{2}$ and $\frac{3\pi}{2}$, but from $\frac{3\pi}{2}$ to 2π, r increases from 0 to 1 and the circle is completed. (If you take increasing values of θ from here, you just go round the circle again and again with suitable gaps every π radians, in the same way that you do from $\theta = 0$ to 2π). Alternatively, there is a straightforward proof using Cartesian coordinates. Starting with $r = \cos\theta$, multiply both sides by r to get $r^2 = r\cos\theta$. Since $x = r\cos\theta$ and $y = r\sin\theta$, we then see that $x^2 + y^2 = x$. Completing the square gives $\left(x - \frac{1}{2}\right)^2 + y^2 = \frac{1}{4}$. Do you recognize this as a circle?

If you're still worried about $r = \sin\theta$, you should now be able to show that it too is a circle. This one, though, has center at $\left(\frac{1}{2}, \frac{\pi}{2}\right)$ in polar coordinates and again the radius is $\frac{1}{2}$.

• • • **BREAK**

There is an easy way of obtaining the graph of $r = \sin\theta$ from that of $r = \cos\theta$. Recall that $\cos\left(\frac{\pi}{2} - \theta\right) = \sin\theta$? How does that help you sketch $r = \sin\theta$? What is the effect on the curve $r = \cos\theta$ of changing θ to $\frac{\pi}{2} - \theta$?

Looking at trigonometric functions opens up floodgates. You should find a lot of interesting shapes of the form $r = \cos 2\theta$,

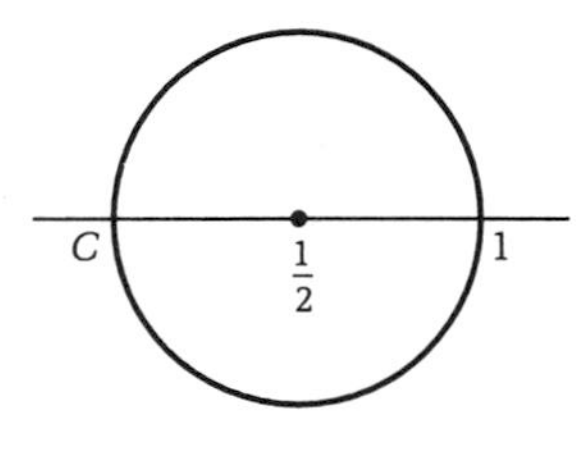

FIGURE 16

$r = \cos 3\theta$, and so on. Something like $r = 1 - \sin\theta$ is interesting too. If you're hooked on these polar curves, we suggest you try to graph a few more of them.

If we can remind you of your spider days in the previous section, remember that we came up with a polar equation of the form $r = ka^{\theta}$. (In actual fact k was 5 and a was $\left(\frac{\sqrt{3}}{2}\right)^{\frac{6}{\pi}}$.) Now this curve has an interesting property. Look at the values of r for two values of θ.

If $\theta = \theta_1$, $r_1 = ka^{\theta_1}$ and if $\theta = \theta_2$, $r_2 = ka^{\theta_2}$.

"So what?" we hear you ask. OK, so take the ratio $r_2 : r_1$. Then

$$\frac{r_2}{r_1} = \frac{ka^{\theta_2}}{ka^{\theta_1}} = a^{\theta_2 - \theta_1}. \tag{3}$$

So here's the insight. If we take any two values of θ which differ by a given amount ($(\theta_2 - \theta_1)$ is constant), the resulting ratio $\frac{r_2}{r_1}$ is the *same*, no matter where you are on the spiral. Because of this property, curves with polar equation $r = ka^{\theta}$ are called ***equiangular spirals*** (for more details see [2] and [3]).

The property also means that, in some sense, the curve is self-similar. The distance from the origin increases by the same amount (by the same ratio) for every constant angle that the spiral goes through. Every section of the spiral is then a replication of the previous section. Zooming in (or out) on the spiral you see the same shape. Hence the spiral is much like a fractal (see Chapter 8).

1.4 FIBONACCI STRIKES

We have constructed spirals with cards on cobwebs but there are other methods of construction. Take your favorite sequence—the Fibonacci sequence 1, 1, 2, 3, 5, 8, 13, 21, ... comes to mind (see Chapter 3)—and use a polar grid with rays at angles of $\frac{\pi}{6}$ for a start. When $\theta = 0$, let $r = F_1 = 1$. When $\theta = \frac{\pi}{6}$, let $r = F_2 = 1$. Keep going so that when $\theta = n\frac{\pi}{6}$, $r = F_{n+1}$.

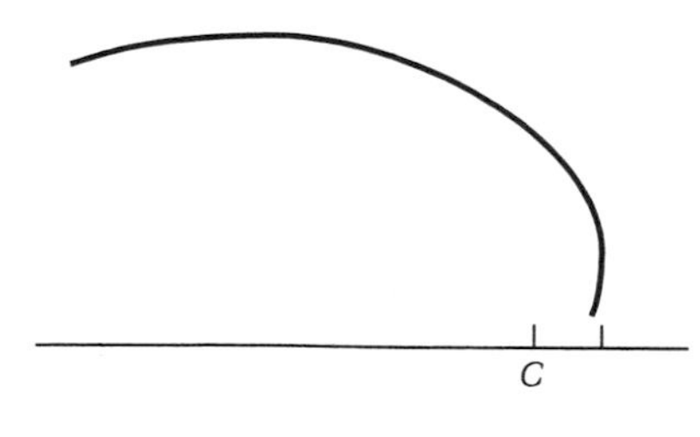

FIGURE 17

• • • BREAK

Plot the Fibonacci points as indicated above and draw a smooth curve between them. You should get a respectable spiral. Check it out.

You have probably ended up with something like the graph of Figure 17. Actually, with a little work, you can give a relation between r and θ for this curve. Have a go and see what you come up with.

But you don't *have* to use the Fibonacci sequence. Something like 1, 2, 4, 8, 16, 32, ... or 1, 4, 9, 16, 25, ..., will give you a spiral too.

• • • BREAK

Experiment with different sequences of numbers and see what your spirals look like.

Just to wrap this one up, let's find the relation between r and θ for the curve formed from 1, 2, 4, 8, 16, ... Using rays $\frac{\pi}{6}$ apart, when $\theta = n\frac{\pi}{6}$, r must be 2^{n-1}. So $r = 2^{\frac{6\theta}{\pi}-1} = \frac{1}{2}\left(2^{\frac{6\theta}{\pi}}\right)$. It looks as if we've ended up with another equiangular spiral. Did you get an equiangular spiral for the Fibonacci curve?

• • • BREAK

Use the Binet formula (see Chapter 3) to express the Fibonacci curve by an equation in polar coordinates.

1.5 DÉNOUEMENT

We started off this chapter asking what there is in common between the sunflower and the snail. The answer is that the seeds of the sunflower and the shell of the snail both exhibit a spiral structure. If you look at the snail's shell, you'll see a clear spiral. What may not be obvious, at first, is that the spiral is equiangular. "The whorls continually increase in breadth and do so in a steady and unchanging ratio" (see [1, Volume 2, p. 753]).

The same kind of behavior is to be found in the Nautilus shell and in many other shells, too. But it is not to be found in sunflowers. Instead, sunflowers exhibit the Fibonacci spiral behavior. This is illustrated in Figure 18.

In this chapter we have only skimmed the surface of the study of spirals and polar curves. There is a lot more out there to investigate. You might actually like to do some of that investigating. If you do, don't forget about three-dimensional spirals. The common or garden helix not only occurs in circular staircases and in bed-

FIGURE 18

springs but also seems to have something to do with DNA too. But what on earth is an Archimedean screw?

Oh! It's suddenly occurred to us that we haven't mentioned drains and that was in the title of the chapter. What's the path traced out by a fleck of fat as it goes down the drain? And does it matter whether the fat is in Sydney, Southampton, or Seattle?

• • • FINAL BREAK

Here are a few problems for you to try out your new skills on.

1. In Section 1, when using the card construction of a spiral we found that 75° (or $\frac{5\pi}{12}$) was a critical angle. Would the same angle be critical if the angle between the rays was changed?
2. Draw four equally spaced equiangular spirals on a piece of card. Pin the point C to the center of a turntable. What effect do you get when the turntable rotates?
3. Consider the polar curve whose equation is $r = \sin\theta$. What would happen if θ were allowed to be negative?
4. Find the polar equation for the circle, center (a, α) and radius b. [*Hint*: First get the Cartesian equation.]

REFERENCES

1. Thompson, D'Arcy W. *On Growth and Form*, Cambridge University Press, London, 1952.
2. Lockwood, E.H. *A Book of Curves*, Cambridge University Press, London, 1963.
3. Coxeter, H.S.M. *Introduction to Geometry*, Wiley, New York, 1961.

ANSWERS FOR FINAL BREAK

1. Suppose the angle between rays is θ. Then the critical case occurs when $\pi - \theta - \alpha = \alpha$. Hence the critical value of α is $\frac{1}{2}(\pi - \theta)$, which certainly depends on the angle θ between the rays.
2. You should get an interesting optical effect.
3. You get the same circle again and again.
4. In Cartesian coordinates the center is $(a \cos \alpha, a \sin \alpha)$ and the radius is b. So the equation is

   $$(x - a \cos \alpha)^2 + (y - a \sin \alpha)^2 = b^2.$$

   Simplifying, this becomes

   $$x^2 + y^2 - 2xa \cos \alpha - 2ya \sin \alpha = b^2 - a^2.$$

   Converting to polar coordinates, we obtain

   $$r^2 - 2ra \cos \theta \cos \alpha - 2ra \sin \theta \sin \alpha = b^2 - a^2,$$

   or

   $$r^2 - 2ra \cos(\theta - \alpha) = b^2 - a^2.$$

   The moral is that, when dealing with circles whose centers are not at the origin, it's easier to use Cartesian coordinates than polar coordinates.

2 CHAPTER A Far Nicer Arithmetic

2.1 GENERAL BACKGROUND: WHAT YOU ALREADY KNOW

Suppose you have to do an addition, say $357 + 586$. How much do you need to know in order to know the last digit of the answer? Would the last digit of each summand be enough? Suppose, instead, that it's a subtraction problem? A multiplication problem? A division problem?

We think you will immediately see that, to know the last digit of the answer to the *addition* problem you only need to know the last digit of each of the summands. (In the above example you would simply add 7 to 6, ignore the 1 in the tens position and get 3.) Likewise, to know the last two digits of the answer, you only need the last two digits of the summands (so that, above, you would add 57 to 86, ignore the 1 in the 100 position, and get 43), and so on. Indeed, the traditional algorithm for adding a column of figures exploits this fundamental fact. And what goes for addition is true of subtraction and multiplication, too, though not of division, even where the divisor is an exact factor of the dividend (think of $12 \div 2$, $22 \div 2$).

Let us concentrate for the moment on the last digit. The last digit of the integer a may be described as the remainder when a is divided by 10. Then, as we have said above, the remainder when $a + b$ is divided by 10 may be obtained from the remainders when a and b are divided by 10; and the same goes for the remainder when $a - b$ or ab is divided by 10. And what goes for remainders when you divide by 10 is just as true for remainders when you divide by 100, 1000, and so on.

Now comes the big, but obvious step—the same goes when you divide by *any integer* m. Those of you who have done arithmetic in various bases will already have met this fact, but we feel that none of our readers will have any real difficulty in understanding this. Thus, for example, if $a = 44$ and $b = 23$, we see that 44 has a remainder of 4 and 23 has a remainder of 3 when divided by 5. Thus we know that the remainder when $a + b (= 67)$ is divided by 5 is the same as the remainder when $4 + 3$ is divided by 5, that is, 2; and $a - b$ $(= 21)$ leaves a remainder of $4 - 3$, or 1, when divided by 5. In just the same way we know that 44×23 leaves the same remainder as 4×3 when divided by 5, that is, a remainder of 2. It is very striking that we don't need to calculate 44×23 to get this last result!

• • • **BREAK**

Check the results of the last paragraph. Then try some examples with a value of m different from 5.

Mathematicians like to use short, pithy phrases,[1] so, instead of the long-winded "the remainder when you divide by m" they say "the remainder mod m" and the set of all integers leaving the same remainder mod m is called a ***residue class*** mod m; here "mod" is short for "modulo," and m is called the "modulus." Thus the remainder mod m is a specially chosen member of its residue class, namely, that integer r in the class which satisfies $0 \leq r < m$. Notice that there are only finitely many residue classes mod m (in fact, there are precisely m such classes), but that a residue class contains infinitely many integers.

[1]Another example of a short, pithy phrase is "the triangle ABC" instead of "the triangle with vertices A, B, C." Can you think of further examples?

Now we also need a word for the *process* which, working mod m, starts with the residue classes of a and b and produces the residue class of $(a + b)$. The word we use is, of course, the ***addition*** of residue classes. Likewise, we speak of the ***subtraction*** and ***multiplication*** of residue classes; moreover, we use the usual symbols to denote these three operations on residue classes.

Of course, the addition of remainders is not quite like the addition of integers. Certainly, if the modulus is 10, then the sum of the remainders 3 and 5 is 8; but the sum of the remainders 6 and 7 is 3, not 13, because a remainder mod m must lie between 0 and $(m - 1)$. We now introduce some notation to overcome this slight awkwardness.

Let $[a]_m$ stand for ***the set of all integers which leave the same remainder as a when you divide by m***; thus $[a]_m$ consists of all integers $(a + km)$ where k is any integer. As we have said, we call $[a]_m$ a ***residue*** class, more precisely, the ***residue class of a* mod *m***. Thus if $a = 0$, $m = 2$, then $[a]_m$ is just the set of *even* integers; if $a = 1$, $m = 2$, then $[a]_m$ is the set of *odd* integers; and $[1]_3$ is the set $\{\ldots, -5, -2, 1, 4, 7, \ldots\}$.

Now we may write the basic arithmetical facts as

$$\left.\begin{aligned} [a]_m + [b]_m &= [a + b]_m, \\ [a]_m[b]_m &= [ab]_m, \end{aligned}\right\} \tag{1}$$

strictly speaking, we may regard the relations (1) as *defining* the addition and multiplication of residue classes mod m. Notice that, in (1), there is *no* restriction on the possible integer values of a and b; they could be remainders but they need not be. Indeed, a or b could be negative.

Notice, too, that we omitted subtraction from (1). We had two reasons for doing this. First, we cannot just *define* subtraction any way we please; subtraction must fit with addition. More formally, $a - b$ is the solution x to the equation $x + b = a$. Of course, a moment's reflection will convince you that the rule

$$[a]_m - [b]_m = [a - b]_m \tag{2}$$

does produce a result compatible with our definition above of the sum of residue classes; moreover, it is obvious that $[0]_m$ is the zero for our addition of residue classes, that is,

$$[a]_m + [0]_m = [a]_m,$$

so that

$$-[a]_m = [-a]_m. \tag{3}$$

However, the second reason for our delaying the statement of the subtraction rule is more important in practice. We have talked so far in this chapter of *integers* and may have left the impression (though we didn't actually say so!) that we had *positive* integers in mind. But, for (2) or (3) to make sense we must allow negative integers and zero, since $a - b$ may well fail to be positive, even if a and b are positive. You may wonder if there is any difficulty in talking of the residue class $[a]_m$ if a is negative. The answer is: No, there isn't. For recall that the residue class $[a]_m$ consists of all integers $(a + km)$, where k is *any* integer, and so contains positive and negative integers, whatever value we give to a. The usual remainder, which number theorists also call the ***residue***, is just the smallest nonnegative integer in the collection $(a + km)$; but, quite often, another representative of the residue class is more natural or more useful, as in (3). Notice that the modulus m is always a *positive* integer. For we never allow $m = 0$ (why not?) and there would be no difference between $[a]_m$ and $[a]_{-m}$ if we did allow negative m. To understand this last statement take a careful look at the definition of $[a]_m$ and $[a]_{-m}$. For example, if $a = 2$, $m = 3$, we see that, by definition, $[2]_3$ is the set of integers $2 + 3k$, where k is *any* integer; so if k is

$$\ldots, -3, -2, -1, 0, 1, 2, 3, \ldots,$$

then the corresponding elements of $[2]_3$ are

$$\ldots, -7, -4, -1, 2, 5, 8, 11, \ldots.$$

Likewise, we see that, by definition, $[2]_{-3}$ is the set of integers $2 - 3k$, where k is *any* integer; so k is $\ldots, -3, -2, -1, 0, 1, 2, 3, \ldots,$ then the corresponding elements of $[2]_{-3}$ are

$$\ldots, 11, 8, 5, 2, -1, -4, -7, \ldots,$$

that is, the same set of integers as above.

Now we can be quite precise. We allow $a, b, \ldots$ to range over the set of integers $\mathbb{Z}$; m is a fixed positive integer; $[a]_m$ is the set of integers $(a + km, k \in \mathbb{Z})$; (1) gives the rules for adding and multiplying residue classes mod m; (2) shows how to subtract residue classes to achieve the usual properties of subtraction; and (3) shows how to take the additive inverse.

We have, in our discussion so far, been skirting round one of the most fundamental ideas in mathematics, that of an ***equivalence relation*** and the associated set of ***equivalence classes***. Those of you who are familiar with this idea may have recognized that our residue classes $[a]_m$ are just special cases of equivalence classes. To "level the playing field," let's take a moment to make all our readers familiar with this idea.

Let S be a set and let $\sim$ be a ***relation*** on the set S. Thus it may happen that $a \sim b$ for certain pairs (a, b) of elements of S. As an example, think of the set $\mathbb{Z}$ and let $a \sim b$ mean that m exactly divides $(a - b)$; we write this as $m \mid (a - b)$. We say that the relation $\sim$ on S is an ***equivalence*** relation if it satisfies three conditions:

I. (Reflexivity) $a \sim a$, for all $a \in S$;

II. (Symmetry) If $a \sim b$, then $b \sim a$, for all $a, b \in S$; and

III. (Transitivity) If $a \sim b$, $b \sim c$, then $a \sim c$, for all $a, b, c \in S$.

Given an equivalence relation $\sim$ on a set S, the set is partitioned into ***equivalence classes*** by the rule that a, b go into the same class if and only if $a \sim b$. You may check that the three conditions above are precisely what we need to make sense of this rule. For example, condition I guarantees that every element of S belongs to *some* class.

If we revert to our example, then we see that the equivalence class (subset of $\mathbb{Z}$) containing a is precisely the residue class of a mod m, or $[a]_m$. We call the equivalence relation of our example ***congruence*** mod m, and write $a \equiv b \bmod m$ for $a \sim b$.

• • • **BREAK**

Consider the following relations and decide which of conditions I, II, and III they satisfy:

(i) $S = \mathbb{Z}$, $a \sim b$ means $a \geq b$;

(ii) $S = \mathbb{Z}$, $a \sim b$ means $a > b$;

(iii) $S = \mathbb{Z}$, $a \sim b$ means $a \neq b$;

(iv) $S =$ set of straight lines in the plane,

$$\ell \sim m \quad \text{means "}\ell \text{ is parallel to } m.\text{"}$$

(v) S is as in (iv), $\ell \sim m$ means "ℓ is perpendicular to m."

Because of the simple nature of the rules (1), (2), and (3), it follows immediately that the addition and multiplication of residue classes have all the nice, convenient properties of the addition and multiplication of integers (with one important exception, described below). Thus (writing $[a]$ for $[a]_m$ for simplicity), we have

$$\text{(Commutative laws)} \quad \begin{cases} [a] + [b] = [b] + [a], \\ [a][b] = [b][a], \end{cases} \tag{4}$$

$$\text{(Associative laws)} \quad \begin{cases} ([a] + [b]) + [c] = [a] + ([b] + [c]), \\ ([a][b])[c] = [a]([b][c]), \end{cases} \tag{5}$$

$$\text{(Distributive law)} \quad [a]([b] + [c]) = [a][b] + [a][c], \tag{6}$$

$$\text{(Zero)} \quad [a] + [0] = [a], \quad [a][0] = [0], \tag{7}$$

$$\text{(Unity)} \quad [a][1] = [a]. \tag{8}$$

Rules (1) to (8) are said to be the rules of ***modular arithmetic***.

• • • **BREAK**

Check the results of (4) through (8) by example and by definition.

The only rule of ordinary arithmetic that you might expect to hold, but which actually fails for modular arithmetic, is the one which states, for integers[2]. a, b, that, if $ab = 0$, then $a = 0$ or

[2]Or, indeed, for any real numbers a, b! This rule is the basis of our method of solving equations. Division cannot even be *defined* (let alone carried out) without it

$b = 0$. The corresponding statement for residue classes is false; for example,

$$[2]_{10}[5]_{10} = [10]_{10} = [0]_{10},$$

but, of course, $[2]_{10} \neq [0]_{10}$, $[5]_{10} \neq [0]_{10}$. We will have much more to say about this phenomenon later—it is very closely related to the reason why, in division, the last digit of the dividend and of the divisor do *not* give us the last digit of the quotient. However, by way of compensation for the loss of this *division rule* for integers, the arithmetic of residue classes has many great advantages. Perhaps the most obvious is that the set of residue classes mod m is *finite*, consisting of the m elements $[0]_m, [1]_m, \ldots, [m-1]_m$. Thus the addition and multiplication tables can be completely described by a square array. Let us illustrate this for the moduli (plural of "modulus"!) 5 and 6; we emphasize that these tables, unlike the traditional tables, are not *part* of the process of adding and multiplying—they actually tell you everything! In these tables—and often when doing modular arithmetic—we write the remainder instead of the residue class containing it (in accordance with our principle of using *simplified notation*—see Chapter 9).

+	0	1	2	3	4
0	0	1	2	3	4
1	1	2	3	4	0
2	2	3	4	0	1
3	3	4	0	1	2
4	4	0	1	2	3

Addition mod 5

×	0	1	2	3	4
0	0	0	0	0	0
1	0	1	2	3	4
2	0	2	4	1	3
3	0	3	1	4	2
4	0	4	3	2	1

Multiplication mod 5

(9)

+	0	1	2	3	4	5
0	0	1	2	3	4	5
1	1	2	3	4	5	0
2	2	3	4	5	0	1
3	3	4	5	0	1	2
4	4	5	0	1	2	3
5	5	0	1	2	3	4

Addition mod 6

×	0	1	2	3	4	5
0	0	0	0	0	0	0
1	0	1	2	3	4	5
2	0	2	4	0	2	4
3	0	3	0	3	0	3
4	0	4	2	0	4	2
5	0	5	4	3	2	1

Multiplication mod 6

(10)

• • • **BREAK**

Construct some tables of your own choosing—of course, with some prime and some non-prime moduli.

We close this section by drawing your attention to the important notational innovations we have introduced: if the integers a and b belong to the same residue class mod m, then we may write

$$a \equiv b \bmod m. \tag{11}$$

Thus (11) means exactly the same as $[a]_m = [b]_m$. We read (11) as "a is congruent to b mod m." Notice that (11) is equivalent to

$$a - b \equiv 0 \bmod m$$

or

$$m \text{ exactly divides } (a - b). \tag{12}$$

For example, $38 \equiv 12 \bmod 13$. In the phrase "m exactly divides n" we will always suppose in this chapter that m is a *positive* integer, but we allow n to be *any* integer. Instead of writing the phrase we will usually use the abbreviated form

$$m \mid n. \tag{13}$$

Then (13) is true if and only if there is an integer k such that $n = mk$.

• • • **BREAK**

Some things to think about: What can you infer if you know that:

(1) $m \mid n$ and $n \mid m$?

(2) $m \mid n$ and $n \mid q$?

(3) $a \mid n$ and $b \mid n$ and a and b are ***coprime***, that is, $\gcd(a, b) = 1$.[3]

[3]gcd means greatest common divisor, so "$\gcd(a, b) = 1$" means the only common factor of a and b is 1.

2.2 SOME SPECIAL MODULI: GETTING READY FOR THE FUN

From what you'll learn in this section, you'll get a reputation as a lightning calculator—but you won't, in fact, have done the calculations your audience thinks you've done. Read on!

Of course, so long as we continue to write numbers in base 10, arithmetic mod 10 (i.e., the arithmetic of residue classes mod 10) will continue to hold a special place among the various modular arithmetics. However, there are other modular arithmetics which are also of special interest and which we will now describe. The first two, arithmetic mod 9 and arithmetic mod 11, also derive their special interest from the fact that we write numbers in base 10.

***Arithmetic* mod 9** Here the special interest, as we have said, derives not from any particular properties of the residue classes mod 9 but from the fact that remainders mod 9 can be quickly calculated for numbers written in base 10, as we will suppose they are. Thus let n be any positive integer and let $s(n)$ be the integer obtained by adding the digits of n. Thus,

$$\text{if } n = 3857, \quad \text{then } s(n) = 3 + 8 + 5 + 7 = 23.$$

We now have an important result, which we state as a theorem.

Theorem 1 $[n]_9 = [s(n)]_9$.

Of course, this only means that $n - s(n)$ is divisible by 9. The general argument is clear if we deal with our particular case above. Then

$$n = 3000 + 800 + 50 + 7,$$
$$s(n) = 3 \quad + 8 \quad + 5 \ + 7,$$

so that

$$n - s(n) = 3(1000 - 1) + 8(100 - 1) + 5(10 - 1)$$
$$= 3(999) + 8(99) + 5(9).$$

It thus suffices to observe that for any positive integer q, $10^q - 1$ is divisible by 9. This last fact comes immediately from our rule of multiplication (1); for

$$\text{since } [10]_9 = [1]_9, \text{ it follows that } [10^q]_9 = [1^q]_9 = [1]_9.$$

(Of course, you may also observe that $10^q - 1$ is written in base 10 as a sequence of q 9's.)

Of course, we can iterate the s-function, that is, we can calculate $s(s(n))$ which we naturally write as $s^2(n)$, and so on. Eventually, we must reach a number lying between 1 and 9, and this must be the remainder when n is divided by 9, with the small modification that if we eventually reach the number 9 itself, then there was no remainder and our number n is divisible by 9. So, for our example of $n = 3857$, $s(n) = 23$, $s^2(n) = 5$, so the remainder is 5. If $n = 49826$, then $s(n) = 29$, $s^2(n) = 11$, $s^3(n) = 2$, so the remainder is 2. If $n = 5247$, then $s(n) = 18$, $s^2(n) = 9$, so 5427 is divisible by 9. (You should check these.)

Now let us consider this question:

***What is the remainder when* 3857 × 49826 *is divided by* 9?**

This question appears at first sight to involve either long, tedious calculation or appeal to a hand calculator or computer. Actually, however, it is easily answered without any mechanical aids. For Theorem 1 tells us that

$$[n]_9 = [s(n)]_9 = [s^2(n)]_9 = \dots .$$

Thus

$$[3857]_9 = [5]_9, \quad [49826]_9 = [2]_9,$$

so

$$[3857 \times 49826]_9 = [5 \times 2]_9 = [10]_9 = [1]_9,$$

and the remainder is 1!

We will see in Section 5 how these ideas can be used very effectively to check calculations. However, there is already an interesting point revealed by this calculation. Those for whom arithmetic is merely a skill would believe that, to find the remainder when a number expressed in a complicated way is divided by 9, one must (a) first express the number in traditional (base 10) form; and (b) then carry out the division. We have shown that this belief is wrong.

Notice that, since 3 is a factor of 9, it follows from Theorem 1 that

Corollary 2 $[n]_3 = [s(n)]_3$.

Thus we may use the same technique as that above to find the remainder when n is divided by 3; we simply apply the s-function repeatedly until we achieve a number between 1 and 9. Then

if we achieve 1, 4, or 7, the remainder is 1,

if we achieve 2, 5, or 8, the remainder is 2,

if we achieve 3, 6, or 9, the remainder is 0.

Notice that this gives us an easy check for divisibility by 3 (and by 6).

***Arithmetic* mod 11.** A similar technique may also be used to do arithmetic mod 11. Once again the interest derives from the ease with which remainders mod 11 can be calculated. Admittedly, it is not as easy as calculating remainders mod 9, but it is very easy compared, say, with calculating remainders mod 7.

Let us again take 3857 as our example. If $n = 3857$, we again write

$$n = 3000 + 800 + 50 + 7.$$

Now, however, we exploit the (obvious?) fact that $10 \equiv -1 \bmod 11$. Then (by (1)), since $(-1)^1 = -1$, $(-1)^2 = 1$, $(-1)^3 = -1, \ldots$

$$10^1 \equiv (-1)^1 = -1 \bmod 11, \text{ so we see that } 11 \text{ divides } 10 + 1\ (= 11);$$

$$10^2 \equiv (-1)^2 = 1 \bmod 11, \text{ so we see that } 11 \text{ divides } 100 - 1\ (= 99);$$

$$10^3 \equiv (-1)^3 = -1 \bmod 11, \text{ so we see that } 11 \text{ divides } 1000 + 1\ (= 1001);$$

$$\vdots$$

We may express all these facts in one formula

$$10^q \equiv (-1)^q \bmod 11, \tag{14}$$

noting that

$$(-1)^q = \begin{cases} 1 & \text{if } q \text{ is even,} \\ -1 & \text{if } q \text{ is odd.} \end{cases} \tag{15}$$

From (14) and (15) we see that what we need now, instead of the function s which we used in calculating mod 9, is the function σ which gives us the *alternating sum* of the digits. Thus if

$$n = 3857 = 3000 + 8000 + 50 + 7,$$

then

$$\sigma(n) = -3 + 8 - 5 + 7.$$

(Notice that the signs in $\sigma(n)$ alternate, *ending* with a positive sign.) Then the result for the modulus 11 which corresponds to Theorem 1 is

Theorem 3 $[n]_{11} = [\sigma(n)]_{11}$.

Once again we can iterate the function σ when necessary (it wasn't necessary in our example, since $\sigma(3857) = 7$, so the remainder is 7). However, a small difficulty arises because $\sigma(n)$ need not be positive. What then should we do if, as is possible, $\sigma(n)$ is a large negative number? Well, suppose $\sigma(n) = -n_1$ and $\sigma(n_1) = n_2$, where n and n_1 are positive, but n_2 is not necessarily positive. Then

$$n_1 \equiv n_2 \bmod 11,$$

$$\text{so } -n_1 \equiv -n_2 \bmod 11, \text{ and } n \equiv \sigma(n) = -n_1 \equiv -n_2 \bmod 11.$$

Thus our basic result that $n \equiv \sigma(n) \bmod 11$ remains valid

$$\text{if we interpret } \sigma(-n), \text{ for } n \text{ positive, to mean } -\sigma(n);$$

and we can then iterate automatically. Let us given an example. If $n = 908172$, then

$$\sigma(n) = -9 - 8 + 1 - 7 + 2 = -21$$

and $\sigma(-21) = -\sigma(21) = -(-2 + 1) = 1$. Thus $908172 \equiv 1 \bmod 11$, so 1 is the remainder when 908172 is divided by 11.

Another example of the application of Theorem 3 is this:

$$\text{if } n = 13464, \text{ then } \sigma(n) = 1 - 3 + 4 - 6 + 4,$$

so 0 is the remainder when 13464 is divided by 11, that is, 13464 is exactly divisible by 11. In Section 5 we will show how to exploit our results on the calculation of remainders on division by 9 or 11.

• • • **BREAK**

Before reading the next section we suggest that you check your own understanding of this section by showing, without doing any division, that:

2037618 is divisible by 9;
30207618 is divisible by 9 (Why is this now especially easy?);
9456238 is divisible by 11;
35343 is divisible by 99;
194601 is divisible by 3;
and 190614 is divisible by 6.

You might also like to prove that any number which reads the same forward and backward (called a ***palindromic number*** with an *even* number of digits is divisible by 11.

2.3 ARITHMETIC mod P: SOME BEAUTIFUL MATHEMATICS

If you look back at the multiplication tables for multiplication mod 5 and mod 6 you will notice an important difference. In fact, this difference becomes even more obvious if, in each table, the row and column headed by 0 are omitted. Then (9) and (10) become, respectively,

×	1	2	3	4
1	1	2	3	4
2	2	4	1	3
3	3	1	4	2
4	4	3	2	1

Multiplication mod 5

×	1	2	3	4	5
1	1	2	3	4	5
2	2	4	0	2	4
3	3	0	3	0	3
4	4	2	0	4	2
5	5	4	3	2	1

Multiplication mod 6

Thus each row (and column) of the mod 5 table is just a permutation of 1, 2, 3, 4; but not every row of the mod 6 table is a permutation of 1, 2, 3, 4, 5; and those which are not contain zeros.

Of course, the essential difference between 5 and 6 which is responsible for the distinctive properties of the multiplication tables is that 5 is prime, whereas 6 is not. If m is not prime, then m may be factored as $m = k\ell$, with neither k nor ℓ divisible by m, so that (with $[\ ]$ meaning $[\ ]_m$)

$$[0] = [m] = [k\ell] = [k][\ell], \quad [k] \neq [0], \quad [\ell] \neq [0].$$

We now show why the multiplication table mod p (where p is prime) must have the features of our example when $p = 5$. Recall that we write $m \mid n$ to mean "m divides n (exactly)." The key fact is the following:

Theorem 4 *If p is prime and $p \mid k\ell$, then $p \mid k$ or $p \mid \ell$.*

This is regarded by some mathematicians as the definition of a prime, provided we insist that $p \neq 1$. Alternatively, Theorem 4 follows easily from the basic result on the factorization of integers as a product of primes.

For the rest of this section we suppose that p is a prime different from 2, since the case $p = 2$ is trivial for our discussion, as you will see.

Now let a be any number prime to p (i.e., p does not divide a, or $p \nmid a$), and consider the sequence of integers

$$a, 2a, 3a, \ldots, (p-1)a. \tag{16}$$

We claim that no two of the integers on the list (16) are congruent mod p. For suppose we were wrong; then we would have numbers k, ℓ, with $1 \leq k < \ell \leq p-1$, such that $ka \equiv \ell a \bmod p$. This means that $p \mid a(\ell - k)$. But $p \nmid a$, by hypothesis; and $p \nmid \ell - k$ since $1 \leq \ell - k < p$. Thus, by Theorem 4, $p \nmid a(\ell - k)$. This contradiction shows that, after all, we were right—no two of the integers in (16) are congruent mod p.

A very similar argument shows that no integer in the list (16) is divisible by p. Thus the list (16) consists of representatives of $(p-1)$ distinct nonzero residue classes mod p. But there are only $(p-1)$ distinct nonzero residue classes mod p and they are represented by $1, 2, 3, \ldots, (p-1)$. Therefore, the list of residue classes $[a], [2a], [3a], \ldots, [(p-1)a]$ is just a reordering, or permutation, of the list $[1], [2], [3], \ldots, [p-1]$.

• • • **BREAK**

Check this last assertion for various (appropriate) values of p and a.

We may draw some very important conclusions from this observation. First, we now know that, in the list (16), some number congruent to 1 must occur. This means that, given $a \neq 0$, there exists b such that $ba \equiv 1 \bmod p$ (you may check various instances of this on the multiplication table mod 5).

We then say that b is the ***(multiplicative) inverse*** of a mod p. Of course, strictly speaking, it is the residue class of b which is the inverse of the residue class of a, but it is quite intolerable to be forced always to be so precise. So long as we *know* what we mean we can (and should) express ourselves informally—this is our *Principle of Licensed Sloppiness* in action (see Chapter 9).

It is easy to check that, when an inverse exists, it is unique; here, of course, the uniqueness refers to the *residue class*. So *every nonzero residue modulo p has an inverse when p is prime.* This shows that arithmetic mod p is easier in an important sense than ordinary arithmetic—we don't have to introduce fractions. In fact, with $p \nmid a$, we can obviously solve for x any congruence $xa \equiv c \bmod p$. For if $ba \equiv 1 \bmod p$, then $(bc)a \equiv c \bmod p$. And Theorem 4 immediately implies that the solution is unique. So if we used a *prime base* for our numeral system (instead of base 10), we could determine the last digit in a division problem, too, from the last digit in the divisor and dividend, provided, of course, that the divisor *is* a factor of the dividend, and that its last digit is not zero. (This is not, let us add, a good enough reason for abandoning base 10!)

• • • **BREAK**

You may like to debate the relative merits of the bases 10 and 12. What are the advantages and disadvantages of base 2?

Our second conclusion, based on the observation immediately preceding the first break of this section, will lead us to a famous theorem. Since the numbers in the list (16) are congruent, in some order, to the numbers $1, 2, 3, \ldots, (p-1)$, the overall products must be congruent. Thus

$$a^{p-1}(p-1)! \equiv (p-1)! \bmod p.$$

where $n!$ is the ***factorial function*** $1 \times 2 \times 3 \times \cdots \times n$. But we may now invoke Theorem 4 again. For certainly $p \nmid (p-1)!$, as an elementary application of Theorem 4 shows. Thus we conclude from the congruence above a famous theorem of Fermat[4] (1601–1665):

Theorem 5 *Let p be a prime and a an integer such that $p \nmid a$. Then*

$$a^{p-1} \equiv 1 \bmod p.$$

(You might like to know that Theorem 5 is a special case of a fundamental result in group theory, which itself follows from a celebrated theorem due to the great French mathematician Lagrange (1736–1813). Look out for Lagrange's Theorem when you study group theory!)

Notice how Fermat's Theorem enables you to answer apparently very difficult questions quickly. Here's an example. What is the remainder when 2^{1000} is divided by 7? Now 1000, on division by 6, leaves a remainder of 4 (so 1000 may be expressed as $6q + 4$ for some q). Thus $2^{1000} = 2^{6q+4} = 2^{6q} \times 2^4$ for some q. But $2^6 \equiv 1 \bmod 7$. Thus $2^{6q} \equiv 1 \bmod 7$, so $2^{1000} \equiv 2^4 = 16 \equiv 2 \bmod 7$. Thus, finally, $2^{1000} \equiv 2 \bmod 7$. So the answer is 2. Of course, here we've given you the justification of the method. The procedure itself is very much quicker since we only need the *remainder* when 1000 is divided by 6.

Theorem 4 has yet another very interesting consequence. Suppose $n^2 \equiv 1 \bmod p$. Then $p \mid (n^2 - 1)$ or $p \mid (n-1)(n+1)$. Theorem 4 tells us that $p \mid (n-1)$ or $p \mid (n+1)$, so that $n \equiv \pm 1 \bmod p$. This gives us an important property of prime numbers p, namely,

$$\boxed{\textbf{\textit{If } } \boldsymbol{n^2 \equiv 1 \bmod p}\textbf{\textit{, then }} \boldsymbol{n \equiv \pm 1 \bmod p}.} \tag{17}$$

[4]This is sometimes called Fermat's Little Theorem, in contrast to the famous Fermat's Last Theorem (FLT) which he never proved. FLT asserts that, if $n \geq 3$, then the equation $a^n + b^n = c^n$ cannot be solved in non-zero integers a, b, c. At the time of writing this chapter, we are hoping that FLT is being proved by Andrew Wiles, of Princeton University. (At the time of proofreading it had!)

Thus, since, by hypothesis, $p \neq 2$, there are precisely two residue classes mod p whose squares are $[1]$, namely $[\pm 1]$. Of course, $[-1]$ is the residue class containing $p - 1$ (and -1). No surprises yet.

Now, as we have pointed out, to every residue class $[a]$ there is a residue class $[b]$ such that $[a][b] = 1$; and what we have proved above tells us that $[a] \neq [b]$ unless $[a] = [1]$ or $[p - 1]$. It follows that if we consider the totality of nonzero residue classes mod p, namely, $[1], [2], [3], \dots, [p - 1]$, and (temporarily!) throw away $[1]$ and $[p - 1]$, then the rest fall into pairs of distinct residue classes $([a], [b])$ such that their product is $[1]$. See (9) for the verification of these facts in the simple case $p = 5$.

It now follows (this is the "blinding light") that if we multiply together *all* the nonzero residue classes

$$[1], [2], [3], \dots, [p - 1],$$

then their product is $[1][1][p - 1]$, that is, $[-1]$. In other words,

$$[(p - 1)!] = [-1],$$

or, as we prefer to express it:

Theorem 6 (Known as Wilson's Theorem[5]) *If p is a prime,* $(p - 1)! \equiv -1 \bmod p$.

[5]Oystein Ore (in his book *Number Theory and Its History*, McGraw-Hill, 1948) gives the following historical account of this theorem:

> In the *Meditationes Algebraicae* by Edward Waring, published in Cambridge in 1770, one finds, as we have already mentioned, several announcements on the theory of numbers. One of them is the following. For any prime p the quotient
>
> $$\frac{1 \cdot 2 \cdot \cdots \cdot (p - 1) + 1}{p}$$
>
> is an integer.
>
> This result Waring ascribes to one of his pupils John Wilson (1741–1793). Wilson was a senior wrangler at Cambridge and left the field of mathematics quite early to study law. Later he became a judge and was knighted. Waring gives no proof of Wilson's theorem until the third edition of his *Meditationes*, which appeared in 1782. Wilson probably arrived at the result through numerical computations. Among the posthumous papers of Leibniz there were later found similar calculations on the remainder of $n!$, and he seems to have made the same conjecture. The first proof of the theorem of Wilson was given by J.L. Lagrange in a treatise that appeared in 1770.

It is usual nowadays to state the theorem as a congruence.

Let's just check this for $p = 7$. Then $6! = 6 \times 5 \times 4 \times 3 \times 2 \times 1 = 720$ and $720 \equiv -1 \bmod 7$ since 7 is, clearly, a factor of 721.

Of course, Theorems 5 and 6 are true (but trivial!) if $p = 2$.

• • • **BREAK**

Try this for some other primes p. Does it work for non-prime moduli? Does (17) work for non-prime moduli?

Now some final thoughts on arithmetic mod p. We gave a proof above that there are precisely *two* residue classes of integers n satisfying $n^2 \equiv 1 \bmod p$, namely, $n \equiv \pm 1 \bmod p$. What is the situation if we consider a fixed but arbitrary integer a, with $p \nmid a$, and seek to *solve* the congruence $n^2 \equiv a \bmod p$? A slightly subtler argument shows us that there are always either *two* solutions or no solutions. For suppose we square all the nonzero residues. Then it is easy to see that $b^2 \equiv c^2 \bmod p$ if and only if $b \equiv \pm c \bmod p$ (Theorem 4 again). Thus, the $p-1$ nonzero residues $[1]$, $[2]$, $[3], \ldots, [p-1]$ may be arranged in pairs $[b]$, $[c]$ such that $[b]^2 = [c]^2$. This means that there are $\frac{p-1}{2}$ residues which are squares (and they are squares of exactly two residues), and $\frac{p-1}{2}$ residues which are not. Let us look at the example $p = 11$. Then, squaring the residues $[1], [2], \ldots, [10] \bmod 11$, we get

$$[1], [4], [9], [5], [3], [3], [5], [9], [4], [1],$$

just as the theory predicted. Thus

$$[1], [3], [4], [5], [9] \text{ are squares, while } [2], [6], [7], [8], [10] \text{ are not.}$$

How can one tell, without squaring all the nonzero residues, whether a given residue is a square or not? There is a beautiful theorem which answers this question, namely, Theorem 7 below. Before stating it, we introduce the term ***quadratic residue*** for a residue (such as $[1]$, $[3]$, $[4]$, $[5]$, $[9] \bmod 11$, above) which *is* the square of another. Remember also that p is always an *odd* prime in this section. (Do you notice where we have already used this fact? *Hint*: Look at our proof of Wilson's Theorem.)

Theorem 7 *Suppose $p \mid a$. Then a is a quadratic residue if and only if*

$$a^{\frac{p-1}{2}} \equiv 1 \bmod p.$$

(If a is not a quadratic residue then $a^{\frac{p-1}{2}} \equiv -1 \bmod p$.)

One part of this is easy to prove. If a is a quadratic residue mod p, say $a \equiv b^2 \bmod p$, then

$$a^{\frac{p-1}{2}} \equiv (b^2)^{\frac{p-1}{2}} = b^{p-1} \equiv 1 \bmod p,$$

the last congruence being Fermat's Theorem. Now, certainly, $a^{\frac{p-1}{2}} \equiv \pm 1 \bmod p$, since $\left(a^{\frac{p-1}{2}}\right)^2 = a^{p-1} \equiv 1 \bmod p$. But how do we know that we must have a quadratic residue a if $a^{\frac{p-1}{2}} \equiv 1 \bmod p$?

★ Let us explain the answer without going into every detail. Congruences mod p behave, in many ways, just like ordinary equations. Now we know that an equation of the form

$$c_0x^k + c_1x^{k-1} + \cdots + c_{k-1}x + c_k = 0, \quad \text{of degree } k,$$

cannot have more than k roots. In the same way, the congruence

$$c_0x^k + c_1x^{k-1} + \cdots + c_{k-1}x + c_k \equiv 0 \bmod p$$

cannot have more than k solutions. In particular, the congruence

$$x^{\frac{p-1}{2}} - 1 \equiv 0 \bmod p$$

cannot have more than $\frac{p-1}{2}$ solutions. But we know by our previous arguments that all quadratic residues satisfy this congruence, and there are $\frac{p-1}{2}$ quadratic residues. Hence the quadratic residues provide the full complement of solutions of the congruence

$$x^{\frac{p-1}{2}} - 1 \equiv 0 \bmod p,$$

and so none of the other residues, that is, the nonquadratic residues, can satisfy this congruence; thus, by (17), they must satisfy the congruence $x^{\frac{p-1}{2}} + 1 \equiv 0 \bmod p$. ★

If we try this with $p = 11$, then $\frac{p-1}{2} = 5$ and

$1^5 = 1,$ $\qquad$ $2^5 = 32 \equiv -1 \bmod 11,$

$$\begin{aligned} 3^5 &= 243 \equiv 1 \bmod 11, & 6^5 &= 2^5 \cdot 3^5 \equiv -1 \bmod 11, \\ 4^5 &= 1024 \equiv 1 \bmod 11, & 7^5 &\equiv 18^5 = 3^5 \cdot 6^5 \equiv -1 \bmod 11, \\ 5^5 &= 3125 \equiv 1 \bmod 11, & 8^5 &= (2^5)^3 \equiv -1 \bmod 11, \\ 9^5 &= (3^5)^2 \equiv 1 \bmod 11, & 10^5 &\equiv (-1)^5 = -1 \bmod 11, \end{aligned}$$

as predicted.

2.4 ARITHMETIC mod NON-PRIMES: THE SAME BUT DIFFERENT

We know things cannot be as nice with arithmetic mod m, if m is not a prime, as they are with arithmetic mod p. For, as we have said, if m is not prime, then the following situation is bound to arise for certain numbers a, b:

$$ab \equiv 0 \bmod m, \quad \text{although } a \not\equiv 0 \bmod m \text{ and } b \not\equiv 0 \bmod m. \tag{18}$$

You can see this because if m is *not* a prime we can express m as $m = ab$, where a, b are *proper* factors of m (remember our example based on $10 = 2 \times 5$). However, we can salvage something. For there is a generalization of Theorem 4 which we can use. Recall that we say m is ***prime to k***, or that m and k are ***coprime***, if the greatest common divisor of m and k is 1, that is, $\gcd(m, k) = 1$. Then we have

Theorem 8 *If $m \mid k\ell$ and m is prime to k, then $m \mid \ell$.*

This is proved just as in the special case when m is prime (i.e., Theorem 4), using the factorization of $k\ell$ as a product of primes.

• • • **BREAK**

Try writing out the proof of Theorem 8 which we have sketched. Test the statement of the theorem with some examples.

From Theorem 8 we may proceed as in the previous section, but confining our attention to *those residues which are prime to m.*

Notice that this terminology makes sense, since

$$a \text{ is prime to } m \text{ if and only if } a + km \text{ is prime to } m$$

(Why?). Thus we suppose that we have written out the remainders

$$r_1\ (= 1), r_2, \ldots, r_{q-1}, r_q\ (= m - 1) \tag{19}$$

prime to m. We multiply by any number a prime to m, to get

$$ar_1, ar_2, \ldots, ar_{q-1}, ar_q. \tag{20}$$

We then argue, as we did before, that the residue classes represented in the list (20) simply form a permutation of those represented in (19). From this we first conclude that the list (20) contains a number congruent to 1 mod m, so that every residue prime to m has an inverse. Second, we observe that, multiplying all the numbers in (19) together and all those in (20) together, we have

$$r_1 r_2 \cdots r_q \equiv ar_1 ar_2 \cdots ar_q \equiv a^q r_1 r_2 \cdots r_q \bmod m. \tag{21}$$

But $r_1 r_2 \cdots r_q$ is prime to m so that, by Theorem 8,

$$a^q \equiv 1 \bmod m. \tag{22}$$

The number q deserves a special name. It depends on m, of course, and is called the ***Euler Φ-function***[6] and is written $q = \Phi(m)$; it is the number of residue classes mod m which are prime to m. Then (22) is Euler's Theorem, so-called since it was discovered by the great Swiss mathematician Leonhard Euler (1707–1783).

Theorem 9 $a^{\Phi(m)} \equiv 1 \bmod m$ *if a is prime to m.*

This, of course, generalizes Fermat's Theorem (Theorem 5); for you should have no difficulty in seeing that

$$\text{if } p \text{ is prime, then } \Phi(p) = p - 1.$$

Notice that we have, in Theorem 9 (or Theorem 5), a situation in which we know that the number $a^{\Phi(m)} - 1$ is divisible by m without having to do any dividing. We show below that $\Phi(5040) = 1152$, so we know that (for example) $13^{1152} - 1$ is divisible by 5040. How long

[6] Φ is the capital Greek letter "phi."

would it take a high-speed computer to show this by straightforward arithmetic?

Theorem 9 has the important consequence.

Corollary 10 *Let a be prime to m. Then there exists a positive integer n, called the order of a mod m, such that*

$$a^s \equiv 1 \bmod m \quad \text{if and only if} \quad n \mid s.$$

Proof Let n be the smallest positive integer such that

$$a^n \equiv 1 \bmod m.$$

Such an n exists by Theorem 9, which tells us that there is a positive integer, namely, $\Phi(m)$, with

$$a^{\Phi(m)} \equiv 1 \bmod m.$$

Now let s be a positive integer. If $s = nk$, then

$$a^s = (a^n)^k \equiv 1 \bmod m.$$

Conversely, suppose $a^s \equiv 1 \bmod m$. We may divide s by n, getting a quotient q and a remainder r, so that

$$s = qn + r, \quad 0 \leq r < n.$$

Then $a^r = a^s(a^n)^{-q} \equiv 1 \bmod m$. By the minimality of n, we must have $r = 0$, so $n \mid s$. □

• • • **BREAK**

Find the orders of 3 mod 10; 3 mod 7; 6 mod 11.

How easy is it to calculate $\Phi(m)$? It is not too difficult if we use some clever counting. First, if m is a power of a prime, say $m = p^n$, then

$$\Phi(p^n) = p^{n-1}(p-1). \tag{23}$$

To see this, consider the residues that are *not* prime to p^n; these are represented by those numbers up to p^n which *are* divisible by p. Which are these representative numbers? They are just $p, 2p, 3p, \ldots, p^n$, and there are obviously p^{n-1} of them. Thus, there are $p^n - p^{n-1}$ residues which *are* prime to p, as (23) asserts.

★ Now suppose $m = k\ell$, where k, ℓ are coprime. We write the numbers from 1 to $k\ell$ in a rectangular array as follows:

$$\begin{array}{llll} 1 & 2 & \cdots & \ell \\ \ell+1 & \ell+2 & \cdots & 2\ell \\ 2\ell+1 & 2\ell+2 & \cdots & 3\ell \\ \vdots & \vdots & \ddots & \vdots \\ (k-1)\ell+1 & (k-2)\ell+2 & \cdots & k\ell \end{array} \qquad (24)$$

(Try this with $k = 9$, $\ell = 8$, say.) We now make a series of claims or assertions.

Claim 1 If the entry at the head of a column is (is not) prime to ℓ, then all the entries in that column are (are not) prime to ℓ. This is just a restatement of (18). Thus, if we wish to strike out numbers in (24) which are not prime to $k\ell$, we may first strike out whole columns consisting of numbers none of which is prime to ℓ, leaving just $\Phi(\ell)$ columns, which consist entirely of numbers prime to ℓ. (Carry your example $k = 9$, $\ell = 8$ along with our argument.)

Claim 2 The final column of (24) consists of representatives of the residue classes

$$[1]_k, [2]_k, \ldots, [k]_k,$$

in some order. This is our old argument (based on Theorem 7 and used in the proof of Euler's Theorem); it is here that we make vital use of the fact that k, ℓ are coprime.

Claim 3 *Every* column of (24) consists of representatives of the residue classes

$$[1]_k, [2]_k, \ldots, [k]_k,$$

in some order. For how do we pass to the column immediately to the left of the last column? We just subtract 1. So we subtract 1 from the residue classes $[1]_k$, $[2]_k, \ldots, [k]_k$ (in some order). This, of course, is just a further (cyclic) permutation of these classes. So we continue, stepping to the left and subtracting 1, across the whole array (24).

Claim 4 Thus, in every column of (24), there are $\Phi(k)$ residues prime to k.

Claim 5 From Claims 1 and 4 we conclude that there are $\Phi(k)\Phi(\ell)$ entries in (24) which are prime to both k and ℓ.

Claim 6 A number is prime to $k\ell$ if and only if it is prime to k and to ℓ. (Prove this!)

From Claims 5 and 6 we obtain the key result

$$\Phi(k\ell) = \Phi(k)\Phi(\ell) \quad \text{if } k, \ell \text{ are mutually prime.} \tag{25}$$

Now you should have no difficulty in deducing from (25) that if

$$k_1, k_2, \ldots, k_n$$

is any (finite) collection of mutually prime integers, then

$$\Phi(k_1 k_2 \cdots k_n) = \Phi(k_1)\Phi(k_2) \cdots \Phi(k_n). \tag{26}$$

★

• • • **BREAK**

Exemplify the whole argument for $k = 9$, $\ell = 8$; and any other pairs of values k, ℓ you care to choose. Illustrate (25) with examples. Illustrate (26) for the case $n = 3$.

It is easy, using (23) and (26), to calculate $\Phi(m)$ for any m. Let us give an example. What is Φ (one million)? Now you surely agree that it would be very tedious to enumerate all the numbers up to 1,000,000 which are prime to 1,000,000, and then count them. All we need to do, in fact, is this:

$$1{,}000{,}000 = 10^6 = 2^6 \cdot 5^6.$$

Thus, by (25),

$$\begin{aligned}\Phi(1{,}000{,}000) = \Phi(2^6)\Phi(5^6) &= 2^5 \cdot 1 \cdot 5^5 \cdot 4, \quad \text{by (23)}\\ &= 4 \cdot 10^5\\ &= 400{,}000.\end{aligned}$$

If you think just of the units digits, you may see why exactly $\frac{4}{10}$ of the numbers from 1 to 1,000,000 are prime to 1,000,000, and feel you don't need the complicated argument we've given in this case. But now try $\Phi(5040)$. We know of no easier method than

$$5040 = 2^4 \cdot 3^2 \cdot 5 \cdot 7, \quad \text{so} \quad \Phi(5040) = 2^3 \cdot 3 \cdot 2 \cdot 4 \cdot 6 = 1152.$$

• • • BREAK

Try calculating Φ for yourself for some nice, juicy numbers! What is $\Phi(600)$, $\Phi(1728)$, $\Phi(90)$? Here's a place where man (and woman!) beats the machine!

Show by an example that the relation $\Phi(k\ell) = \Phi(k)\Phi(\ell)$ fails if k, ℓ are *not* coprime.

You may think, at this point, that, after all, there isn't so much difference between mod p arithmetic (p a prime) and mod m arithmetic (m not a prime). Let's just give one example to show that there really is. We saw in formula (17) of Section 3 that there are exactly two residues n mod p such that $n^2 \equiv 1 \bmod p$, namely, $n \equiv \pm 1 \bmod p$. But there is no such result for general residues mod m. Thus, for example,

$$1^2 \equiv 5^2 \equiv 7^2 \equiv 11^2 \equiv 1 \bmod 12,$$

so that there are *four* residues n mod 12 such that $n^2 \equiv 1 \bmod 12$.

Why do you think this difference appears between arithmetic mod p and arithmetic mod m?

2.5 PRIMES, CODES, AND SECURITY

One of the simplest questions and, as we have seen, one of the most important questions that can be asked about an integer is: Is it prime? For most integers the surprising answer is, we don't know—at least we don't know how to determine whether or not an integer is prime in reasonable time. Consequently, the question, "What are the factors of a number N?" is even more unreasonable. We will see that the reluctance of numbers to be factorized can, surprisingly, be quite useful. We will also see that the reason we want to find primes is so that we can then multiply a couple of them together to get a composite number.

So why is it so difficult to determine whether or not a given number N is prime? All you have to do is to see if 2 divides N (if it does, and $N \neq 2$, clearly N isn't a prime), then if 3 divides N, then if 5 divides N, and so on, until you've checked all the primes up to the largest prime number p such that $p \leq \sqrt{N}$.

• • • **BREAK**

Of course, if $p = \sqrt{N}$ it means N is the square of the prime p, and hence composite. Why don't you need to check any primes bigger than $\sqrt{N}$?

Thus we see that, in theory, there is no difficulty—but, in practice, the difficulty is *time*. We know how to decide whether or not a number is prime, but the test that we have just mentioned can involve many steps. This shows that there is sometimes a difference in mathematics between knowing something is possible and actually making it happen.

The challenge, then, is to find some practical method for determining the primality of a number. One way of attempting it might be to use Fermat's Little Theorem (Theorem 5). In the last section we discovered that if p was a prime number and $p \nmid a$, then $a^{p-1} \equiv 1 \bmod p$. Is the converse true? We might ask less. If $a^{N-1} \equiv 1 \bmod N$ for some $a \not\equiv 1 \bmod N$, does this mean that N is a prime?

You might like to test this using, say, $a = 2$. If you do, we suggest that you use a computer, although when, in 1819, Sarrus discovered that $2^{340} \equiv 1 \bmod 341$, he had to do it by hand.

In fact, the situation is very complicated. There are composite numbers N for which $a^{N-1} \equiv 1 \bmod N$ for certain values of a satisfying $1 < a < N$. Because the N with this latter property are behaving like primes, we call them ***pseudoprimes in base a***. You shouldn't have trouble finding a pseudoprime in base 7.

• • • **BREAK**

Find two numbers N which are *not* pseudoprime in base 7.

★ Let's push pseudoprimes a step further. Suppose $N - 1 = 2^s d$, with d odd and $s \geq 1$. (We can easily get the $2^s d$ factorization of $(N - 1)$; simply keep dividing by 2 until the result is no longer even.) Take $a < N$, such that a and N are relatively prime. Then we call N a ***strong pseudoprime*** in base a if $a^d \equiv 1 \bmod N$ or $a^{2^r d} \equiv -1 \bmod N$ for some r with $0 \leq r < s$. You should check that all primes p are strong pseudoprimes for every base a such that $1 < a < p$.

And now let's look at a test for primality that's a bit "iffy." It won't be up to our usual high mathematical standards of exactness. Start with N, the number we want to test for primality and choose at random k (bigger than one) small numbers a $(1 < a < N)$ and such that a and N have no common factors. (The Euclidean Algorithm gives an efficient way of checking this—no time difficulties there.) Now test to see whether N is a strong pseudoprime in base a.

Two things can happen. It is possible that N may not be a strong pseudoprime for some base a, in which case N is composite. On the other hand, if N passes the test for each one of the k values of a, then the probability that N is a prime is greater than $1 - 4^{-k}$. Even for relatively small a this gives a high probability that N is prime. Admittedly it does not give an ironclad guarantee that N is prime but you might be happy to balance certainty against efficiency.

Just in case you are worried by this, be assured that there are efficient tests of primality that come with no element of doubt. Unfortunately, they require a knowledge of some deep results in algebraic number theory so we'll skip them. ★

Let's suppose we've found ourselves a couple of nice big prime numbers p, q, say, with 200 digits each. Now form the composite number $m = pq$ and let's go into the coding business. To do this, choose some e which is relatively prime to $(p - 1)$ and $(q - 1)$, we'll see why in a minute. If someone wants to send us a message, they first convert it into a number. This can be done in many ways. They could, for example, let $A = 01$, $B = 02$, and so on, so long as they told us what they planned in advance. Running all the numbers for all letters of the message together gives them a number M. Now we want M to satisfy $M \leq m - 1$, but we also want it to be prime to m. We can be sure of both these requirements if $M < \min(p, q)$, the smaller of p and q. If M doesn't already satisfy $M < \min(p, q)$, they would break it into separate parts $M_1, M_2, \ldots$, so that each M_i would satisfy $M_i < \min(p, q)$. Each M_i would then be handled just like M, in the following way. They calculate the remainder E of $M^e \bmod m$. This new number E, the encoded message, they then send off to us by carrier pigeon—or e-mail.

While we've been waiting for the message we haven't been idle. We've calculated d such that $1 \leq d \leq (p-1)(q-1) = \Phi(m)$ and $ed \equiv 1 \bmod \Phi(m)$. Again the Euclidean Algorithm comes in handy here. Then we take the message E we've just received and calculate the remainder F of $E^d \bmod m$. Somewhat surprisingly, $F = M$.

To see why, let's go through the whole arithmetic process from the end to the beginning. Now

$$
\begin{aligned}
F &\equiv E^d \bmod m \\
&\equiv (M^e)^d \bmod m.
\end{aligned}
$$

But $ed \equiv 1 \bmod \Phi(m)$. Hence $ed = 1 + k\Phi(m)$. So

$$
\begin{aligned}
F &\equiv M^{1+k\Phi(m)} \bmod m \\
&\equiv M(M^{\Phi(m)})^k \bmod m.
\end{aligned}
$$

But, by Euler's Theorem (Theorem 9), $M^{\Phi(m)} \equiv 1 \bmod m$, since M is prime to m. Hence

$$F \equiv M \bmod m.$$

Since F, M both lie between 1 and $m - 1$, we conclude that $F = M$. And we've decoded the original message!

So what? If we can do it, so can anyone else, surely! After all, the person who was sending us the message knew m and e. Surely they, and any other Tomasina, Dick, or Harriet who knew the encoding procedure, together with m and e, could find d and do the decoding.

Ah, but we didn't spend all that time finding 200 digit primes for nothing. For example, suppose we have an adversarial situation in which we desperately wish to keep secrets from an enemy. Then, to find d, the enemy would have to know $\Phi(m) = (p-1)(q-1)$ and this requires them to find p and q starting with m. And factorization, especially of 400 digit numbers, can take quite some time—even with today's speedy computers. The best factorization algorithm, on the fastest computer, or even on linked computers, would take a long long time. Too long for the message to be of any interest to the enemy by the time they'd done the decoding.

This method of encryption is generally called a ***public key crypto-system*** because the key (m and e) can be made public without endangering the security of the system. You can put your key in the classified pages of the paper for everyone to see. Despite this, only you can decode messages sent to you this way. Because an effective implementation of this was produced by Rivest, Shamir, and Adleman in 1978, this public key system is usually known as the RSA system. Is this the only public key crypto-system?

Those of you interested in reading more about primes should consult Paulo Ribenboim's well-written account, "Prime Number Records," in *The College Mathematics Journal*, Vol. 25, No. 4 (1994), pp. 280–290. In this article Ribenboim discusses how many primes exist, how one can produce them, how one can recognize them, and how they are distributed among the natural numbers. The article also records the "biggest sizes reached so far—the prime number records."

2.6 CASTING OUT 9'S AND 11'S: TRICKS OF THE TRADE

In Section 2 we showed that it was particularly easy to compute remainders mod 9 and mod 11. We will now see how to take advantage of this to check a calculation involving the operations of addition, subtraction, and multiplication.

Consider, for example, the assertion

$$46 \times 28 - 65 \times 17 = 183. \tag{27}$$

Let us ***cast out 9***'s from the two sides of (27), that is, calculate mod 9. On the left we get $1 \times 1 - 2 \times 8$, which is congruent mod 9 to $1 - 7$ and hence (since $-6 + 9 = 3$) to 3. On the right we also get 3. So we say that the result (27) satisfies our "check" (and this simply means that both sides, when divided by 9, leave a remainder of 3). If we had believed that $46 \times 28 - 65 \times 17 = 173$, this check would have proved us wrong, since the right side would have produced 2, while the left produced 3.

We must emphasize that the check by casting out 9's can never *prove* that the result of a calculation is correct—but it can very often detect error. Some errors would, obviously, escape this check. The commonest error of this type is that of (accidentally) interchanging digits. Thus, for example, if we had misread our answer as 138, the check would not have revealed this, because the remainders mod 9 of 183 and 138 are, of course, the same (since we obtain them by adding the digits). Similarly, if we had calculated the left-hand side using 56 instead of 65 we could not have detected this.

A very important principle emerges from this technique of checking by casting out 9's.

You may know an answer is wrong without knowing the right answer!

This is an important principle in life itself, of which many people (e.g., parents, priests, and politicians) seem to be unaware. In fact, it is customary to deny to people who do not claim to know the right answer to a problem the right to make any comment at all on the problem, and especially on the correctness of an expert's proposed solution. (Think of examples from your own experience!)

We can make doubly sure of our answer in (27) by casting out 11's. Going back to (27) and casting out 11's (i.e., calculating mod 11), as explained in Section 2, we get on the left

$$(2 \times 6) - ((-1) \times 6) \equiv 1 + 6 = 7,$$

while on the right, we get

$$1 - 8 + 3 = -4 \equiv 7 \bmod 11,$$

and so the result (27) satisfies this second, independent check. It is now highly unlikely that (27) is wrong if we did a calculation (by hand or machine) to get the result. If, on the other hand, we had merely guessed the answer, we might still be wrong. Notice, in particular, that this check would have detected the error of writing 138 instead of 183 on the right of (27), but it would have failed to detect the error of writing 381 instead of 183.

You may wonder why we speak of "casting out 9's" rather than just saying that we calculate mod 9. The reason is that often the most efficient technique to calculate mod 9 is, literally, to cast out 9's. Suppose, for example, that we wish to check the calculation

$$67 + 24 + 86 = 177.$$

Calculating mod 9, we get on the left $4 + 6 + 5$. We now notice that $4 + 5 = 9$, so we ***cast out*** the pair 4, 5 and are left with 6. Here, of course, we also get 6 as the residue of 177 mod 9, so we suspect our calculation is correct.

We could be even cleverer. We could cast out the pair 7, 2 from the *original* digits on the left, so we would be left with $6 + 4 + 8 + 6$; we could then cast out $6 + 4 + 8$ since this is 18 (we would be casting out a pair of 9's!), and we would be left with 6.

Of course, there is no fixed rule as to how to use this beautiful technique; you should simply try to make the process as short and easy as possible.

• • • BREAK

1. See if you can figure out the following diagram:

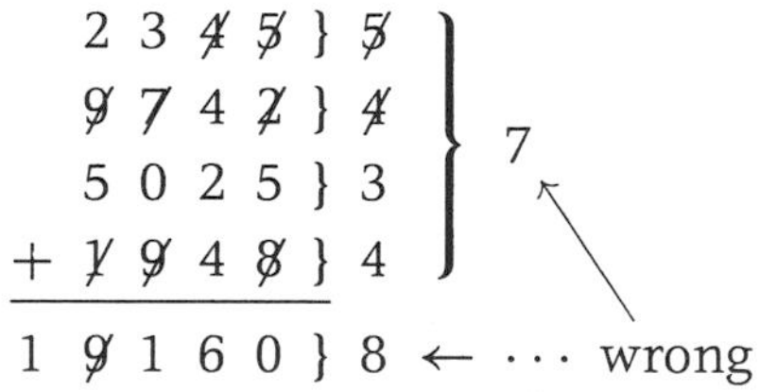

$$\begin{array}{rrrrrlll}
 & 2 & 3 & \not{4} & \not{5} & \} \; \not{5} & \left.\begin{array}{l} \\ \\ \\ \\ \end{array}\right\} & 7 \\
 & \not{9} & \not{7} & 4 & \not{2} & \} \; \not{4} & & \nwarrow \\
 & 5 & 0 & 2 & 5 & \} \; 3 & & \\
+ & \not{1} & \not{9} & 4 & \not{8} & \} \; 4 & & \\
\hline
1 & \not{9} & 1 & 6 & 0 & \} \; 8 & \leftarrow \cdots & \text{wrong}
\end{array}$$

2. Can you see how one may, in a similar way, literally "cast out 11's?"

Can we check division problems by casting out 9's and 11's? Let us first discuss casting out 9's. If the result of a division problem is announced in the form of (quotient, remainder), then the answer is "Yes." To check whether the result of dividing a by b is to produce a quotient q and a remainder r is just to check whether $a = bq + r$ and we know we can check this. Indeed, this shows that we can check any calculation involving fractions the same way. For example, the statement

$$\frac{a}{b} + \frac{c}{d} = \frac{e}{f} \quad \text{is equivalent to the statement} \quad f(ad + bc) = bde.$$

However, as you will see, such a check may be rather a waste of time. If, for instance, our number b above is divisible by 9 (say, $b = 9$), then all we are checking is whether fad is divisible by 9! So it is best to confine attention to *fractions whose denominators are prime to* 3, in using the technique of casting out 9's to check a calculation. But then we have, in fact, a far quicker method than that suggested by the argument above. For we saw, in Section 4, that, for any m, every residue mod m prime to m has an inverse. In fact, with $m = 9, 11$, the tables are

Residue	1	2	4	5	7	8
Inverse	1	5	7	2	4	8

$m = 9$

Residue	1	2	3	4	5	6	7	8	9	10
Inverse	1	6	4	3	9	2	8	7	5	10

$m = 11$

Thus, let us suppose we want to check the calculation

$$\frac{2}{31} + \frac{7}{26} = \frac{269}{806}. \tag{28}$$

Casting out 9's, that is, calculating mod 9, we find $31 \equiv 4$, $26 \equiv 8$. Now $\frac{2}{4} = \frac{1}{2}$ and, from the table above,

$$\frac{1}{2} \equiv 5, \quad \text{while} \quad \frac{1}{8} \equiv 8.$$

Thus

$$\frac{2}{31} + \frac{7}{26} \equiv \frac{1}{2} + \frac{7}{8} \equiv 5 + 7 \times 8 = 61 \equiv 7.$$

Turning to the right-hand side of (28), $806 \equiv 14 \equiv 5$ and $\frac{1}{5} \equiv 2$, so

$$\frac{269}{806} \equiv \frac{8}{5} \equiv 8 \times 2 = 16 \equiv 7.$$

The calculation passes the mod 9 test!

Obviously, the same arguments apply to casting out 11's, so let us now apply the mod 11 test. Then

$$\frac{2}{31} \equiv \frac{2}{-3+1} \equiv -1,$$

and

$$\frac{7}{26} \equiv \frac{7}{-2+6} = \frac{7}{4} \equiv 7 \times 3 = 21 \equiv -2+1 = -1,$$

so

$$\frac{2}{31} + \frac{7}{26} \equiv -2 \bmod 11.$$

On the other hand,

$$\frac{269}{806} \equiv \frac{2-6+9}{8+6} = \frac{5}{14} \equiv \frac{5}{-1+4} = \frac{5}{3}$$
$$\equiv 5 \times 4 = 20 \equiv -2 \bmod 11.$$

The calculation also passes the mod 11 test.

So our restriction on applying the mod 9 test to a calculation involving fractions is that none of the denominators have a factor of 3; and we may apply the mod 11 test if no denominator has a factor of 11. (This makes the mod 11 test even more serviceable for calculations involving fractions than the mod 9 test!)

• • • FINAL BREAK

Here are a few problems for you to try out your new skills on.

1. Find the residue mod 9 of the following numbers without using a hand calculator:

   (a) $(2873 + 5915)^2$;

   (b) $(3028 \times 473) - 4629$;

   (c) $144864 \times 3475 \times 84616 \times 2378429$;

   (d) 8^{92};

   (e) $7^{11} + 2^{11}$.

2. Three of the following statements are false. Identify the false statements without using a hand calculator:

(a) $7282 \times 416 = 2913832$;

(b) $4083 + (961 \times 6137) = 25184 \times 290$;

(c) $6184 + (968 \times 39) = 43936$;

(d) $512 \times 8172 \times 903 = 4001216022$.

3. Show that the following statements are false by casting out 9's. Give your arguments:

(a) $\frac{2}{7} + \frac{93}{104} = \frac{12}{13}$;

(b) $\frac{5}{19} \times \frac{8}{23} \div \frac{11}{15} = \frac{29}{209}$.

4. Explain why we may test whether a number is divisible by 99 by casting out 9's and 11's.

*5. Show that $n \equiv 12 \bmod 99$ if and only if $n \equiv 1 \bmod 11$ and $n \equiv 3 \bmod 9$. More generally, show that the residue class of $n \bmod 99$ is determined by its residue classes mod 9 and mod 11; and that these last two residue classes can take *any* values.

6. (For the enthusiastic reader.) Take a look at the article, "Casting Out Nines Revisited," by Peter Hilton and Jean Pedersen, published in *Mathematics Magazine*, Vol. 54, No. 4, September 1981, and set yourself some more problems casting out 9's and 11's.

ANSWERS FOR FINAL BREAK

1. (a) 7.
   (b) 8.
   (c) 0 (Note that it is enough to observe that the first factor is divisible by 9.)
   (d) 1 (Note that $8 \equiv -1 \bmod 9$ and -1 raised to any *even* power is 1.)
   (e) 0 (Note that $7 \equiv -2 \bmod 9$ and $(-2)^{11} = -2^{11}$.)

2. (a) and (b) may be seen to be false by casting out 9's, (c) and (d) may not. However, (d) may be seen to be false by casting out 11's. An even quicker proof of its falsity is to observe that $512 \times 8172 \times 903$ is obviously divisible by 4, whereas 4001216022 fails the test for divisibility by 4 (namely, that the number formed by the last two digits must be divisible by 4).

3. (a) $\frac{2}{7} \equiv 8 \bmod 9$, $\frac{93}{104} \equiv 6 \bmod 9$, $8 + 6 \equiv 5 \bmod 9$; but $\frac{12}{13} \equiv \frac{12}{4} \bmod 9$ and $\frac{12}{4} = 3$.
   (b) $\frac{5}{19} \equiv 5 \bmod 9$, $\frac{8}{23} \equiv 7 \bmod 9$, $\frac{15}{11} \equiv 3 \bmod 9$, $5 \times 7 \times 3 \equiv 6 \bmod 9$; but $\frac{29}{209} \equiv \frac{29}{29} = 1 \bmod 9$.

4. A number is divisible by 99 if and only if it is divisible by 9 and 11. This is because 9 and 11 are coprime.
5. Obviously if $n = 99k + 12$, then $n \equiv 1 \bmod 11$ and $n \equiv 3 \bmod 9$. In the other direction, suppose $n \equiv 1 \bmod 11$ and $n \equiv 3 \bmod 9$. Then $n = 11k + 1$, so

$$
\begin{aligned}
11k + 1 &\equiv 3 \bmod 9, \\
11k &\equiv 2 \bmod 9, \\
2k &\equiv 2 \bmod 9 \quad \text{(cast out 9's!)}, \\
k &\equiv 1 \bmod 9,\ k = 9\ell + 1,\ n = 99\ell + 12.
\end{aligned}
$$

   In general, if $n \equiv a \bmod 11$, $n \equiv b \bmod 9$, then $n = 11k + a$

$$11k + a \equiv b \bmod 9,$$

$$11k \equiv b - a \bmod 9,$$
$$2k \equiv b - a \bmod 9,$$
$$k \equiv 5(b - a) \bmod 9 \quad \text{(Why?)},$$
$$k = 9\ell + 5(b - a), \;\; n = 99\ell + 55(b - a) + a,$$
$$n \equiv 55(b - a) + a \bmod 99.$$

Here a and b are arbitrary.

3 CHAPTER Fibonacci and Lucas Numbers

3.1 A NUMBER TRICK

Consider the following number trick—try it out on your friends. You ask them to write down the numbers from 0 to 9. Against 0 and 1 they write any two numbers (we suggest two fairly small positive integers just to avoid tedious arithmetic, but all participants should write the *same* pair of numbers). Then against 2 they write the sum of the entries against 0 and 1; against 3 they write the sum of the entries against 1 and 2; and so on. Once they have completed the process, producing entries against each number from 0 to 9, you suggest that, as a check, they call out the entry against the number 6. Thus their table (which, of course, you do not see) might look like the table in the margin. You now ask them to add all the entries in the second column, while you write 341 quickly on a slip of paper.

0	3	
1	2	
2	5	
3	7	
4	12	
5	19	
6	31	← (This is the one used as a check.)
7	50	
8	81	
9	131	

How did you know? Well, let's look at the procedure from an algebraic viewpoint. If you had started with any numbers a and b, your table would have been:

TABLE 1

0	a
1	b
2	$a + b$
3	$a + 2b$
4	$2a + 3b$
5	$3a + 5b$
6	$\boxed{5a + 8b}$
7	$8a + 13b$
8	$13a + 21b$
9	$21a + 34b$

And if we supplement Table 1 with the running sums we get:

TABLE 2

N	u_N	$\sum_N = \sum_{n=0}^N u_n$
0	a	a
1	b	$a + b$
2	$a + b$	$2a + 2b$
3	$a + 2b$	$3a + 4b$
4	$2a + 3b$	$5a + 7b$
5	$3a + 5b$	$8a + 12b$
6	$\boxed{5a + 8b}$	$13a + 20b$
7	$8a + 13b$	$21a + 33b$
8	$13a + 21b$	$34a + 54b$
9	$21a + 34b$	$\boxed{55a + 88b}$

Now we see why the trick works—the running sum $\sum_9$ is actually 11 times u_6, whatever numbers a, b we choose (but the values of a and b themselves are not known when you give the answer)!

But is this really a satisfactory explanation? Why does this strange phenomenon just relate $\sum_9$ and u_6; and why is the multiplying factor 11? If we look at Table 2 more closely we see much more in it. The coefficients of a in u_N are

$$1\ 0\ 1\ 1\ 2\ 3\ 5\ 8\ 13\ 21; \tag{1}$$

and the coefficients of b are almost the same, except that they start with the 0; thus they are

$$0\ 1\ 1\ 2\ 3\ 5\ 8\ 13\ 21\ 34. \tag{2}$$

Moreover, the coefficients of a in $\sum_N$ are again essentially the same sequence, except starting now with the 1 in the second place in (2), namely

$$1\ 1\ 2\ 3\ 5\ 8\ 13\ 21\ 34\ 55. \tag{3}$$

Finally, the coefficients of b in $\sum_N$ are much like the numbers in (3), except that we start in the third place in (2) *and subtract* 1, thus,

$$0\ 1\ 2\ 4\ 7\ 12\ 20\ 33\ 54\ 88. \tag{4}$$

3.2 THE EXPLANATION BEGINS

How can we explain all this? Well, first, we must obviously study carefully this sequence that keeps coming up in some form. We take it in the form (2) and we define it by the rules

$$F_0 = 0, \quad F_1 = 1, \tag{5}$$

$$F_{n+2} = F_{n+1} + F_n, \quad n \geq 0. \tag{6}$$

We call (5) the ***initial conditions*** and (6) the ***recurrence relation***. In fact, (5) and (6) together determine a very famous sequence (i.e., (2)) called the ***Fibonacci sequence***,[1] named after Leonardo of Pisa, who lived from 1180 to 1250 and who was called Fibonacci, meaning "the son of Bonacci." The Fibonacci sequence turns out to have all sorts of applications in nature, but we will concentrate here on its remarkable arithmetical properties.

We notice that our sequence $\{u_n,\ n = 0, 1, 2, \ldots\}$ satisfies the same recurrence relations as the Fibonacci sequence, but it has different initial conditions, namely,

$$u_0 = a, \quad u_1 = b, \tag{7}$$

[1]The numbers in the sequence are often called the ***Fibonacci numbers***. For example, we may say that 13 is a Fibonacci number, but 7 is not a Fibonacci number.

$$u_{n+2} = u_{n+1} + u_n, \quad n \geq 0. \tag{8}$$

Among the (infinitely) many sequences satisfying (8) there is another of particular interest besides the Fibonacci sequence, namely, the sequence given by

$$L_0 = 2, \quad L_1 = 1, \tag{9}$$

$$L_{n+2} = L_{n+1} + L_n, \quad n \geq 0. \tag{10}$$

This is the ***Lucas sequence***,[2] named after the French mathematician, Edouard Lucas who worked at the end of the 19th century. Table 3 is a table of the values of F_n and L_n up to $n = 13$. You may wish to extend it and study it. You will surely notice many number patterns involving the Fibonacci and Lucas numbers.

TABLE 3

n	0	1	2	3	4	5	6	7	8	9	10	11	12	13
F_n	0	1	1	2	3	5	8	13	21	34	55	89	144	233
L_n	2	1	3	4	7	11	18	29	47	76	123	199	322	521

Thus, for example, you may conjecture that

$$F_{n-1} + F_{n+1} = L_n, \quad n \geq 1, \tag{11}$$

and

$$F_{2n} = F_n L_n, \quad n \geq 0. \tag{12}$$

However, you may have some difficulty in proving such identities right now—but, don't despair, we are going to show you how to do it.

It may appear that we have strayed from our original problem in looking at the Fibonacci and Lucas numbers. It turns out, in fact, that by doing so, we will be able to provide a more satisfactory, more comprehensive explanation of our number trick, and show that it is just one example of a whole pattern of similar phenomena. And we hope to produce for you some more surprises!

[2]The numbers in this sequence, which begins 2, 1, 3, 4, . . ., are known as the ***Lucas numbers***. You may notice that there are some numbers which are *both* Fibonacci and Lucas. How many such numbers are there? It is true, but not easy to prove, that there are only three such numbers. In fact, we recently discovered that an even stronger statement is true—*no* Fibonacci number greater than 3 is a factor of *any* Lucas number. Who would have thought it!

• • • **BREAK**

Equation (12) could have been conjectured by simply multiplying together F_n and L_n for some successive values of n and looking at the answers. Experiment with other operations (e.g., add together F_n and L_n for some successive values of n) and see what other conjectures you can make. Make a record of these conjectures and try to prove them after you've read the next section.

Obviously, the values of u_n, for every n, are determined by (7) and (8). We call these values the ***solution*** of the recurrence relation (8) subject to the initial conditions (7). Now the recurrence relation (8) is a special case of what are called ***linear recurrence relations of the second order***, and we may just as well discuss the method of solution in the general case. Thus we study the sequence

$$\{u_n,\ n = 0, 1, 2, \ldots\},$$

given by

$$u_0 = a, \quad u_1 = b \quad (\textit{initial conditions}), \tag{13}$$

$$u_{n+1} = pu_{n+1} + qu_n \tag{14}$$

(linear recurrence relation of the second order),

where p, q are any real numbers. Note that (14) is ***linear*** because u_{n+2}, u_{n+1}, u_n only occur to the first power, and ***second order*** because it involves $n + 2, n + 1, n$. We want to find an expression for u_n as a function of n (involving, of course, a and b). We first make the obvious but important remark that, since u_n *is* determined by (13) and (14), ***any function of n satisfying*** (13) and (14) ***must be*** u_n.

Our second remark is also of great importance. If we find any two functions of n, say v_n and w_n satisfying (14), then, for any constants r and s, the function $rv_n + sw_n$ also satisfies (14). This is because the relation (14) is ***linear***; but we do advise you to check this important consequence of linearity for yourselves.

We now look for such functions v_n, w_n. Our experience (perhaps not yet yours!) tells us to try a function x^n as a solution of (14); notice that n is here the variable, not x which is an unknown, to be determined. Then x^n will satisfy (14) if, for all $n \geq 0$,

$$x^{n+2} - px^{n+1} - qx^n = 0.$$

But this will obviously be true if

$$x^2 - px - q = 0. \tag{15}$$

This is our old friend (or enemy?) the quadratic equation. We know that (15) has two distinct solutions (which may not be real, but that won't bother us here!), provided that

$$p^2 + 4q \neq 0. \tag{16}$$

So we assume (16) and we call the solutions of (15) α and β; it is a *very important* matter of technique here that we do *not* look for explicit expressions for α and β in terms of p and q unless we are politely asked to do so by someone who gives us a reason for making the request! What *is* important is that α, β satisfy the key relations (derived from (15))

$$\alpha + \beta = p, \quad \alpha\beta = -q. \tag{17}$$

So if α, β satisfy (17), then α^n and β^n are solutions of the relation (14). These then will serve as our functions v_n and w_n. As already pointed out, we conclude that, for any constants r, s,

$$u_n = r\alpha^n + s\beta^n \tag{18}$$

is a solution of (14).

It only remains to take account of the initial conditions (13). We use these to determine r and s in (18). Thus, substituting $n = 0$ and then $n = 1$ into (18), we have

$$\left.\begin{aligned} r + s &= a \\ r\alpha + s\beta &= b. \end{aligned}\right\} \tag{19}$$

These equations *do* determine r and s, since, by imposing condition (16), we have ensured that $\alpha \neq \beta$. Notice that, once again, we do *not* write down explicit formulae for r and s. We sum up our conclusion in the following theorem.

Theorem 1 *Suppose $p^2 + 4q \neq 0$. Then the solution of the linear recurrence relation $u_{n+2} = pu_{n+1} + qu_n$, subject to the initial conditions $u_0 = a$, $u_1 = b$, is given by $u_n = r\alpha^n + s\beta^n$, where α,β*

are the roots of the quadratic equation $x^2 - px - q = 0$ *and* r, s *are given by* (19).

• • • **BREAK**

Take the recurrence relation $u_{n+2} = 3u_{n+1} - 2u_n$, with initial conditions $u_0 = 0$, $u_1 = 2$. Go through the steps of our argument to find α, β in this case, and then to find r, s. Check that you have, in fact, found the right formula for u_n, by comparing the values you get for u_n, from the recurrence relation and from your formula, up to $n = 4$.

Let's take stock of what we have found out about the Fibonacci and Lucas numbers. First, in both cases we are dealing with the recurrence relation $u_{n+2} = u_{n+1} + u_n$, so that $p = 1$, $q = 1$; certainly $p^2 + 4q \neq 0$. Thus α, β are the roots of the equation $x^2 - x - 1 = 0$ and

$$\alpha + \beta = 1, \quad \alpha\beta = -1. \tag{20}$$

For the Fibonacci numbers we have, from (19), since $a = 0, b = 1$,

$$r + s = 0, \quad r\alpha + s\beta = 1,$$

so that

$$r = \frac{1}{\alpha - \beta}, \quad s = -\frac{1}{\alpha - \beta}, \tag{21}$$

and we conclude from Theorem 1 that

$$\boxed{F_n = \frac{\alpha^n - \beta^n}{\alpha - \beta}}. \tag{22}$$

This is really a very remarkable formula when you come to think of it! We are dealing with positive integers F_n generated in a very simple fashion; yet it turns out that F_n is given by a formula, namely (22), which, though very neat from an algebraic viewpoint, involves numbers α, β whose actual values (as roots of $x^2 - x - 1 = 0$) are $\frac{1 \pm \sqrt{5}}{2}$. You might try checking (22) for $n = 2, 3, 4$, say, so that you're comfortable with the formula, and appreciate how extraordinary it is.

Now for the Lucas numbers. Of course, α, β remain the same as for the Fibonacci numbers, but r and s will now be different from their values in (21) above. For now $a = 2$, $b = 1$, so that we must find the (unique) solution of the simultaneous equations

$$r + s = 2, \quad r\alpha + s\beta = 1. \tag{23}$$

However, we know from (20) that

$$\alpha + \beta = 1,$$

so that it is easy to see that equations (23) are satisfied by $r = 1$, $s = 1$. We conclude that

$$\boxed{L_n = \alpha^n + \beta^n} \tag{24}$$

Let us straightaway remark that (22) and (24) together immediately imply the truth of our formula (12) $F_{2n} = F_n L_n$, which we guessed just by looking at small values of n. At the risk of boring you, we again emphasize how easy it is to see this precisely because we *haven't* replaced α, β by their actual numerical expressions. In fact, we see that if we take the ***general*** linear recurrence relation of the second order (14), instead of the special case (8), and then define the ***generalized*** Fibonacci numbers f_n by the initial conditions $f_0 = 0$, $f_1 = 1$, and the ***generalized*** Lucas numbers by the initial conditions $\ell_0 = 2$, $\ell_1 = p$, then we again get

$$f_n = \frac{\alpha^n - \beta^n}{\alpha - \beta}, \quad \ell_n = \alpha^n + \beta^n, \tag{25}$$

where α, β are the roots of $x^2 - px - q = 0$, and the relation $f_{2n} = f_n \ell_n$ holds just as obviously. A triumph for algebra over mere arithmetic!

• • • **BREAK**

Prove what we have just said, that is, (25) and the relation $f_{2n} = f_n \ell_n$. Check the latter in some special case (not $p = q = 1$).

Before you take your next break we want to make two further points, of which the first is much more important from our point of view in this chapter.

We have discussed sequences $\{u_n\}$ where n takes the values $0, 1, 2, \ldots$. But there is no reason why we should not consider negative values of n as well. We simply have to insist that the recurrence relation (8)—or, more generally,[3] (14)—holds for *all* integers n. Thus, for example, by starting with $F_0 = 0$, $F_1 = 1$ and working *backward* with $F_n = F_{n+2} - F_{n+1}$ we obtain

$$F_{-1} = 1, \quad F_{-2} = -1, \quad F_{-3} = 2, \ldots;$$

and, likewise, starting with $L_0 = 2$, $L_1 = 1$, we obtain

$$L_{-1} = -1, \quad L_{-2} = 3, \quad L_{-3} = -4, \ldots.$$

It is particularly important to note that the formulas (22) and (24) *continue* to hold for negative as well as nonnegative n. However, we can do even better. For, remembering that $\alpha\beta = -1$, we get

$$\left.\begin{aligned} F_{-n} &= \frac{\alpha^{-n} - \beta^{-n}}{\alpha - \beta} = \frac{1}{\alpha^n\beta^n}\frac{\beta^n - \alpha^n}{\alpha - \beta} = (-1)^{n+1}F_n, \\ &\text{and} \\ L_{-n} &= \alpha^{-n} + \beta^{-n} = \frac{1}{\alpha^n\beta^n}(\beta^n + \alpha^n) = (-1)^n L_n. \end{aligned}\right\} \tag{26}$$

Our second point is to satisfy the curiosity of those who ask what happens in the general case if $p^2 + 4q = 0$. In this case, the conclusions of Theorem 1 break down because $\alpha = \beta$ and so the equations (19) don't determine the numbers r, s. What we find is that, in this case, if the function α^n satisfies the recurrence relation $u_{n+2} = pu_{n+1} + qu_n$, then so does the function $n\alpha^n$. For, given that $\alpha^2 = p\alpha + q$, we have

$$(n+2)\alpha^{n+2} - p(n+1)\alpha^{n+1} - qn\alpha^n = n\alpha^n(\alpha^2 - p\alpha - q) + \alpha^{n+1}(2\alpha - p). \tag{27}$$

But the unique solution α of the equation $x^2 - px + \frac{p^2}{4} = 0$ (remember that $p^2 + 4q = 0$) is given by $\alpha = \frac{p}{2}$, so it follows from (27) that

$$(n+2)\alpha^{n+2} - p(n+1)\alpha^{n+1} - qn\alpha^n = 0.$$

[3]In the general case, we need $q \neq 0$ so that u_n becomes a linear function of u_{n+1} and u_{n+2}; but this is reasonable since, if $q = 0$, equation (14) is really a *first-order* recurrence relation.

Hence in this ***singular*** case (to use the correct mathematical term), the solution of (14), subject to the initial conditions (13), is given by

$$u_n = \alpha^n(r + sn), \tag{28}$$

where $\alpha = \frac{p}{2}$ and r, s are determined by

$$r = a, \quad \alpha(r + s) = b. \tag{29}$$

As an example, suppose $a = 0$, $b = 1$, $p = 4$, $q = -4$ in (13) and (14). Then $p^2 + 4q = 0$, $\alpha = \frac{p}{2} = 2$, $r = 0$, $s = \frac{1}{2}$, so $u_n = 2^n \frac{n}{2} = 2^{n-1}n$. You may check from the recurrence relation $u_{n+2} = 4u_{n+1} - 4u_n$ with initial conditions $u_0 = 0$, $u_1 = 1$, that $u_2 = 4$, $u_3 = 12$, $u_4 = 32$, $u_5 = 80, \ldots$.

• • • **BREAK**

Carry out the proposed check.

The ***Binet*** formulae (22) and (24), which we now repeat for your convenience,

$$F_n = \frac{\alpha^n - \beta^n}{\alpha - \beta}, \tag{22}$$

$$L_n = \alpha^n + \beta^n, \tag{24}$$

where

$$\alpha + \beta = 1, \quad \alpha\beta = -1, \tag{20}$$

lead to many interesting results, some of which you may have already observed experimentally. We have explicitly noted that we get a quick proof that $F_{2n} = F_nL_n$. What happens if, instead, we look at F_nL_{n-1}? Well

$$F_nL_{n-1} = \frac{(\alpha^n - \beta^n)(\alpha^{n-1} + \beta^{n-1})}{\alpha - \beta} = \frac{\alpha^{2n-1} - \beta^{2n-1}}{\alpha - \beta} + \alpha^{n-1}\beta^{n-1}$$
$$= F_{2n-1} + (-1)^{n-1},$$

using (20). We may reexpress this as

$$F_nL_{n-1} - F_{2n-1} = (-1)^{n-1}. \tag{30}$$

Similarly,

$$F_{n-1}L_n = \frac{(\alpha^{n-1} - \beta^{n-1})(\alpha^n + \beta^n)}{\alpha - \beta} = \frac{\alpha^{2n-1} - \beta^{2n-1}}{\alpha - \beta} - \alpha^{n-1}\beta^{n-1}$$
$$= F_{2n-1} + (-1)^n,$$

so that

$$F_{n-1}L_n - F_{2n-1} = (-1)^n. \tag{31}$$

You should check (30) and (31) against Table 3.

• • • BREAK

Take time now to try to prove some of your own conjectures. But we should tell you that, although using (22) and (24) is often useful, it is not always necessary.

For example, if you followed our earlier hint you might have conjectured that $F_n + L_n = 2F_{n+1}$. In this case, *because the relation is linear*, all you need to do is check that it works for two consecutive values of n and you can be assured that it will continue to work from then on—and back. (To see why this is true, think of the first place ahead where it might fail and work your way backward—and the first place back and work your way forward!) This principle applies also to relation (11).

On the other hand, if your conjectures involved *nonlinear* relations, you may find (22) and (24) helpful as we did in obtaining (30) and (31) (don't forget that $\alpha + \beta = 1$ and $\alpha\beta = -1$). Perhaps you'd like to try to prove that

$$F_{n+2}L_n = F_{2n+2} + (-1)^n.$$

What do you think F_nL_{n+2} might equal?

3.3 DIVISIBILITY PROPERTIES

Formulae like (12), (30), and (31) are, for obvious reasons, called ***quadratic*** formulae. Many other such quadratic formulae are available, of course. An important result of a different kind comes from the following algebraic considerations.

Although α, β are not themselves integers, it follows from (20)—we will not give you a detailed proof—that:

Any symmetric polynomial in α and β with integer coefficients is, in fact, an integer.

Here a polynomial in two variables x, y, say $p(x, y)$ is said to be ***symmetric*** if $p(x, y) = p(y, x)$, as polynomials. Thus

$$x^2 + y^2, \quad x^3 - 3xy + y^3,$$

are symmetric polynomials. For these two polynomials you may then verify the boxed statement in the following way:

$$\begin{aligned} \alpha^2 + \beta^2 &= (\alpha + \beta)^2 - 2\alpha\beta = 1 + 2 = 3, \\ \alpha^3 - 3\alpha\beta + \beta^3 &= (\alpha + \beta)(\alpha^2 - \alpha\beta + \beta^2) - 3\alpha\beta \\ &= (\alpha + \beta)((\alpha + \beta)^2 - 3\alpha\beta) - 3\alpha\beta \\ &= (1 + 3) + 3 \\ &= 7. \end{aligned}$$

In fact, these two cases give us the key to the argument in the general case—it is a true statement, not easy to prove, that a symmetric polynomial in α and β with integral coefficients may always be expressed as a polynomial in $\alpha + \beta$ and $\alpha\beta$ with integer coefficients, and so is itself an integer.

• • • **BREAK**

Verify that $\alpha^4 + \alpha^2\beta^2 + \beta^4$ is an integer; and express it as a polynomial in $\alpha + \beta$ and $\alpha\beta$.

Now suppose that m exactly divides n; remember that, in Chapter 2, we adopted the notation $m \mid n$ for this. Then

$$\frac{F_n}{F_m} = \frac{\alpha^n - \beta^n}{\alpha^m - \beta^m}.$$

But the polynomial $\alpha^m - \beta^m$ exactly divides the polynomial $\alpha^n - \beta^n$ if $m \mid n$. For if $n = mk$, then

$$\frac{x^n - y^n}{x^m - y^m} = \frac{X^k - Y^k}{X - Y}, \quad \text{where} \quad X = x^m, \ Y = y^m,$$

and $X - Y$ obviously divides $X^k - Y^k$. Thus, if $m \mid n$, then $\frac{\alpha^n-\beta^n}{\alpha^m-\beta^m}$ is a polynomial in α, β, with integer coefficients. Moreover, it is symmetric; for interchanging α and β, we get $\frac{\beta^n-\alpha^n}{\beta^m-\alpha^m}$, which is the same as $\frac{\alpha^n-\beta^n}{\alpha^m-\beta^m}$. Thus, if $m \mid n$, $\frac{F_n}{F_m}$ is an integer, that is, $F_m \mid F_n$. Look at some examples, to feel comfortable about this remarkable result. Let us write it down formally.

Theorem 2 *If $m \mid n$, then $F_m \mid F_n$.*

Actually, even more is true. Suppose $d = \gcd(m, n)$, the greatest common divisor of m and n. Then we have, even more surprisingly,

Theorem 3 *Let $d = \gcd(m,n)$. Then $\gcd(F_m,F_n) = F_d$.*

For a proof, which only involves ideas already discussed, see [1].

Is there a similar result for Lucas numbers? Strictly, the answer is no. Thus $L_2 = 3$, $L_4 = 7$, and $2 \mid 4$, but $3 \nmid 7$. The difference lies in the fact that $m \mid n$ does *not* guarantee that $\alpha^m + \beta^m \mid \alpha^n + \beta^n$.

Consider the argument above, adapted to Lucas numbers. We have

$$\frac{L_n}{L_m} = \frac{\alpha^n + \beta^n}{\alpha^m + \beta^m};$$

this is certainly symmetric—but when is it a polynomial? Well, if $n = mk$, then

$$\frac{x^n + y^n}{x^m + y^m} = \frac{X^k + Y^k}{X + Y}, \quad \text{where} \quad X = x^m, \; Y = y^m,$$

but $X + Y$ does not always divide $X^k + Y^k$. However, *it does if k is odd.* Let us say that ***m divides n oddly*** (written "$m \mid n$ oddly") if $n = mk$ with k odd. Then we conclude (compare Theorem 2)

Theorem 4 *If $m \mid n$ oddly, then $L_m \mid L_n$.*

Does Theorem 3 also have an analogue? Yes, it does but, as you might expect, it is a little more complicated. As a preparation, let us talk a little about the Fibonacci and Lucas numbers mod 2, that is, the residue class mod 2 of the Fibonacci and Lucas numbers. Since the Fibonacci sequence begins 0, 1, and the Lucas sequence begins 2, 1, they *begin in the same way* mod 2 and therefore *stay the same* mod 2, that is,

$$F_n \equiv L_n \bmod 2, \quad \text{for all } n. \tag{32}$$

Here we used a type of argument which will recur—if we are dealing with a *linear* phenomenon, it is sufficient (as we pointed out in the break at the end of Section 2) to verify it for *two successive values* of n. Returning to F_n mod 2 (or L_n mod 2) we see that the sequence is (for $n = 0, 1, 2, \ldots$)

$$0\,1\,1\,0\,1\,1\,\cdots,$$

that is, the triple 0 1 1 simply repeats indefinitely (forward and backward). Thus we have proved

Proposition 5 *F_n is even $\Leftrightarrow$ L_n is even $\Leftrightarrow$ $3 \mid n$.*

In the light of Theorem 4, Proposition 5 (and Theorem 3), the following result (Theorem 6) is not so surprising. However, to state it, we need one more concept. For any integer n we say that its 2-***value*** is k if k is the largest nonnegative integer such that $2^k \mid n$, and we write $(n)_2 = k$. Thus $(n)_2 = 0$ if and only if n is odd, $(100)_2 = 2$, $(240)_2 = 4$, and $(ab)_2 = (a)_2 + (b)_2$, for any positive integers a, b.
Then we have

Theorem 6 *Let $d = \gcd(m,n)$. Then*

$$\gcd(L_m, L_n) = \begin{cases} L_d & \text{if } (m)_2 = (n)_2, \\ 2 & \text{if } (m)_2 \neq (n)_2 \text{ and } 3 \mid d, \\ 1 & \text{if } (m)_2 \neq (n)_2 \text{ and } 3 \nmid d. \end{cases}$$

Again, for the really curious (and fairly ambitious) a proof may be found in [1].

• • • **BREAK**

Check Theorems 3 and 6 by taking some numerical examples.

Now let's get back to this business of *linearity*. Here's a relationship we have already conjectured (see (11)):

$$L_n = F_{n+1} + F_{n-1} = F_n + 2F_{n-1}. \tag{33}$$

Since this is a ***linear*** *relationship* (unlike $F_{2n} = F_n L_n$), we have only to verify it for two successive values of n. Let's take $n = 0, 1$, remembering that $F_{-1} = 1$. Then $L_0 = 2$, $F_1 = 1$, $F_{-1} = 1$, so $L_0 = F_1 + F_{-1}$, and $L_1 = 1$, $F_2 = 1$, $F_0 = 0$, so $L_1 = F_2 + F_0$. We have now *proved* (33)—or (11).

We would still, however, like to regard (33) as a special case of what we get from the general case (14). We recall that we have already characterized the generalized Fibonacci numbers $\{f_n\}$ as the solution of (14) determined by the initial conditions

$$f_0 = 0, \quad f_1 = 1.$$

We now prove

Theorem 7 *If $\{u_n\}$ is given by*

$$u_0 = a, \quad u_1 = b,$$

$$u_{n+2} = pu_{n+1} + qu_n,$$

then

$$u_n = bf_n + qaf_{n-1}.$$

In the special case $p = q = 1$, we have $u_n = bF_n + aF_{n-1}$.

Proof We have only to verify this for two successive values of n. Again we choose $n = 0, 1$. So we must first determine f_{-1}; but, from (14),

$$qf_{-1} = f_1 - pf_0,$$

so, by (34), $f_{-1} = \frac{1}{q}$. Substituting $n = 0$ into the conjectured equality

$$u_n = bf_n + qaf_{n-1} \quad \text{yields} \quad u_0 = qa\left(\frac{1}{q}\right) = a,$$

which is true. Substituting $n = 1$ yields $u_1 = b \cdot 1 = b$, which is also true, so *the equality is proved.* □

Notice that not only does (33) appear as a special case of this theorem, but we have also now explained one of the features of our number trick—the appearance of the Fibonacci numbers as coefficients of a and b in the u_N column of Table 1. Perhaps we should now get on with the job of explaining everything about that trick!

3.4 THE NUMBER TRICK FINALLY EXPLAINED

First, what about the coefficients in the $\sum_N$ column of Table 1? We must prove

Theorem 8 *If the sequence* $\{u_n\}$ *is given by*

$$u_0 = a, \quad u_1 = b,$$

$$u_{n+2} = u_{n+1} + u_n, \quad n \geq 0,$$

and if $\Sigma_N = \Sigma_{n=0}^{N} u_n$, *then*

$$\Sigma_N = aF_{N+1} + b(F_{N+2} - 1) = u_{N+2} - b. \tag{34}$$

Proof Notice that the second equality above was already proved in Theorem 7, so we have only to prove that

$$\Sigma_N = u_{N+2} - b. \tag{35}$$

This statement cries out for a proof by induction! If $N = 0$, (35) asserts that $a = (a + b) - b$, which is obviously true. Then, if (35) is true for a particular N, we have

$$\begin{aligned} \Sigma_{N+1} &= \Sigma_N + u_{N+1} \\ &= u_{N+2} + u_{N+1} - b \\ &= u_{N+3} - b, \end{aligned}$$

which gives us the inductive step. □

Now for the trick itself. Notice that the multiplying factor 11 is, in fact, L_5. We prove

Theorem 9 $\sum_{4k+1} = L_{2k+1}u_{2k+2}$.

(Our number trick was just the case $k = 2$ of this theorem.)

Proof We are going to put together a number of facts already proved. We have

$$\begin{aligned} L_{2k+1}u_{2k+2} &= L_{2k+1}(bF_{2k+2} + aF_{2k+1}), \quad \text{by Theorem 7,} \\ &= bL_{2k+1}F_{2k+2} + aL_{2k+1}F_{2k+1} \\ &= b(F_{4k+3} - 1) + aF_{4k+2}, \quad \text{by (30) and (12),} \\ &= \Sigma_{4k+1}, \quad \text{by Theorem 8.} \end{aligned}$$

□

We hope you agree that we have now provided a *real* explanation of the trick, by showing that it depends on a whole family of relationships, by showing why the Lucas number L_5 (= 11) was involved, and by showing that it could be done with any value of k and not just with $k = 2$. But just to demonstrate that in mathematics explanations are never complete, let's just see whether there is a corresponding trick involving the sums $\sum_N$ where N is *not* of the form $4k + 1$. In fact, we will prove

Theorem 10 $\sum_{4k-1} = F_{2k}(u_{2k} + u_{2k+2})$.

Proof Our method is essentially the same as for Theorem 9. We have, first of all,

$$\begin{aligned} u_{2k} + u_{2k+2} &= b(F_{2k} + F_{2k+2}) + a(F_{2k-1} + F_{2k+1}), \\ &\qquad \text{by Theorem 7,} \\ &= bL_{2k+1} + aL_{2k}, \quad \text{by (33).} \end{aligned}$$

Thus

$$\begin{aligned} F_{2k}(u_{2k} + u_{2k+2}) &= bF_{2k}L_{2k+1} + aF_{2k}L_{2k} \\ &= b(F_{4k+1} - 1) + aF_{4k}, \quad \text{by (31) and (12),} \\ &= \Sigma_{4k-1}, \quad \text{by Theorem 8.} \end{aligned}$$

(Notice that the additional step at the beginning was necessary since we knew, by Theorem 7, how to express u_n in terms of Fibonacci numbers, but not in terms of Lucas numbers.) □

• • • **BREAK**

Try to make a number trick out of Theorem 10. Can you find any nice formulae involving $\sum_{4k}$, $\sum_{4k+2}$?

3.5 MORE ABOUT DIVISIBILITY

In this section we take up a question which might at first seem much too difficult to handle. We have seen that F_n is divisible by 2 if and only if $3 \mid n$. When is F_n divisible by 3? By 4? By 5? ... Of course, if we are looking at any particular example of the question

For what positive n is F_n divisible by q?

we may simply start by writing out the Fibonacci sequence mod q and looking for occurrences of 0. Consider, say, the case $q = 5$. The Fibonacci sequence mod 5 is

$$0\ 1\ 1\ 2\ 3\ 0\ 3\ 3\ 1\ 4\ 0\ 4\ 4\ 3\ 2\ 0\ 2\ 2\ 4\ 1\ 0\ 1\ \cdots . \qquad (36)$$

Notice that we stopped when we saw 0 1 come up again, because we know that the sequence must now repeat in its entirety. We have shown that $5 \mid F_5$ and that 5 is the *smallest* positive integer m such that $5 \mid F_m$. It follows from Theorem 2 that $5 \mid F_n$ if $5 \mid n$. However, if we use the stronger Theorem 3, we find that $5 \mid F_n$ *if and only if* $5 \mid n$. For suppose that $5 \mid F_n$. Since $5 \mid F_5$ it follows from Theorem 3 that $5 \mid F_d$, where $d = \gcd(5, n)$. But 5 is the *smallest* positive integer n such that $5 \mid F_n$, so $d = 5$ and $5 \mid n$.

Thus we may associate with 5 a certain integer m (which happens also to be 5) such that

$$5 \mid F_n \Leftrightarrow m \mid n.$$

We call m the ***Fibonacci Index*** of 5 and write $m = FI(5)$. What we will prove now is that the situation analyzed above in the case $q = 5$ is true for *any* positive integer q (we may as well take $q \geq 2$, of course). The part of the argument above which said, with $q = 5$, that we only had to look at the first occurrence of 0 in (36) *after* the initial occurrence to determine $FI(5)$, plainly generalizes to any q replacing $q = 5$. The subtle part is to show that there *is* a first occurrence. Of course, it is sufficient to show that there is *some occurrence of* 0 *after the initial occurrence*, since there is then certainly a first occurrence!

So that you shouldn't mistake the subtlety of the problem, let us point out that, if we replace the Fibonacci numbers by the Lucas numbers, there may be no L_n such that $q \mid L_n$. Again taking $q = 5$, we find that the Lucas sequence mod 5 is

$$2\ 1\ 3\ 4\ 2\ 1 \cdots. \tag{37}$$

Again we stop because 2 1 has come up again, so the whole sequence will now repeat itself. But there is no zero in (37), so there can be no n such that $5 \mid L_n$.

We now *prove* that, for any q, there does exist a positive n such that $q \mid F_n$. We argue in this way: there are q residues mod q, so there are q^2 ordered pairs of residues mod q. Thus, if we write down the residue classes r_n of F_n mod q, starting with $n = 0$, obtaining, say,

$$r_0\ r_1\ r_2\ r_3 \cdots r_{q^2+1}, \tag{38}$$

stopping where $n = q^2 + 1$, we will have written $q^2 + 1$ *successive pairs* (r_n, r_{n+1}), and so some ordered pair of residues mod q must have been written twice (at least) in (38). This is an application of the ***pigeonhole principle***, which you will find discussed at greater length in Chapter 7. More precisely, it tells us that there must be an integer k (nonnegative) and an integer m (positive) such that the pair (r_k, r_{k+1}) is the same as the pair (r_{m+k}, r_{m+k+1}). In other words,

$$\left.\begin{aligned} F_k &\equiv F_{m+k} \bmod q, \\ F_{k+1} &\equiv F_{m+k+1} \bmod q. \end{aligned}\right\} \tag{39}$$

But, if k is positive,[4] we can go backward from (39). For

$$F_{k-1} = F_{k+1} - F_k \equiv F_{m+k+1} - F_{m+k} = F_{m+k-1},$$

where the congruence is, of course, mod q. Thus, going backward repeatedly, using the recurrence relation, we finally conclude that

$$F_0 \equiv F_m \bmod q.$$

[4]Of course, we could go backward even if $k = 0$, but we would not need to.

But $F_0 = 0$, so $q \mid F_m$. Notice that, in the course of this argument, we have explained what we should have noticed in (36), namely, that the first time a pair repeated it was the original pair (0, 1).

The argument we have given doesn't tell us, for a given q, the smallest m such that $q \mid F_m$ (with $q = 5$ it gives us $m = 20$). However, it does tell us that the m we have found satisfies $m \leq q^2$. Actually, we can do better than this if we use the fact that *consecutive Fibonacci numbers are always coprime.* We then find that, if q can be factorized as

$$q = \prod_i p_i^{n_i}$$

as a product of prime powers, then

$$m \leq \prod_i (p_i^{2n_i} - p_i^{2n_i - 2}). \tag{40}$$

Usually, however, the smallest m will be much smaller than this estimate, though this is a great improvement on q^2. For example, if $q = 60$, then $q^2 = 3600$, but the estimate is 2304.

• • • **BREAK**

Use (6) and (7) to prove our statement that consecutive Fibonacci numbers are coprime. What about consecutive Lucas numbers?

Another interesting question is this:

For what q does there exist n such that $q \mid L_n$?

We believe that, in this generality, the question is open. However, here's one rather extraordinary result in this direction—if p is an odd prime such that p divides some Lucas number, then, for any k, p^k divides some Lucas number. Notice that the conclusion is wrong for the prime 2. Certainly 2 divides some L_n (indeed, precisely those L_n such that $3 \mid n$); also 4 divides some L_n (indeed, precisely those L_n such that $3 \mid n$ but $6 \nmid n$). But the sequence $L_n \bmod 8$ reads

$$2\ 1\ 3\ 4\ 7\ 3\ 2\ 5\ 7\ 4\ 3\ 7\ 2\ 1 \cdots,$$

and thus contains no zero!

• • • **BREAK**

Find a value of q such that the *smallest* positive integer m for which $q \mid F_m$ is given by equality in formula (40). You might also like to show that if $q \geq 3$ and there *is* an n such that $q \mid L_n$, then the collection of those positive n such that $q \mid L_n$ consists of odd multiples of the smallest one. (Use Theorem 6.)

3.6 A LITTLE GEOMETRY!

We have always liked to emphasize, in our work, that the various parts of mathematics are very closely interrelated, and that it is very artificial, and misleading, to keep them rigidly apart. So we are particularly pleased to be able to show you how certain properties of Fibonacci and Lucas numbers may be suggested, and expressed in geometric language.

We consider a path passing through opposite vertices of a rectangle. In a particular example (see Figure 1) the path begins at the origin and passes, by straight-line segments, through the break points (5, 8), (8, 13), and (13, 21), thereby dividing the rectangle

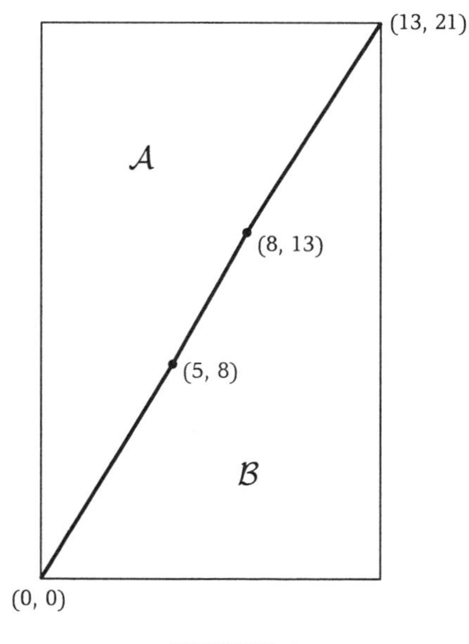

FIGURE 1

into two regions which are labeled $\mathcal{A}$ (above) and $\mathcal{B}$ (below). Do you suppose there is anything interesting about the areas of these regions? Take a few minutes to calculate each of their areas.

Surprised? Do you think we had to choose the coordinates of the break points in any special way to have the areas of $\mathcal{A}$ and $\mathcal{B}$ turn out to be the same? You may have already noticed that the coordinates of the break points (except for the origin) are, in fact, taken from the consecutive Fibonacci numbers 5, 8, 13, 21. Have we found some special sequence of Fibonacci numbers or would the areas of $\mathcal{A}$ and $\mathcal{B}$ turn out to be the same if we adjusted the figure by starting somewhere else in the sequence? Suppose we start at the origin and have the path pass through $(1, 1)$, $(1, 2)$ and terminate at $(2, 3)$—will this produce a similar result? It's easy to see that it does.

We suspect now that we are onto something more general. We are tempted to ask if we will always get the same result with *any* sequence of Fibonacci numbers. That is, if we start at the origin and pass through $P_1(F_{n-1}, F_n)$, $P_2(F_n, F_{n+1})$, and terminate at the opposite corner of the rectangle at $P_3(F_{n+1}, F_{n+2})$, with $n \geq 2$, will the area of the rectangle be bisected by this path? If it is truc, how can we prove it? And what happens if we replace Fibonacci numbers by Lucas numbers?

Before we attempt the proof of our more general statement let's return to the path in Figure 1 and see if we can discover *why* that path should bisect the area of the rectangle. Figure 2 shows the same path with some vertical and horizontal lines added to the diagram. Obviously, the triangles above and below each of the straight-line segments of the path are of equal area. Moreover, it is easy to see that the squares A_1 and B_1 both have sides of length 8, and the squares A_2 and B_2 both have sides of length 5. Thus we see, at a glance, that the path connecting the origin with the point $(13, 21)$ must divide the rectangle into two regions of equal area.

• • • BREAK

There is another nice argument by symmetry to show that the regions $\mathcal{A}$ and $\mathcal{B}$ in Figure 1 have the same area. As a hint, transfer the origin in Figure 1 to the point $(13, 21)$, and look carefully at the new coordinates of the break points. (Actually,

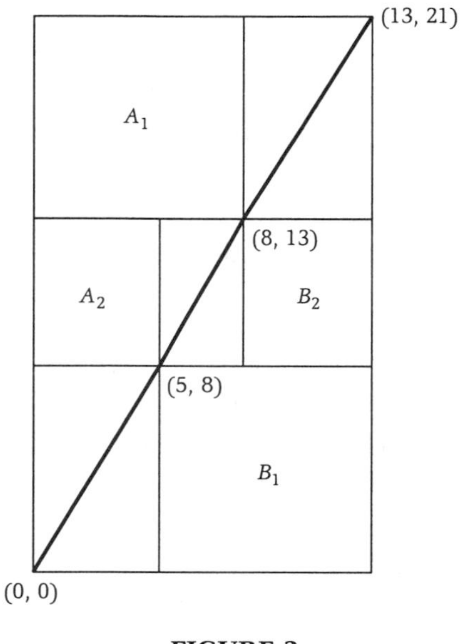

FIGURE 2

the two arguments are really the same, but this variant is even more inspired by geometry!)

But we see much more than just the proof of this particular case. In fact, what makes the areas of the squares A_1, B_1 equal, and the areas of the squares A_2, B_2 equal, is not the fact that the numbers we chose were Fibonacci numbers, but simply the fact that the Fibonacci numbers F_n, $n \geq 1$, are positive numbers satisfying the general recurrence relation

$$u_{n+2} = u_{n+1} + u_n.$$

Thus we find that if we start at the origin and pass along straight-line segments through $P_1(u_n, u_{n+1})$, $P_2(u_{n+1}, u_{n+2})$, and $P_3(u_{n+2}, u_{n+3})$, we will have a path which divides the rectangle into two regions of equal area. In fact, Figure 2, with the points appropriately labeled, serves to prove this much more general statement.

• • • BREAK

Check that the path starting at the origin and passing through the break points (4, 7), (7, 11), and (11, 18) divides the rectangle having (0, 0) and (11, 18) as opposite vertices into regions of equal area.

Extend the path of Figure 1 to include the straight-line segments from (13, 21) to (21, 34) and from (21, 34) to (34, 55). Show that this longer path divides the rectangle, having (0, 0) and (34, 55) as opposite vertices, into regions of equal area.

More generally, show that if we start at (0, 0) and pass, via straight-line segments, through $P_1(u_n, u_{n+1})$, $P_2(u_{n+1}, u_{n+2}) \cdots$, and terminate the path at the opposite corner of the rectangle at $P_{2k-1}(u_{n+2k}, u_{n+2k+1})$ we will have a path which divides the rectangle into regions of equal area.

There is more to this story and the interested reader may wish to consult [2] to find out *why* only paths ending at P_{2k-1}, and not those ending at P_{2k}, bisect the rectangle having the origin and the terminal point as opposite vertices. You may wish to experiment with some paths, adding one point at a time, and see what effect this has on the areas of the regions above and below the path within the rectangle having the first and terminal points as opposite vertices.

• • • FINAL BREAK

1. Verify the following statement by considering values of the pair (m, n) in each of the three cases. Of course, $d = \gcd(m, n)$.

$$\gcd(F_m, L_n) = \begin{cases} L_d & \text{if } (m)_2 > (n)_2, \\ 2 & \text{if } (m)_2 \leq (n)_2 \text{ and } 3 \mid d, \\ 1 & \text{if } (m)_2 \leq (n)_2 \text{ and } 3 \nmid d. \end{cases}$$

Remarks: (i) A proof of this relation is also to be found in [1].

(ii) We have used this relation to prove the unexpected fact that no Fibonacci number greater than 3 can divide a Lucas number.

2. Prove the result quoted in Question 1 in the special case $m = n$, without using Theorem 3. (*Hint*: First show that F_{n-1}, F_n are always coprime.)
3. Let the terms of the sequence $\{u_n\}$ satisfy the recurrence relation

$$u_{n+2} = 3u_{n+1} - 2u_n$$

and suppose

$$u_0 = 1, \ u_1 = 2.$$

Find u_{10}.

REFERENCES

1. Hilton, Peter, and Jean Pedersen, Fibonacci and Lucas numbers in teaching and research, *Journées Mathématiques & Informatique*, **3** (1991–1992), 36–57.
2. Hilton, Peter, and Jean Pedersen, A note on a geometrical property of Fibonacci numbers, *The Fibonacci Quarterly*, **32**, No. 5 (1994), 386–388.

ANSWERS FOR FINAL BREAK

1. Answers will vary.
2. $\gcd(F_n, F_{n-1}) = \gcd(F_{n-1}, F_{n-2})$, since $F_n = F_{n-1} + F_{n-2}$. Hence, $\gcd(F_n, F_{n-1}) = \gcd(F_2, F_1) = 1$.

$$\begin{aligned}\gcd(F_n, L_n) &= \gcd(F_n, F_n + 2F_{n-1}), \quad \text{by (33)}, \\ &= \gcd(F_n, 2F_{n-1}).\end{aligned}$$

Now by our earlier result that $\gcd(F_n, F_{n-1}) = 1$, we infer that

$$\gcd(F_n, 2F_{n-1}) = 1 \quad \text{or} \quad 2.$$

Now if $3 \mid n$, then F_n is even and F_{n-1} is odd, so that $2 \mid 2F_{n-1}$, $4 \nmid 2F_{n-1}$. Thus

$$\gcd(F_n, 2F_{n-1}) = 2.$$

But if $3 \nmid n$, F_n is odd, so

$$\gcd(F_n, 2F_{n-1}) = 1.$$

3. The equation $x^2 - 3x + 2 = 0$ has solutions $x = 1$, $x = 2$. Thus,

$$u_n = r + s2^n.$$

Since $u_0 = 1$,

$$1 = r + s.$$

Since $u_1 = 2$,

$$2 = r + 2s.$$

Hence, $s = 1$, $r = 0$, and $u_n = 2^n$, so $u_{10} = 1024$.

CHAPTER 4

Paper-Folding and Number Theory

4.1 INTRODUCTION: WHAT YOU CAN DO WITH—AND WITHOUT—EUCLIDEAN TOOLS

We begin this story with the Greeks and their fascination with the challeng of constructing regular convex polygons—that is, polygons with all sides of the same length and all interior angles equal. We refer to such N-sided polygons as regular convex N-gons, and we may suppress the word "regular" or the word "convex" if no confusion would result. The Greeks wanted to construct these polygons using what we call ***Euclidean tools***, namely, an unmarked straight edge and a compass. The Greeks (working around 350 B.C.) were successful in devising Euclidean constructions for regular convex polygons having N sides, where

$$N = 2^c N_0, \quad \text{with } N_0 = 1, 3, 5, \text{ or } 15.$$

Of course, we need $N \geq 3$ for the polygon to exist.

Naturally, the Greeks would have liked to answer the question whether or not Euclidean constructions existed for any other values of N. However, it appears that no further progress was made until about 2000 years later, when Gauss (1777–1855) completely settled the question by proving that a Euclidean construction of a regular N-gon is possible *if and only if* the number of sides N is of the form $N = 2^c \Pi \rho_i$, where the ρ_i are distinct Fermat primes—that is, primes of the form $F_n = 2^{2^n} + 1$.

Gauss's discovery was remarkable—and, of course, it tells us precisely which N-gons admit a Euclidean construction, provided we know which Fermat numbers F_n are prime.[1] In fact, we don't know! Certainly not all Fermat numbers are prime. Euler (1707–1783) showed that F_5 $(= 2^{2^5} + 1)$ is not prime and, although many composite Fermat numbers have been identified, to this day the only known prime Fermat numbers are

$$F_0 = 3, \; F_1 = 5, \; F_2 = 17, \; F_3 = 257, \; \text{and } F_4 = 65537.$$

Thus, even with Gauss's contribution, a Euclidean construction of a regular N-gon is known to exist for very few values of N; and even for these N we do not in all cases know an explicit construction.

• • • **BREAK**

Deduce, from Gauss's discovery, for how many odd values of N we know that we can construct a regular N-gon by means of a Euclidean construction. You need not work out the actual values of N.

Despite our knowledge of Gauss's work we still would like to be able somehow to construct *all* regular N-gons. Our approach is to modify the question so that, instead of exact construction, we will ask:

> ***For which $N \geq 3$ is it possible, systematically and explicitly, to construct an arbitrarily good approximation of a regular convex N-gon?***

What we will show is that one can, simply and algorithmically, construct an approximation (to any degree of accuracy) to a convex N-gon for *any* value of N. In fact, we will give explicit (and uncomplicated) instructions involving only the folding of a straight strip of paper in a prescribed periodic manner.

[1] Fermat (1601–1665) must have hoped that all numbers F_n are prime. Probably Abbé Mersenne (1588–1648) hoped that all numbers $2^p - 1$, p prime, are themselves prime. You should look for an example of a Mersenne number which is not prime.

Although the construction of regular convex N-gons would be a perfectly legitimate goal in itself, the mathematics we encounter is generous and we achieve much more. In the process of making what we call the *primary crease lines* on the tape used to construct regular N-gons we obtain tape which can be used to fold certain (but not all) star polygons. It is not too difficult systematically to add *secondary crease lines* to the original tape in order to obtain tape that may be used to construct the remaining regular star polygons; a star polygon is a "polygon" in which the sides are allowed to intersect each other at points different from the vertices. (For a precise definition of a regular star polygon, see the second paragraph of the next section. The pentagram and the Star of David shown here are two well-known examples.)

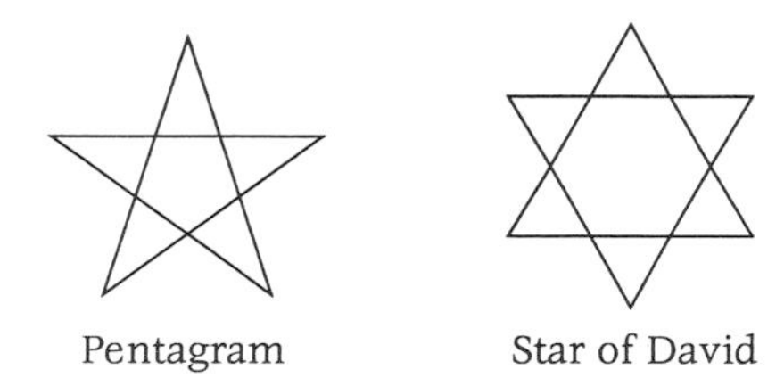

Pentagram Star of David

As it turns out, the mathematics we encounter, using our folding procedure, leads quickly and naturally to ideas in elementary analysis; and the question of which convex or star polygons we can fold by a particular, and particularly simple, procedure leads to new results in number theory. These are the topics covered in Part I.

In Part II we introduce our general folding procedure and show how it enables us to construct arbitrarily good approximations to *any* regular convex polygon. We also discuss the construction of regular star polygons. Moreover, we are once again led into some fascinating number theory, this time very different from that which arose in Part I.

There are many aspects to the topic of Part II (see [3, 5, 6, 8, 9, 10, and 11]), but we will focus most of our attention on the approximation of regular polygons and on one specific new result in number theory, namely, the ***Quasi-Order Theorem***, which we actually use to prove that the Fermat number F_5 is not prime.

We then describe, in an Appendix, how the folded paper may be used to construct some intriguing polyhedral models.

We close this Introduction by stating a key result in the Euclidean geometry of the plane which is absolutely basic to what follows.

Given any polygon (it need not even be convex, let alone regular), we claim that the sum of its exterior angles[2] is 2π radians (see Section 1 of Chapter 1 if you don't know about radians). A nice way of seeing this is, oddly enough, to pretend to be short-sighted. For the diagram

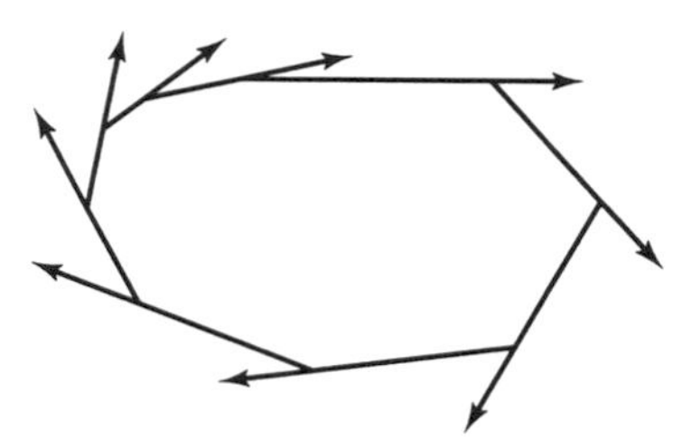

looks to anyone with myopia like this

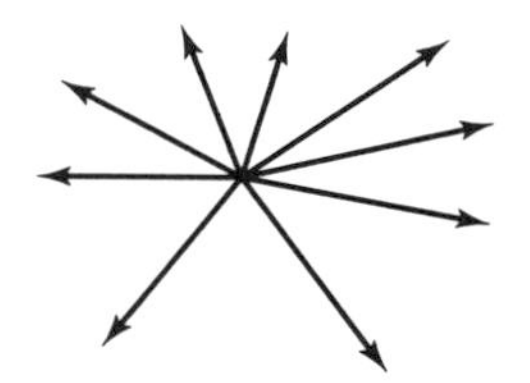

The following facts are immediate consequences:

(1) The sum of the interior angles of an N-gon is $(N - 2)\pi$.

(2) The exterior angles of a regular N-gon are each $\frac{2\pi}{N}$.

(3) The interior angles of a regular N-gon are each $\left(\frac{N-2}{N}\right)\pi$.

• • • **BREAK**

1. Show that the numbers N for which Gauss showed there is a Euclidean construction of a regular N-gon are precisely the numbers N such that $\Phi(N)$ is a power of 2. This fact is the basis of Gauss's proof. (See Chapter 2 for the definition of Euler's Φ-function.)
2. Give a more formal version of our proof that the sum of the exterior angles of a polygon is 2π. Pay special attention to what happens if the polygon is *not* convex.

[2]An exterior angle of a polygon is the supplement of the corresponding interior angle.

I. SIMPLE PAPER-FOLDING

4.2 GOING BEYOND EUCLID: FOLDING 2-PERIOD REGULAR POLYGONS

First we explain a precise and fundamental folding procedure, involving a straight strip of paper with parallel edges that may be used to produce various polygonal shapes which are, in fact, called the regular star $\{\frac{b}{a}\}$-gons (explained in the next paragraph). We suggest that you will find it useful to have a long strip of paper handy. Adding-machine tape or ordinary unreinforced gummed tape work well. Assume that we have a straight strip of paper that has certain vertices marked on its top and bottom edges and which also has *creases* or *folds* along straight lines emanating from vertices at the top edge of the strip. Further assume that the creases at those vertices labeled A_{nk}, $n = 0, 1, 2, \ldots$ (see Figure 1) which are on the top edge, form identical angles of $\frac{a\pi}{b}$ with the top edge,

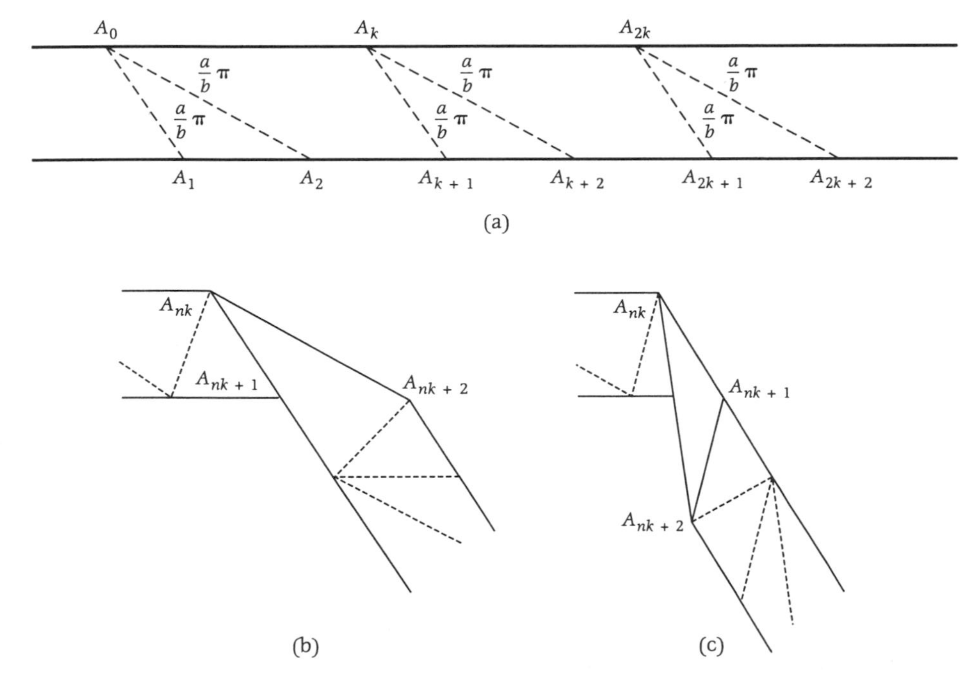

FIGURE 1

with an identical angle of $\frac{a\pi}{b}$ between the crease along the lines $A_{nk}A_{nk+2}$ and the crease along $A_{nk}A_{nk+1}$ (as shown in Figure 1(a)). Suppose further that the vertices A_{nk} are equally spaced. If we fold this strip on $A_{nk}A_{nk+2}$, as shown in Figure 1(b), and then twist the tape so that it folds on $A_{nk}A_{nk+1}$, as shown in Figure 1(c), the direction of the *top edge* of the tape will be rotated through an angle of $2\left(\frac{a\pi}{b}\right)$. We call this process of ***folding and twisting*** the ***FAT***-algorithm (see any of [3, 4, 5, 6, 7, 10, 11]).

Now consider the equally spaced vertices A_{nk} along the top of the tape, with k fixed and n varying. If the FAT-algorithm is performed on a sequence of angles, each of measure $\frac{a\pi}{b}$, at the vertices given by $n = 0, 1, 2, \ldots, b-1$, then the top of the tape will have turned through an angle of $2a\pi$. Thus the vertex A_{bk} will come into coincidence with A_0; and the top edge of the tape will have visited every ath vertex of a bounding regular convex b-gon, thus determining a regular star $\{\frac{b}{a}\}$-gon. As an example, see Figure 5(c) where $a = 2$ and $b = 7$. (In order to fit with our usage of "N-gon" we make a slight adaptation of the Coxeter notation for star polygons (see [1]), so that when we refer to a ***regular star*** $\{\frac{b}{a}\}$***-gon*** we mean a connected sequence of edges that visits every ath vertex of a regular convex b-gon. Thus our N-gon is the special star $\{\frac{N}{1}\}$-gon. When viewing a convex polygon this way we may well use a lowercase letter instead of N.)

• • • BREAK

Which of the two figures in the Introduction is a regular star polygon? And for what values of a, b?

Figure 2 illustrates how a suitably creased strip of paper may be folded by the FAT-algorithm to produce a regular convex p-gon (which may be thought of, for these purposes, as a star $\{\frac{p}{1}\}$-gon). In Figure 2 we have written V_k instead of A_{nk}.

Let us now illustrate how the FAT-algorithm may be used to fold a regular convex 8-gon. Figure 3(a) shows a straight strip of paper on which the dotted lines indicate certain special exact crease lines. In fact, these crease lines occur at equally spaced intervals along the top of the tape so that the angles occurring at the top of each vertical line are (from left to right) $\frac{\pi}{2}, \frac{\pi}{4}, \frac{\pi}{8}, \frac{\pi}{8}$. Figuring

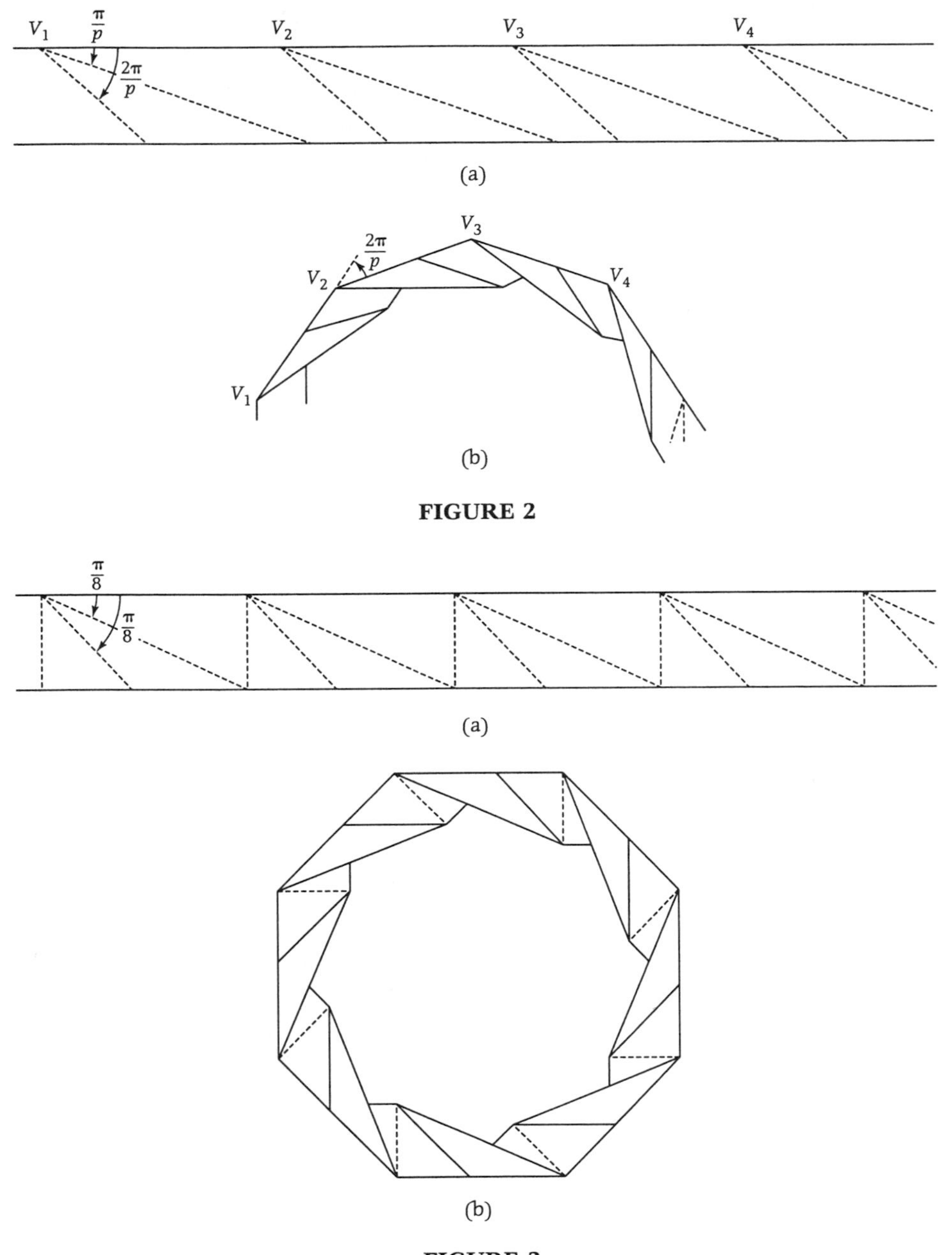

(a)

(b)

FIGURE 2

(a)

(b)

FIGURE 3

out how to fold a strip of tape to obtain this arrangement of crease lines is an interesting exercise for elementary students (complete instructions are given in [7]). Our immediate interest is focused on the observation that this tape has, at equally spaced intervals along the top edge, adjacent angles each measuring $\frac{\pi}{8}$, and we can therefore execute the FAT-algorithm at 8 consecutive vertices along the top of the tape to produce the regular convex 8-gon shown in Figure 3(b). (Of course, in constructing the model one would cut the tape on the first vertical line and glue a section at the end to the beginning so that the model would form a closed polygon.)

Notice that the tape shown in Figure 3(a) also has suitable crease lines that make it possible to use the FAT-algorithm to fold a regular convex 4-gon. We leave this as an exercise for the reader and turn to a more challenging construction, the regular convex 7-gon.

Now, since the regular convex 7-gon is the first polygon we encounter for which we do not have available a Euclidean construction, we are faced with a real difficulty in creating a crease line making an angle of $\frac{\pi}{7}$ with the top edge of the tape. We proceed by adopting a general policy that we say more about in Chapter 9—we call it our ***optimistic strategy***. Assume that we *can* crease an angle of $\frac{2\pi}{7}$ (certainly we can come close) as shown in Figure 4(a). Given that we have the angle of $\frac{2\pi}{7}$ it is then a trivial matter to fold the top edge of the strip DOWN to bisect this angle, producing two adjacent angles of $\frac{\pi}{7}$ at the top edge as shown in Figure 4(b). (We say that $\frac{\pi}{7}$ is the ***putative*** angle on this tape.) Then, since we are content with this arrangement, we go to the bottom of the tape where we observe that the angle to the right of the last crease line is $\frac{6\pi}{7}$—and we decide, as paper-folders, that we will always avoid leaving even multiples of π in the numerator of any angle next to the edge of the tape, so we bisect this angle of $\frac{6\pi}{7}$, by bringing the bottom edge of the tape UP to coincide with the last crease line and creating the new crease line sloping up shown in Figure 4(c). We settle for this (because we are content with an odd multiple of π in the numerator) and go to the top of the tape where we observe that the angle to the right of the last crease line is $\frac{4\pi}{7}$—and, since we have decided against leaving an even multiple of π in any angle next to an edge of the tape, we are forced to bisect this angle twice, each time bringing the top edge of the tape DOWN to coincide

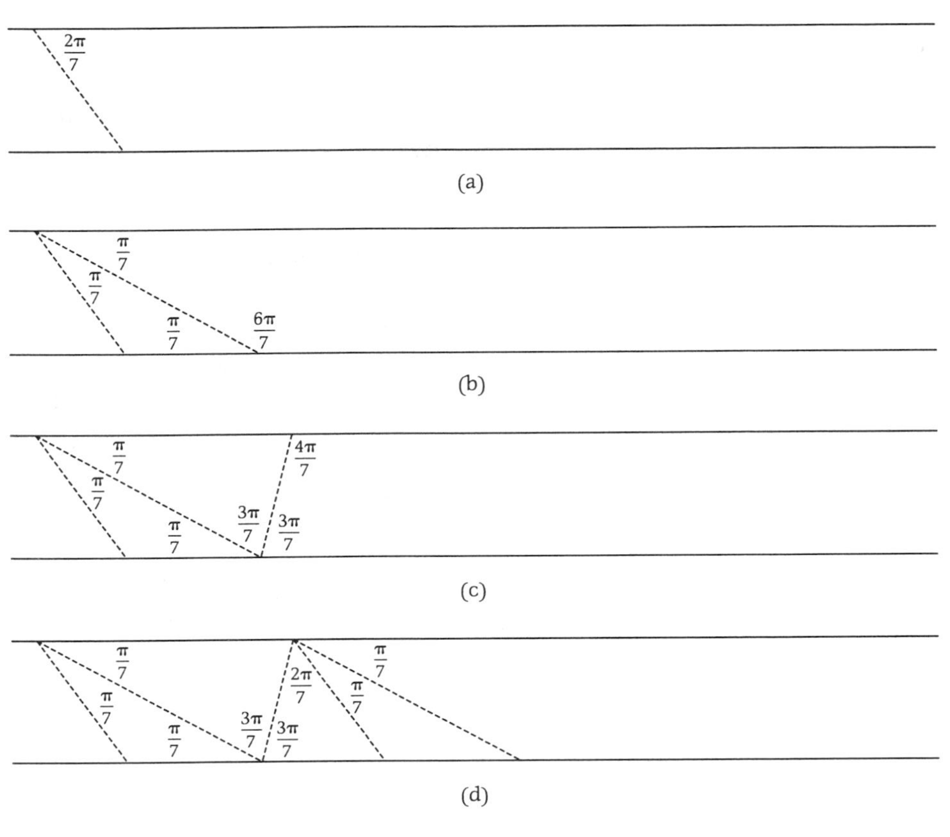

FIGURE 4

with the last crease line, obtaining the arrangement of crease lines shown in Figure 4(d). But now we notice something miraculous has occurred! If we had really started with an angle of exactly $\frac{2\pi}{7}$, and if we now continue introducing crease lines by repeatedly folding the tape DOWN TWICE at the top and UP ONCE at the bottom, we get precisely what we want; namely, pairs of adjacent angles, measuring $\frac{\pi}{7}$, at equally spaced intervals along the top edge of the tape. Let us call this folding procedure the ***D^2U^1-folding procedure*** (or, more simply—and especially when we are concerned merely with the related number theory—the (2, 1)-folding procedure) and call the strip of creased paper it produces ***D^2U^1-tape*** (or, again more simply, **(2, 1)-*tape***). The crease lines on this tape are called the ***primary crease*** *lines*.

• • • **BREAK**

(You'll be sorry if you skip this one!) We suggest that before reading further you get a piece of paper and fold an acute angle which you call an approximation to $\frac{2\pi}{7}$. Then fold about 40 triangles using the D^2U^1-folding procedure as shown in Figure 4 and described above, throw away the first 10 triangles, and see if you can tell that the first angle you get between the top edge of the tape and the adjacent crease line is *not* $\frac{\pi}{7}$. Then try to construct the FAT 7-gon shown in Figure 5(b). You may then *believe* that the D^2U^1-folding procedure produces tape on which the angles approach the values indicated in Figure 5(a).

How do we *prove* that this evident convergence takes place? A very direct approach is to admit that the first angle folded down from the top of the tape in Figure 4(a) might not have been precisely $\frac{2\pi}{7}$. Then the bisection forming the next crease would make the two acute angles nearest the top edge in Figure 4(b) only approximately $\frac{\pi}{7}$; let us call them $\frac{\pi}{7} + \varepsilon$ (where ε may be either positive or negative). Consequently the angle to the right of this crease, at the bottom of the tape, would measure $\frac{6\pi}{7} - \varepsilon$. When this angle is bisected, by folding up, the resulting acute angles nearest the bottom of the tape, labeled $\frac{3\pi}{7}$ in Figure 4(c), would in fact measure $\frac{3\pi}{7} - \frac{\varepsilon}{2}$, forcing the angle to the right of this crease line at the top of the tape to have measure $\frac{4\pi}{7} + \frac{\varepsilon}{2}$. When this last angle is bisected twice by folding the tape down, the two acute angles nearest the top edge of the tape will measure $\frac{\pi}{7} + \frac{\varepsilon}{2^3}$. This makes it clear that every time we repeat a D^2U^1-folding on the tape the error is reduced by a factor of 2^3.

Now it should be clear how our ***optimistic strategy*** has paid off. By blandly *assuming* we have an angle of $\frac{\pi}{7}$ at the top of the tape to begin with, and folding accordingly, we get *what we want*—successive angles at the top of the tape which, as we fold, rapidly get closer and closer to $\frac{\pi}{7}$, whatever angle we had, in fact, started with!

In practice, the approximations we obtain by folding paper are quite as accurate as the *real world* constructions with a straight edge and compass—for the latter are only perfect in the mind. In both cases the real world result is a function of human skill, but our procedure, unlike the Euclidean procedure, is very forgiving in that it tends to reduce the effects of human error—and, for many people (even the not so young), it is far easier to bisect an angle by folding paper than it is with a straight edge and compass.

Figures 5(c) and (d) show the regular $\left\{\frac{7}{2}\right\}$- and $\left\{\frac{7}{3}\right\}$-gons that are produced from the D^2U^1-tape by executing the FAT-algorithm on the crease lines that make angles of $\frac{2\pi}{7}$ and $\frac{3\pi}{7}$, respectively, with an edge of the tape (if the angle needed is at the bottom of the tape, as with $\frac{3\pi}{7}$, simply turn the tape over so that the required angle appears on the top). In Figures 5(c) and (d) the FAT-algorithm was executed on every other suitable vertex along the edge of the tape so that, in (c), the resulting figure, or its flipped version, could be woven together in a more symmetric way and, in (d), the excess could be folded neatly around the points.

We are now in the position of having many *answerable* questions. Here are just two.

(1) Can we use the same general approach used for folding a convex 7-gon to fold a convex N-gon with N odd, at least for certain specified values of N? Can we prove, in general, that the actual angles on the tape really converge to the putative angle we originally sought?

(2) What happens if we consider general folding procedures perhaps with other periods, such as those represented by

$$D^3U^3, \quad D^4D^2, \quad \text{or } D^3U^1D^1U^3D^1U^1?$$

(The ***period*** is determined by the repeat of the exponents, so these examples have periods 1, 2, and 3, respectively.)

It turns out that, in answering these questions, we need to make a straightforward use of the following theorem which is a special case of the well-known Contraction Mapping Principle (see, e.g., [13]).

Theorem 1 *For any three real numbers a, b, and x_0, with $a \neq 0$, let the sequence $\{x_k\}$, $k = 0,1,2,\ldots$, be defined by the recurrence relation*

$$x_k + ax_{k+1} = b, \quad k = 0, 1, 2, \ldots . \qquad (1)$$

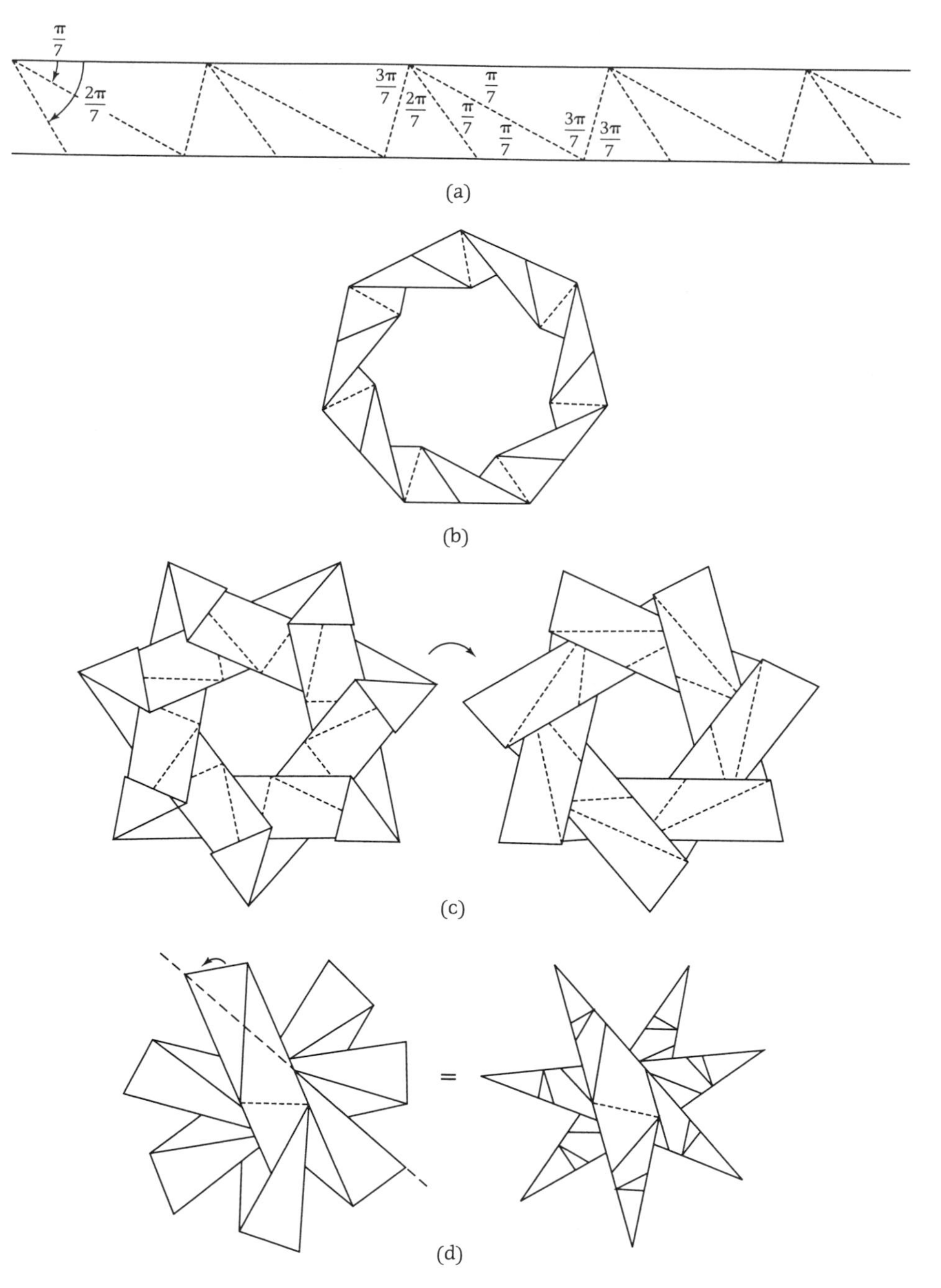
$\frac{\pi}{7}$
$\frac{2\pi}{7}$
$\frac{3\pi}{7}$
$\frac{2\pi}{7}$
$\frac{\pi}{7}$
$\frac{\pi}{7}$
$\frac{\pi}{7}$
$\frac{3\pi}{7}$
$\frac{3\pi}{7}$
(a)
(b)
(c)
(d)

FIGURE 5

Then, if $|a| > 1$, $x_k \to \frac{b}{1+a}$ *as* $k \to \infty$.

Proof Set $x_k = \frac{b}{1+a} + y_k$. Then $y_k + ay_{k+1} = 0$. It follows that $y_k = \left(\frac{-1}{a}\right)^k y_0$. If $|a| > 1$, $\left(\frac{-1}{a}\right)^k \to 0$, so that $y_k \to 0$ as $k \to \infty$. Hence $x_k \to \frac{b}{1+a}$ as $k \to \infty$. Notice that y_k is the *error* at the kth stage, and that the absolute value of y_k is equal to $\frac{1}{|a|^k}|y_0|$. □

You may recognize (1), in the form $x_{k+1} = -\frac{1}{a}x_k + \frac{b}{a}$, as a linear recurrence relation of the *first* order; remember that Chapter 3 was all about linear recurrence relations of the *second order*.

We point out that it is significant in Theorem 1 that neither the convergence nor the limit depends on the initial value x_0. This means, in terms of the folding, that the process will converge to the same limit no matter how we fold the tape to produce the first crease line—this is what justifies our *optimistic strategy*! And, as we have seen in our example, and as we will soon demonstrate in general, the result of the lemma tells us that the convergence of our folding procedure is rapid, since, in all cases, $|a|$ will be a positive power of 2.

Let us now look at the general 1-*period* folding procedure D^nU^n. A typical portion of the tape would appear as illustrated in Figure 6(a). Then if the folding process had been started with an arbitrary angle u_0 we would have, from the kth stage,

$$u_k + 2^n u_{k+1} = \pi, \quad k = 0, 1, 2, \ldots. \tag{2}$$

Equation (2) is of the form (1), so it follows immediately from

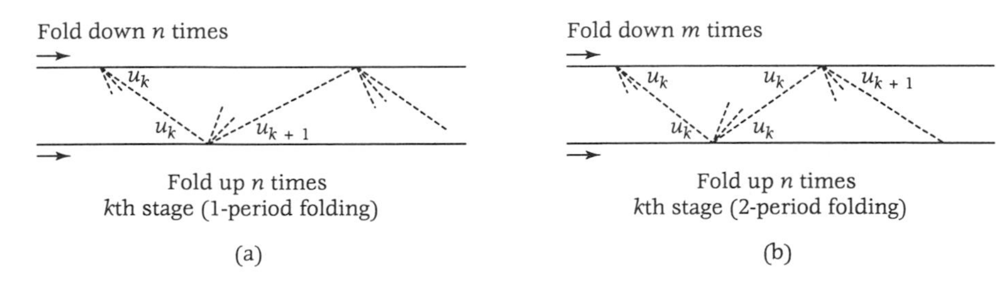

FIGURE 6

Theorem 1 that

$$u_k \to \frac{\pi}{2^n + 1} \quad \text{as } k \to \infty. \tag{3}$$

Furthermore, we can see that, if the original fold differed from the putative angle of $\frac{\pi}{2^n+1}$ by an error E_0, then the error at the kth stage of the $D^n U^n$-folding procedure would be given by

$$E_k = \frac{E_0}{2^{nk}}.$$

We now see from (3) that the $D^n U^n$-folding procedure produces tape from which we may construct regular convex $(2^n + 1)$-gons—and, of course, these include those N-gons for which N is a Fermat number, prime or not. We would like to believe that the ancient Greeks and Gauss would have appreciated the fact that, when $n = 1, 2, 4, 8$, and 16, the $D^n U^n$-folding procedure produces tape from which we can obtain arbitrarily good approximations (by means of the FAT-algorithm) to the regular 3-, 5-, 17-, 257-, and 65537-gon, respectively. Moreover, if $n = 3$, we approximate the regular 9-gon, whose nonconstructibility by Euclidean tools is very closely related to the nontrisectibility of an arbitrary angle.

The case $N = 2^n + 1$ is special, since we may construct $(2^n + 1)$-gons from our folded tape by special methods, in which, however, the top edge does not describe the polygon (as it does in the FAT-algorithm). Figure 7 shows how D^2U^2-tape (see (a)) may be folded along just the short lines of the creased tape to form the *outline* of a regular convex pentagon (see (b)); and along just the long lines of the creased tape to from the *outline* of a slightly larger regular convex pentagon (see (c)); and, finally, we show in (d) the regular convex pentagon formed by an edge of the tape when the FAT-algorithm is executed.

• • • **BREAK**

Fold a length of $D^n U^n$-tape, for various different values of n, and *experiment* with the folded tape to see how many differently sized regular $(2^n + 1)$-gons you can create from each tape.

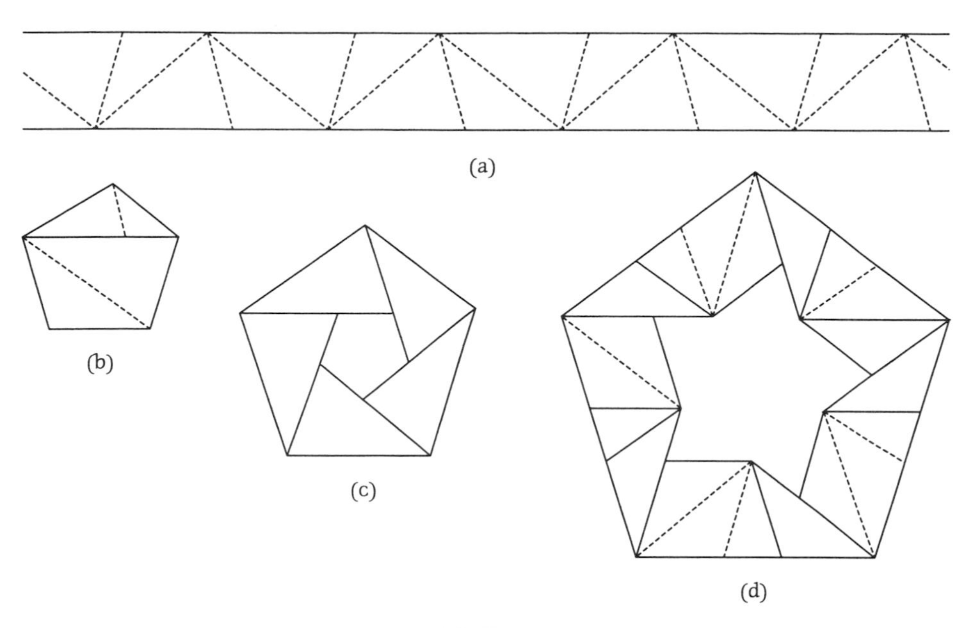

FIGURE 7

We next demonstrate how we construct regular convex polygons with 2^cN sides, N odd, if we already know how to construct regular N-gons. If, for example, we wished to construct a regular 10-gon then we take the D^2U^2-tape (which you may recall produced FAT 5-gons) and introduce a ***secondary crease line*** by bisecting each of the angles of $\frac{\pi}{5}$ next to the top (or bottom) edge of the tape. The FAT-algorithm may be used on the resulting tape to produce the regular convex FAT 10-gons, as illustrated in Figure 8. It should now be clear how to construct a regular 20-gon, a regular 40-gon,. . . .

Now we turn to the general ***2-period*** folding procedure, D^mU^n, which we may abbreviate to (m, n). In this case, a typical portion of the tape would appear as shown in Figure 6(b). If the folding procedure had been started with an arbitrary angle u_0 at the top of the tape we would have, at the kth stage,

$$u_k + 2^n v_k = \pi,$$
$$v_k + 2^m u_{k+1} = \pi,$$

and hence it follows that

$$u_k - 2^{m+n} u_{k+1} = \pi(1 - 2^n), \quad k = 0, 1, 2, \ldots . \tag{4}$$

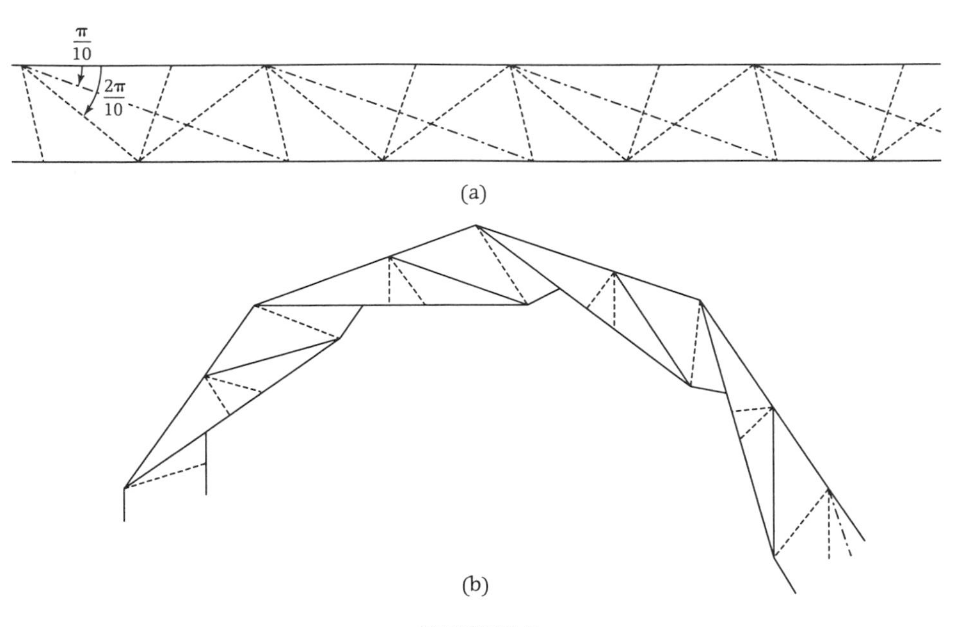

FIGURE 8

Thus, again using Theorem 1, we see that

$$u_k \to \frac{2^n - 1}{2^{m+n} - 1}\pi \quad \text{as } k \to \infty, \tag{5}$$

so that $\frac{2^n-1}{2^{m+n}-1}\pi$ is the putative angle at the top of the tape. Thus the FAT-algorithm will produce, from this tape, a star $\left\{\frac{b}{a}\right\}$-gon, where the fraction $\frac{b}{a}$ may turn out not to be reduced (e.g., when $n = 2$, $m = 4$), with $b = 2^{m+n} - 1$, $a = 2^n - 1$. By symmetry we infer that

$$v_k \to \frac{2^m - 1}{2^{m+n} - 1}\pi \quad \text{as } k \to \infty. \tag{6}$$

Furthermore, if we assume an initial error of E_0, then we know that the error at the kth stage (when the folding $D^m U^n$ has been done exactly k times) will be given by $E_k = \frac{E_0}{2^{(m+n)k}}$. Hence, we see that in the case of our D^2U^1-folding (Figure 4) any initial error E_0 is, as we already saw from our other argument, reduced by a factor of 8 between consecutive stages. It should now be clear why we advised throwing away the first part of the tape—but, likewise, it should also be clear that it is never necessary to throw away very much of the tape. In practice, convergence is very rapid indeed, and if one made it a rule of thumb to always throw away the first 20 crease lines on the tape for any iterative folding procedure, it would turn out to be a very conservative rule.

The above technique can be used to prove the convergence for any given folding sequence of arbitrary period. This is fine, once we have identified the folding sequence; but the question that is more likely to arise, for the paper-folder, is—***How do we know which sequence of folds to make in order to produce a particular regular star polygon with the FAT-algorithm***? We will answer this question in Section 5, but we first discuss in Section 3 some of the number theory results connected with the 2-period folding procedure.

4.3 FOLDING NUMBERS

We have seen in (5) that the (m, n)-folding procedure produces (via the FAT-algorithm) a regular convex s-gon, where

$$s = \frac{2^{m+n} - 1}{2^n - 1}, \quad m \geq 1, \ n \geq 1, \tag{7}$$

provided only that s is an integer; and that, in any case, it will produce a $\left\{ \frac{2^{m+n}-1}{2^n-1} \right\}$-star polygon. We naturally ask as our first *number-theoretic* question

When is $\frac{2^{m+n}-1}{2^n-1}$ ***an integer***?

Second, we ask

How do we recognize integers of this kind?

And, if we are properly ambitious, we further ask,

What rational numbers are of the form $\frac{2^{m+n}-1}{2^n-1}$***, and how do we recognize them***?

Let's be really ambitious! From the *number-theoretic* point of view there is no point in confining attention to the number 2, which we may think of as a *base*, in (7)—this choice is dictated by our insistence on the *bisection* of angles as a basic geometrical operation.

Thus we will fix an integer $t \geq 2$ as "base" and discuss rational numbers of the form $\frac{t^a-1}{t^b-1}$, for given positive integers a, b. Our first question thus becomes

When is $\frac{t^a-1}{t^b-1}$ an integer?

We will find, perhaps surprisingly, that the answer is independent of our choice of base t, thus confirming the correctness of our strategy in making this generalization. In order to deal also with rational numbers of the form $\frac{t^a-1}{t^b-1}$, we prove the following theorem.

Theorem 2 *Let* $d = \gcd(a,b)$. *Then* $\gcd(t^a - 1, t^b - 1) = t^d - 1$.

Proof Let § stand for $t - 1$ in base t. Then, writing numbers in base t, we see that[3]

$$\left.\begin{aligned} t^a - 1 &\overset{(t)}{=} \S\,\S \cdots \S \quad (a \text{ squiggles}),\\ t^b - 1 &\overset{(t)}{=} \S\,\S \cdots \S \quad (b \text{ squiggles}).\end{aligned}\right\} \tag{8}$$

Now there is a beautiful algorithm for calculating gcd's, called the ***Euclidean algorithm***.[4] If we want to calculate $\gcd(a, b)$, where we assume $a > b$, we divide a by b, paying no attention to the quotient but recording the remainder as r_1; of course, $0 \leq r_1 < b$. If $r_1 \neq 0$, we divide b by r_1, recording the remainder as r_2; of course, $0 \leq r_2 < r_1$. If $r_2 \neq 0$, we divide r_1 by r_2, recording the remainder as r_3; and we continue in this way until we get, as eventually we must, a remainder $r_k = 0$. Then $\gcd(a, b) = r_{k-1}$. (If $k = 1$, interpret r_0 as b.) □

• • • **BREAK**

If you are not already familiar with the Euclidean algorithm, follow while we use the above description to calculate gcd(27379, 341), and to show you why the algorithm works. First, we execute the algorithm, writing out the successive

[3]We write $\overset{(t)}{=}$ to indicate that the number on the right is written in base t.

[4]If may be helpful to recall the well-known case $t = 10$ where $\S = 9$. Thus, for example, $10^5 - 1 = 99999$.

steps as

$$\begin{aligned}
27379 &= 341Q_1 + 99 \quad (r_1 = 99),\\
341 &= 99Q_2 \ + 44 \quad (r_2 = 44),\\
99 &= 44Q_3 \ + 11 \quad (r_3 = 11),\\
44 &= 11Q_4 \ + 0 \quad (r_4 = 0).
\end{aligned}$$

So we claim that gcd(27379, 341) = 11. Second, to justify our claim, we observe, from the first equation, that *any* number which is a divisor of both 27379 and 341 must also be a divisor of both 341 and 99; and conversely. Likewise, *any* number which is a divisor of both 341 and 99 must also be a divisor of both 99 and 44; and conversely. And so on.... Now we can easily see that 11 is the greatest common divisor of 44 and 11. Hence the successive steps above show that

$$\begin{aligned}
\gcd(27379, 341) &= \gcd(341, 99) = \gcd(99, 44)\\
&= \gcd(44, 11) = 11.
\end{aligned}$$

See if you can figure out how to use your hand calculator to obtain the above sequence of equations about gcd's. Of course, as soon as you recognize the gcd you can go straight to the final answer.

Use the Euclidean algorithm to calculate the gcd of some consecutive Fibonacci numbers (see Chapter 3). You may notice that these calculations take a very large number of steps, and the answer is always somewhat the same. Can you see why this should be so? Is there a similar result for the Lucas numbers?

Returning to the proof of Theorem 2, it should be plain, from (8) in base t, remembering that r_1 is the remainder when we divide a by b, that, if we divide $t^a - 1$ by $t^b - 1$ we get a remainder of $t^{r_1} - 1$; and that the remainders are indeed successively

$$t^{r_2} - 1,\ t^{r_3} - 1, \dots .$$

Thus if the first zero remainder is r_k when we carry out the Euclidean algorithm on a and b, then the first zero remainder is $t^{r_k} - 1$

when we carry out the Euclidean algorithm on $t^a - 1$ and $t^b - 1$; so, remembering that $r_{k-1} = d$,

$$\gcd(t^a - 1, t^b - 1) = t^{r_{k-1}} - 1 = t^d - 1.$$

From Theorem 2 we derive the consequence which enables us to answer our first question. Consequences of theorems are called, by mathematicians, ***corollaries***.

Corollary 3 $t^b - 1 \mid t^a - 1 \Leftrightarrow b \mid a$.

Proof We claim that it is obvious from simple algebra that if $b \mid a$ then $t^b - 1 \mid t^a - 1$ (this is, in fact, just the case $r_1 = 0$ of the argument above). Conversely, suppose $t^b - 1 \mid t^a - 1$. Then

$$\gcd(t^a - 1, t^b - 1) = t^b - 1,$$

so $t^b - 1 = t^d - 1$ by Theorem 2. But it is plain that

$$t^b - 1 = t^d - 1 \quad \Rightarrow \quad b = d.$$

Notice that, as claimed, the condition that $t^b - 1 \mid t^a - 1$ is independent of t. □

Corollary 4 $t^n - 1 \mid t^{n+m} - 1 \Leftrightarrow n \mid m$.

You shouldn't need us to prove this!

Since we now know that for $\frac{t^{n+m}-1}{t^n-1}$ to be an integer we require $n \mid m$, it will be to our advantage, when considering integers of this form, to change notation, writing instead

$$s = \frac{t^{xy} - 1}{t^x - 1} \tag{9}$$

so that

$$n = x, \quad m + n = xy. \tag{10}$$

However, we should note that the condition $m \geq 1$, $n \geq 1$ of (7) translates into the condition

$$x \geq 1, \quad y \geq 2, \tag{11}$$

on the integers x, y. We now prove

Theorem 5 *The integer s determines the values of x,y.*

Proof If we write $\frac{t^{xy}-1}{t^x-1}$ in base t, then, using the fact that

$$\frac{t^{xy}-1}{t^x-1} = t^{x(y-1)} + t^{x(y-2)} + \cdots + t^x + 1,$$

we see that

$$s = \frac{t^{xy}-1}{t^x-1} \overset{(t)}{=} \underbrace{10\cdots0}\,\underbrace{10\cdots0}\cdots\underbrace{10\cdots0}\,1, \tag{12}$$

where the block $\underbrace{10\cdots0}$, of length x, is repeated and there are y 1's. Thus x and y are determined by (9). □

We call (x, y) the ***coordinates*** of s and may sometimes (as with points in the plane) even write

$$s = (x, y). \tag{13}$$

We write $\mathcal{F}_t$ for the set of (***integral***) ***folding numbers s*** (***with base t***), given by (9), abbreviating simply to ***folding numbers*** if no confusion would occur (see [9] for a discussion on generalizing the concept of folding numbers). Remember the restrictions $x \geq 1$, $y \geq 2$; however, it is sometimes useful to allow $y = 1$. Of course, if $y = 1$, then $s = 1$, independently of the choice of x, so we lose uniqueness there. We may write $\overline{\mathcal{F}_t}$ for $\mathcal{F}_t \cup (1)$, $\mathcal{F}_t$ with the degenerate folding number 1 adjoined. Figure 9_t shows how a table containing values of (x, y), for $1 \leq x, y \leq 4$, would appear *in*

y				
4	1111	1010101	1001001001	1000100010001
3	111	10101	1001001	100010001
2	11	101	1001	10001
1	1	1	1	1
↑				
y				
$x \rightarrow$	1	2	3	4

FIGURE 9_t

base t. In Figure 9_2 we have written the entries in base 10 to display the actual magnitudes of the folding numbers in $\mathcal{F}_2$.

Formula (12) tells us how to recognize folding numbers; and (10) then tells us the folding procedure to use in order to construct a regular polygon with s sides. For example, reverting to the base $t = 2$, we observe that

$$341 \overset{(2)}{=} 1\,0\,1\,0\,1\,0\,1\,0\,1.$$

Thus $x = 2$, $y = 5$, so $n = 2$, $m + n = 10$, $m = 8$; we conclude that the D^8U^2-procedure allows us to construct a regular convex 341-gon. On the other hand,

$$11 \overset{(2)}{=} 1\,0\,1\,1,$$

and hence is not a folding number (in base 2). Thus there is no 2-period folding procedure to fold a regular convex 11-gon.

We know now which integers belong to $\mathcal{F}_t$ and how to recognize them. What about rational numbers of the form $\frac{t^a-1}{t^b-1}$? Let us assume that $a > b$ as before. First, we reduce the fraction, and Theorem 2 tells us how to do it—we have

$$\frac{t^a - 1}{t^b - 1} = \frac{t^a - 1}{t^d - 1} \bigg/ \frac{t^b - 1}{t^d - 1}, \quad \text{where } d = \gcd(a, b) \tag{14}$$

and the fraction on the right of (14) is reduced. Moreover, $\frac{t^a-1}{t^d-1} \in \mathcal{F}_t$ and $\frac{t^b-1}{t^d-1} \in \overline{\mathcal{F}}_t$; in fact, $\frac{t^b-1}{t^d-1} = 1$ precisely when our original number $\frac{t^a-1}{t^b-1}$ is an integer. Suppose this is not the case.

Let $a = a'd$, $b = b'd$. Then $\gcd(a', b') = 1$ and

$$\frac{t^a - 1}{t^d - 1} = (d, a'), \quad \frac{t^b - 1}{t^d - 1} = (d, b'). \tag{15}$$

Thus the reduced fraction (on the right of (14)) has a folding number (d, a') as numerator and a ***prime section*** of it, namely (d, b'), as denominator. Here we say that (x, y') is a ***prime section*** of (x, y) if y' is coprime to y and less than y.

These then are the rational numbers expressible as $\frac{t^a-1}{t^b-1}$. If we have such a rational number $\frac{t^a-1}{t^b-1}$, how do we fold tape to produce the star $\left\{\frac{t^a-1}{t^b-1}\right\}$-gon? Well, given a folding number (d, a') and a prime section (d, b'), we have $a = a'd$, $b = b'd$, so $m + n = a'd$, $n = b'd$,

y	$2^y - 1$													
26	67708863													
25	33554431													
24	16777215													
23	8389607													
22	4194303													
21	2097151													
20	1048575													
19	524287													
18	262143													
17	131071													
16	65535													
15	32767													
14	16383													
13	8191	22369621												
12	4095	5592405												
11	2047	1398101												
10	1023	349525												
9	511	87381	19173961											
8	255	21845	2396745											
7	127	5461	299593	17895697										
6	63	1365	37449	1118481	34636833									
5	31	341	4681	69905	1082401	17043521								
4	15	85	585	4369	33825	266305	2113665	16843009						
3	7	21	73	273	1057	4161	16513	65793	26257	1049601	4196353	16781313	67117057	
2	3	5	9	17	33	65	129	257	513	1025	2049	4097	8193	$2^x + 1$
1	1	1	1	1	1	1	1	1	1	1	1	1	1	1
y / x	1	2	3	4	5	6	7	8	9	10	11	12	13	$14 \leq x \leq 26$

FIGURE 9_2

and $m = (a' - b')d$. Thus, for example, we can fold tape to produce a regular $\left\{ \frac{341}{21} \right\}$-gon since, in base 2,

$$341 \overset{(2)}{=} 101010101,$$

$$21 \overset{(2)}{=} 10101.$$

Hence $d = 2$, $a' = 5$, $b' = 3$, so $n = b'd = 6$, $m = (a' - b')d = 4$, and we conclude that the D^4U^6-folding procedure produces a $\left\{ \frac{341}{21} \right\}$-gon; notice that

$$\frac{2^{m+n} - 1}{2^n - 1} = \frac{2^{10} - 1}{2^6 - 1} = \frac{1023}{63} = \frac{341}{21},$$

as it should!

Suppose now that we really want to produce an 11-gon. In the course of this chapter we will give you two procedures. The first, which we now describe, is simpler, but messier in practice. Although 11 is not a folding number, there are folding numbers having 11 as a factor; for example, 33 is a folding number. If we produce a regular 33-gon, we can easily (in principle!) produce a regular 11-gon simply by visiting consecutively every third vertex of our regular 33-gon.

We will prove a theorem which tells us that every odd number a appears as a factor of some $(x, y) \in \mathcal{F}_2$; and that we may, in fact, choose x arbitrarily. Which would be the best folding number to choose as a multiple of a? The number theorist would surely say "the smallest," while the paper-folder would say "that involving the smallest total number of folds"—this number being $m + n$, or xy. Miraculously, these two are the same (Theorem 8 below). We call this a miracle because, of course, there is no guarantee that, of two folding numbers, the smaller requires the fewer folds. Thus

$$1023 = \frac{2^{10} - 1}{2^1 - 1} \quad \text{requires 10 folds,}$$

$$257 = \frac{2^{16} - 1}{2^8 - 1} \quad \text{requires 16 folds.}$$

As usual, we work with an arbitrary base t and consider a fixed but arbitrary integer[5] a coprime to t. We prove

Theorem 6 *Every column x of Figure 9_t contains a first entry at height $y_0 = y_0(x)$ such that $a \mid (x,y_0)$ and the entries (x,y) in column x such that $a \mid (x,y)$ are precisely those for which y is a multiple of y_0.*

Proof Now t is coprime to a and to $t^x - 1$, hence to $a(t^x - 1)$. Thus (see Corollary 10 of Chapter 2) there exists a positive integer z_0 such that $t^z \equiv 1 \bmod a(t^x - 1)$ if and only if z is a multiple of z_0. It immediately follows that $t^x - 1 \mid t^{z_0} - 1$ so that, by Corollary 3, $x \mid x_0$ and we may write $z_0 = xy_0$. Hence any multiple of z_0 is of the form xy, where y is a multiple of y_0. We conclude that

$$a \left| \frac{t^{xy} - 1}{t^x - 1} \right. \quad \Leftrightarrow \quad y \text{ is a multiple of } y_0,$$

as claimed. □

Let $h(a) = y_0(1)$; we call $h(a)$ the ***height*** of a. It is the height, in the table, of the first element of the form $(1, y)$ divisible by a.

Proposition 7 *Every folding number (x,y), with base t, such that $a \mid (x,y)$ satisfies $h(a) \mid xy$.*

Proof Now $a \mid \frac{t^{xy}-1}{t^x-1}$, so $a \mid \frac{t^{xy}-1}{t-1}$; thus $h(a) \mid xy$ by Theorem 6. □

We come now to our key result.

Theorem 8 *The smallest folding number s having a as a factor is of the form (x,y) with*

$$xy = h(a).$$

Proof Suppose not. Then if $s = (x, y)$ is the smallest folding number having a as a factor, we have (by Proposition 7) $h \mid xy$, but

[5]We may assume $a > 1$. This use of the symbol "a" is, of course, not to be confused with previous ones.

$h \neq xy$ (where $h = h(a)$). Thus $xy \geq 2h$. Now remember from (11) that $y \geq 2$. Thus (you should check each step in the line below)

$$s = \frac{t^{xy} - 1}{t^x - 1} > \frac{t^{xy}}{t^x} = t^{x(y-1)} \geq t^{\frac{1}{2}xy} \geq t^h > \frac{t^h - 1}{t - 1}. \qquad (16)$$

But, by the definition of h, the folding number $\frac{t^h-1}{t-1}$ *is* divisible by a, so that (16) contradicts the minimality of s. We are thus forced to conclude that $h = xy$, as claimed. □

We draw attention to an immediate and remarkable consequence of Theorem 8. We think of $s = (x, y)$ *as a function of* t and prove

Corollary 9 *Let $s = (x,y)$, $s' = (x',y')$. Then if, for some t, $s \mid s'$, it follows that $xy \mid x'y'$.*

Proof We fix a value of t such that $s \mid s'$. We let s play the role of a in Theorem 8; then s is obviously the smallest folding number divisible by s, so that $h(s) = xy$. But Proposition 7 guarantees that $h(s) \mid x'y'$. □

This result is remarkable since the conclusion is independent of t, and only requires one value of t such that $(x, y) \mid (x', y')$. As we will see, there may well be only one such value of t. Notice also that our proof of this striking result of pure number theory was inspired by very practical considerations of efficient paper-folding.

We also see clearly in the statement of Corollary 9 the great advantage of well-conceived generalization. If we had stuck to base 2 we would have had a vastly inferior result. In fact, we are now able to raise a question which is natural but would have been quite beyond us. Notice that, with this question, we leave the realm of geometry, and firmly enter that of number theory; but it is the mathematics itself which has compelled us to make this move. Our question is:

Given $s = (x, y)$, $s' = (x', y')$, for which t is it true that $s \mid s'$?

We will describe the answer to this question, which may take a form which, at first glance, will disappoint you.

Of course, we may suppose that $xy \mid x'y'$; otherwise, as Corollary 9 tells us, the answer is that $s \mid s'$ for no value of t. Now let $\gcd(xy, x') = d$, so that $xy = ad$, $x' = bd$, with $\gcd(a, b) = 1$. Since $xy \mid x'y'$, it follows that $ad \mid bdy'$, so that $a \mid by'$. But $\gcd(a, b) = 1$, so that $a \mid y'$. This prepares the way for our answer which is contained in the following theorem.

Theorem 10 *Suppose* $xy \mid x'y'$ *and let* $d = \gcd(xy, x')$, $h = \gcd(x, x')$. *Then*

$$\frac{t^{xy}-1}{t^x-1} \,\Bigg|\, \frac{t^{x'y'}-1}{t^{x'}-1} \quad \Leftrightarrow \quad \frac{t^d-1}{t^h-1} \,\Bigg|\, \frac{y'}{a}.$$

This is proved in [4] or [6] and we will not repeat the proof here. Rather we want to show why the theorem is very useful. First notice that

$$h = \gcd(x, x') = \gcd(x, xy, x') = \gcd(x, d).$$

Next, notice that $\frac{y'}{a}$ is independent of t. There are thus two possibilities:

Case 1 $h = d$, that is, $d \mid x$. Then $\frac{t^d-1}{t^h-1} = 1$, whatever the value of t and, of course, $1 \mid \frac{y'}{a}$. Thus, in this case,

$$s \mid s' \quad \text{for all } t.$$

An example of this case is provided by

$$s = \frac{t^{45}-1}{t^9-1}, \quad s' = \frac{t^{90}-1}{t^6-1}.$$

Then $d = \gcd(45, 6) = 3$ and $x = 9$ so $d \mid x$. We may easily verify the truth of our claim in this case since

$$\frac{s'}{s} = \frac{(t^{90}-1)}{(t^{45}-1)}\frac{(t^9-1)}{(t^6-1)} = \frac{(t^{45}+1)}{(t^3+1)}\frac{(t^9-1)}{(t^3-1)},$$

which is obviously a polynomial in t with integer coefficients.

Case 2 $h \neq d$, that is, $d \nmid x$. Then $h < d$, so that $\frac{t^d-1}{t^h-1}$ is a polynomial in t of positive degree $d - h$ and leading term t^{d-h}. It follows that $\frac{t^d-1}{t^h-1}$ tends to infinity with t and can thus be less than or equal to $\frac{y'}{a}$ (which, as we have already pointed out, is a constant independent of t) for only finitely many integer values of t. Thus, in this case,

$$s \mid s' \quad \text{for only finitely many values of } t.$$

We immediately infer the striking result:

Theorem 11 *If $s \mid s'$ for infinitely many t, then $s \mid s'$ for all t.*

In fact, we have a stronger statement. Remember that we may think of s and s' as polynomials in t with integer coefficients and leading coefficient 1.

Theorem 12 *The following statements are equivalent:*

(i) $s \mid s'$ *for all t;*

(ii) $s \mid s'$ *for infinitely many t; and*

(iii) $s \mid s'$ *as polynomials with integer coefficients.*

Proof We know that (ii) $\Rightarrow$ (i), and it is obvious that (iii) $\Rightarrow$ (i) (see our analysis of the example given in Case 1). Thus it remains to show that (ii) $\Rightarrow$ (iii) for, of course, (i) $\Rightarrow$ (ii). In fact, we prove a more general result which is no harder.

Let $f(t)$, $g(t)$ be two polynomials with integer coefficients, and let g have leading coefficients 1. Suppose that $g(t)$ divides $f(t)$ for infinitely many (positive integer) values of t. *Then $g(t)$ divides $f(t)$ as a polynomial.*

To see this, first divide $f(t)$ by $g(t)$, getting a quotient $q(t)$ and a remainder $r(t)$; both of these will also be polynomials with integer coefficients. We want to show that $r(t)$ is the zero polynomial. If not, it is a polynomial of degree less than that of $g(t)$, and

$$f(t) = q(t)g(t) + r(t). \tag{17}$$

Suppose that $g(t)$ divides $f(t)$ for the increasing sequence of integers $t = t_1, t_2, t_3, \ldots$. Since $r(t)$ has only finitely many zeros, we may drop these from the list (if they occurred) and still have an infinite sequence. Then (17) shows that $g(t)$ divides $r(t)$ for these same values of t. Now degree $g(t)$ > degree $r(t)$. This implies that $\frac{r(t)}{g(t)}$ tends to 0 as t tends to infinity, so that, for t sufficiently large,

$$\left|\frac{r(t)}{g(t)}\right| < 1.$$

But, however large we take t, there is some t_N in our sequence of integers which is larger than t and for which $g(t_N)$ divides $r(t_N)$. We are in a hopeless contradiction—How can $g(t_N)$ divide $r(t_N)$ as integers when the ratio

$$\left|\frac{r(t_N)}{g(t_N)}\right|$$

is less than 1 and $r(t_N) \neq 0$? Thus $r(t)$ must be the zero polynomial, as our statement claimed. □

We close with two examples, treated in some detail. The first example is beyond the scope of a modern computer, if tackled head-on.

Example 13 Let $s = (2, 4) = \frac{t^8-1}{t^2-1}$, $s' = (4, 40) = \frac{t^{160}-1}{t^4-1}$. Then $d = \gcd(8, 4) = 4$, $h = \gcd(2, 4) = 2$. Moreover, $8 = 4a$, so $a = 2$, and $y' = 40$. Thus, by Theorem 10,

$$s \mid s' \quad \Leftrightarrow \quad \frac{t^4 - 1}{t^2 - 1} \Big| 20 \quad \Leftrightarrow \quad t^2 + 1 \mid 20.$$

It is easy to see that this holds for precisely $t = 2, 3$ (remember that $t \geq 2$). We are in Case 2 and $s \mid s'$ only when $t = 2, 3$.

Example 14 We consider the innocent statement $21 = 7 \times 3$. Now 21 and 7 are folding numbers,

$$21 = \frac{2^6 - 1}{2^2 - 1}, \quad 7 = \frac{2^3 - 1}{2^1 - 1}, \quad \text{or} \quad 21 = (2, 3), \ 7 = (1, 3).$$

We look at a general base t and apply Theorem 10. It is easy to see that we are in Case 1, so that

$$\frac{t^3-1}{t-1} \,\Big|\, \frac{t^6-1}{t^2-1} \quad \text{for } all \text{ bases } t.$$

Thus, for instance,

$$(t=3) \quad 13 \mid 91,$$
$$(t=4) \quad 21 \mid 273,$$
$$(t=5) \quad 31 \mid 651 \; \ldots .$$

If we write 7 and 21 in base 2 we get 111 and 10101; and we may restate our conclusion by saying that $111 \mid 10101$, interpreted in *any* base. Thus the statement $7 \mid 21$ is seen as just one case of a completely general fact. Of course, we know by Theorem 11 that

$$\frac{t^3-1}{t-1} \quad \text{must divide} \quad \frac{t^6-1}{t^2-1} \quad \text{as polynomials;}$$

in fact,

$$\frac{(t^6-1)(t-1)}{(t^3-1)(t^2-1)} = t^2 - t + 1.$$

Now consider 21 and 3; 3 is also a folding number,

$$3 = \frac{2^2-1}{2^1-1} = (1,2).$$

We look at a general base t and ask when

$$\frac{t^2-1}{t-1} \,\Big|\, \frac{t^6-1}{t^2-1}.$$

Applying Theorem 10, we conclude that

$$\frac{t^2-1}{t-1} \,\Big|\, \frac{t^6-1}{t^2-1} \quad \Leftrightarrow \quad \frac{t^2-1}{t-1} \,\Big|\, 3 \quad \Leftrightarrow \quad (t+1) \mid 3.$$

Since $t \geq 2$, we conclude that

$$\frac{t^2-1}{t-1} \,\Big|\, \frac{t^6-1}{t^2-1} \quad only \text{ when } t = 2.$$

We are in Case 2 with a vengeance. We now know that $11 \mid 10101$ in base 2 *and in no other base*. Thus the statement $3 \mid 21$ is seen as a highly singular phenomenon!

We close this section with the remark that there is one further generalization which we might have introduced at almost no cost in complication; instead of looking at numbers

$$s(x, y) = \frac{t^{xy} - 1}{t^x - 1},$$

we could have looked at numbers

$$r(x, y) = \frac{t^{xy} - u^{xy}}{t^x - u^x},$$

where t, u are coprime (we may assume $t > u$). All our results, suitably modified, hold in this generality—and, of course, then become even more striking. You can find the details in [9].

• • • **BREAK**

You may wish to experiment and try to *guess* some theorems and results for numbers of the form

$$r(x, y) = \frac{t^{xy} - u^{xy}}{t^x - u}, \quad \text{where } t, u \text{ are coprime and } t > u.$$

For example, verify with a hand-calculator (comparing the results with Example 14) that

$$\text{(a)} \quad \frac{5^3 - 3^3}{5 - 3} \,\Big|\, \frac{5^6 - 3^6}{5^2 - 3^2} \quad \text{but} \quad \frac{5^2 - 3^2}{5 - 3} \nmid \frac{5^6 - 3^6}{5^2 - 3^2};$$

$$\text{(b)} \quad \frac{7^3 - 2^3}{7 - 2} \,\Big|\, \frac{7^6 - 2^6}{7^2 - 2^2} \quad \text{but} \quad \frac{7^2 - 2^2}{7 - 2} \nmid \frac{7^6 - 2^6}{7^2 - 2^2};$$

Express the quotient when $\frac{t^6 - u^6}{t^2 - u^2}$ is divided by $\frac{t^3 - u^3}{t - u}$ as a polynomial in t and u.

4.4 SOME MATHEMATICAL TIDBITS

We insert here three attractive pieces of mathematics related to the set of folding numbers $\mathcal{F}_t$. The material of this section is in no way essential for what follows in later sections, so you may postpone it (or even ignore it!) if you choose; but we think many of you will enjoy these tidbits.

Tidbit 1 A nice number trick which you can try on your friends is to exploit the identity

$$(x, yz) = (xy, z)(x, y). \tag{T1}$$

Of course, (T1) simply reflects the obvious identity

$$\frac{t^{xyz} - 1}{t^x - 1} = \frac{t^{xyz} - 1}{t^{xy} - 1} \cdot \frac{t^{xy} - 1}{t^x - 1};$$

but it is very impressive if you happen to be at a party where a table of $\mathcal{F}_2$ (see Figure 9_2) is prominently displayed. What you do is (appearing to think very hard) divine that the entry in position (3, 6), that is, 37449, is the product of the entry in position (9, 2), that is, 513, and the entry in position (3, 3), that is, 73. Try some other examples for yourself.

• • • **BREAK**

1. Practice this number trick using the entries in Figure 9_2.
2. Write out some of the entries of $\mathcal{F}_3$ expressed in base 10 and try the number trick again.

Tidbit 2 We proved (Theorem 2 of this section) that

$$\gcd(t^a - 1, t^b - 1) = t^d - 1, \quad \text{where } d = \gcd(a, b).$$

It is natural to ask whether the similar relationship holds true for the lcm,[6] that is, whether

$$\operatorname{lcm}(t^a - 1, t^b - 1) = t^\ell - 1, \quad \text{where } \ell = \operatorname{lcm}(a, b). \tag{T2}$$

Now there is an obvious case in which (T2) does hold. If $b \mid a$, then $t^b - 1 \mid t^a - 1$, and $a = \operatorname{lcm}(a, b)$, so

$$\operatorname{lcm}(t^a - 1, t^b - 1) = t^a - 1 = t^\ell - 1.$$

[6]lcm, of course, stands for "least common multiple."

Likewise (T2) holds if $a \mid b$. The remarkable conclusion which we can draw, however, is that (T2) holds *only* in these two very special cases!

The argument runs as follows. We first prove a lemma, that is, a helpful result.

Lemma 15 *The expression of a rational number different from* 1 *as*

$$\frac{t^p - 1}{t^q - 1}$$

(if it exists) is unique.

Proof We must show that, if

$$\frac{t^p - 1}{t^q - 1} = \frac{t^r - 1}{t^s - 1},$$

then either $p = q$, $r = s$ or $p = r$, $q = s$. Now if

$$\frac{t^p - 1}{t^q - 1} = \frac{t^r - 1}{t^s - 1},$$

then

$$t^{p+s} - t^p - t^s = t^{r+q} - t^r - t^q.$$

Suppose, as we may, that q is the *least* of the numbers p, q, r, s. Then, dividing by t^q, we get an obvious contradiction (remember $t \geq 2$) if q is *uniquely* the least, so we must suppose it is not. Thus $q = p$, $q = s$, or $q = r$. We are happy with $q = p$, $q = s$, since $q = p$ implies $r = s$ and $q = s$ implies $p = r$. But if $q = r$ and $q \neq p$, $q \neq s$, we have another obvious contradiction, since then

$$\frac{t^p - 1}{t^q - 1} > 1, \quad \frac{t^r - 1}{t^s - 1} < 1.$$

This proves the lemma. □

Now suppose that (T2) holds. Then since, for any two numbers m, n,

$mn = \gcd(m, n)\ \mathrm{lcm}(m, n)$ (Can you see why this is true?)

we have, by Theorem 2,

$$(t^a - 1)(t^b - 1) = (t^d - 1)(t^\ell - 1)$$

so that

$$\frac{t^a - 1}{t^d - 1} = \frac{t^\ell - 1}{t^b - 1}.$$

Lemma 15 now comes into play to tell us that

$$d = a, \quad b = \ell, \quad \text{that is,} \quad a \mid b,$$

or

$$d = b, \quad a = \ell, \quad \text{that is,} \quad b \mid a.$$

It is amusing to see that (T2) cannot hold precisely because Theorem 2 *does* hold.

• • • **BREAK**

Do you see how you could have avoided the use of Lemma 15 (take a look at Theorem 5)? But we thought Lemma 15 was nice too!

Tidbit 3 We have pointed out in the previous section the important fact that, given any number a prime to t, there are numbers in $\mathcal{F}_t$ having a as a factor; and we have further emphasized the importance of finding the smallest such number s in $\mathcal{F}_t$. Is there a nice algorithm for finding s? The answer is yes!

First, we must determine $h(a)$, the *height* of a (see the definition following the proof of Theorem 6). Now we know (see Chapter 2) that

$$t^{\Phi(a(t-1))} \equiv 1 \bmod a(t-1),$$

so

$$h(a) \mid \Phi(a(t-1)).$$

In fact, we are looking for the *smallest* factor h of $\Phi(a(t-1))$ such that

$$t^h \equiv 1 \bmod a(t-1).$$

Suppose

$$\Phi(a(t-1)) = p_1^{m_1} p_2^{m_2} \cdots p_k^{m_k}$$

is a factorization of $\Phi(a(t-1))$ as a product of primes. We test

$$h_1 = p_1^{m_1-1} p_2^{m_2} \cdots p_k^{m_k}$$

to see if $t^{h_1} \equiv 1 \bmod a(t-1)$. If so, we test

$$h_2 = p_1^{m_1-2} p_2^{m_2} \cdots p_k^{m_k}$$

and continue to lower the exponent of p_1 by 1 until *either* we reach

$$h_i = p_1^{m_1-i} p_2^{m_2} \cdots p_k^{m_k}$$

such that $t^{h_i} \not\equiv 1 \bmod a(t-1)$ *or* the exponent reaches zero, and still $t^{h_i} \equiv 1 \bmod a(t-1)$. In the former case we fix the exponent of p_1 at $m_1 - i + 1$ $(= n_1)$ and start again with h_{i-1} which we now call

$$h_{i+1} = p_1^{n_1} p_2^{m_2} \cdots p_k^{m_k}.$$

Now, however, we operate on the exponent of p_2 just as we did above for the exponent of p_1. In the latter case, we eliminate p_1 and start with

$$h_i = p_2^{m_2} \cdots p_k^{m_k},$$

now operating on the exponent of p_2 just as we did above for the exponent of p_1. Then, after handling p_2, we handle p_3, $p_4, \ldots, p_k$. At the last step q for which

$$h_q \equiv 1 \bmod a(t-1),$$

we will have found $h(a)$. Notice that we can choose the order of the prime factors ourselves.

Let us give an example; we take $t = 3$, $a = 14$. Now

$$\Phi(28) = \Phi(4)\Phi(7) = 2 \times 6 = 12,$$

so we know that $3^{12} \equiv 1 \bmod 28$. Since $12 = 2^2 \cdot 3$, we first try $h_1 = 2 \cdot 3$. Then $3^6 - 1 = 728$, which is certainly divisible by 28. Next we try $h_2 = 3$, but $3^3 \not\equiv 1 \bmod 28$. Thus $h_3 = h_1 = 2 \cdot 3 = 6$ and we must finally try $h_4 = 2$, but again $3^2 \not\equiv 1 \bmod 28$, so $h(14) = h_3 = h_1 = 6$.

• • • BREAK

1. Produce a flow chart to describe the procedure for finding $h(a)$.
2. With $t = 2$, calculate $h(11)$, $h(33)$, and $h(99)$.

Now we know how to calculate $h(a)$, we also know that s will be the smallest integer expressible as

$$\frac{t^{h(a)} - 1}{t^x - 1}$$

which is a multiple of a. (Certainly, if $x = 1$, then $\frac{t^{h(a)}-1}{t^x-1}$ is a multiple of a.) Thus an easy algorithm is to examine the proper factors of $h(a)$ in descending order, stopping the first time we find such a factor x that $\frac{t^{h(a)}-1}{t^x-1}$ is a multiple of a.

Let us return to our example $t = 3$, $a = 14$. We found $h(a) = 6$ so we consider numbers (in $\mathcal{F}_3$) of the form $\frac{3^6-1}{3^x-1}$ with x a proper factor of 6. We first try $x = 3$. Then $\frac{3^6-1}{3^3-1} = 3^3 + 1 = 28$ which is a multiple of 14, so we have found the required s.

As a second example, take $t = 2$, $a = 21$. Then $\Phi(a) = 12$ and we quickly find that $h(a) = 6$. We consider numbers of the form $\frac{2^6-1}{2^x-1}$, with x a proper factor of 6. With $x = 3$, we get 9 which is no good. With $x = 2$ we get 21, which works. Indeed, in this case, our algorithm has obtained for us the fact we already know, namely, that 21 is itself a folding number.

A particularly easy case is that in which a is prime. For let x be the *largest* proper factor of $h(a)$. Then we claim that

$$a \left| \frac{t^{h(a)} - 1}{t^x - 1} \right.,$$

so that

$$s = \frac{t^{h(a)} - 1}{t^x - 1}.$$

For

$$a \left| \frac{t^{h(a)} - 1}{t^x - 1} \cdot \frac{t^x - 1}{t - 1} \right.,$$

but

$$a \nmid \frac{t^x - 1}{t - 1}, \quad \text{by the minimality of } h(a).$$

Thus, since a is prime, it follows (see Theorem 4 of Chapter 2) that

$$a \mid \frac{t^{h(a)} - 1}{t^x - 1}.$$

Let us apply this to the case $t = 2$, $a = 13$. Then

$$\Phi(a) = 12 = 2^2 \cdot 3.$$

We try $h_1 = 2 \cdot 3 = 6$, but $2^6 \not\equiv 1 \bmod 13$; we try $h_2 = 2^2$, but $2^4 \not\equiv 1 \bmod 13$. Thus $h(a) = 12$. It follows that the folding number s we require is given by

$$s = \frac{2^{12} - 1}{2^6 - 1} = 2^6 + 1 = 65.$$

• • • **BREAK**

With $t = 2$, carry out the algorithm described above to calculate s when $a = 11$, $a = 33$, and $a = 99$.

II. GENERAL PAPER-FOLDING

4.5 GENERAL FOLDING PROCEDURES

Suppose we want to construct a regular star $\left\{\frac{11}{3}\right\}$-gon. Then, of course, $b = 11$, $a = 3$ in the notation used at the beginning of Section 2, and we proceed as we did when we wished to construct the regular convex 7-gon in that section—we adopt our ***optimistic strategy*** (which means that we *assume* we've got what we want and, as we will show, we then actually *get* an arbitrarily good approximation to what we want!) Thus we assume we can fold the desired putative angle of $\frac{3\pi}{11}$ at A_0 (see Figure 10(a)) and we adhere to the same principles that we used in constructing the regular 7-gon, namely, we adopt the following rules.

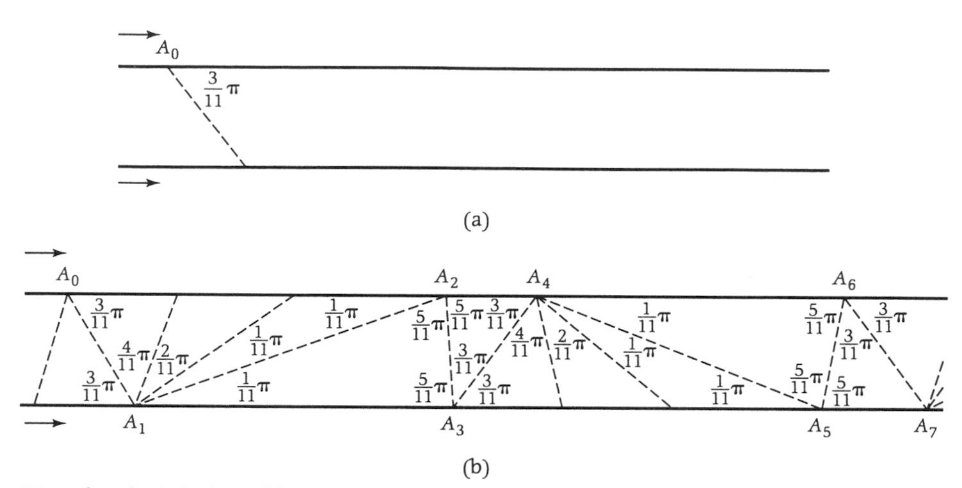

(Note that the indexing of the vertices is *not* the same as that in Figure 1.)

FIGURE 10

(1) Each new crease line goes in the forward (left to right) direction along the strip of paper.

(2) Each new crease line always *bisects* the angle between the last crease line and the edge of the tape from which it emanates.

(3) The bisection of angles at any vertex continues until a crease line produces an angle of the form $\frac{a'\pi}{b}$ where a' is an *odd* number; then the folding stops at that vertex and commences at the intersection point of the last crease line with the other edge of the tape.

Once again the ***optimistic strategy*** works; and following this procedure results in tape whose angles converge to those shown in Figure 10(b). We could denote this folding procedure as $D^1U^3D^1U^1D^3U^1$, interpreted in the obvious way on the tape—that is, the first exponent "1" refers to the one bisection (producing a line in a downward direction) at the vertices A_{6n} (for $n = 0, 1, 2, \ldots$) on the top of the tape; similarly, the "3" refers to the 3 bisections (producing creases in an upward direction) made at the bottom of the tape through the vertices A_{6n+1}; etc. However, since the folding procedure is *duplicated* halfway through, we can abbreviate the notation and write simply $\{1, 3, 1\}$, with the understanding that we alternately fold from the top and bottom of the tape as described, with the *number* of bisections at each vertex running, in order, through the values 1, 3, 1, ... We call this a ***primary folding procedure of period 3*** or a ***3-period folding***.

A proof of convergence for the general folding procedure of arbitrary period may be given that is similar to the one we gave for the primary folding procedure of period 2. Or one could revert to an error-correction type of proof like that given for the 7-gon in Section 1. We leave the details to the reader, and explore here what we can do with this (1, 3, 1)-tape. First, note that, starting with the putative angle $\frac{3\pi}{11}$ at the top of the tape, we produce a putative angle of $\frac{\pi}{11}$ at the bottom of the tape, then a putative angle of $\frac{5\pi}{11}$ at the top of the tape, then a putative angle of $\frac{3\pi}{11}$ at the bottom of the tape, and so on. Hence we see that we could use this tape to fold a star $\left\{\frac{11}{3}\right\}$-gon, a convex 11-gon, and a star $\left\{\frac{11}{5}\right\}$-gon. More still is true; for, as we see, if there are crease lines enabling us to fold a star $\left\{\frac{11}{a}\right\}$-gon, there will be crease lines enabling us to fold star $\left\{\frac{11}{2^k a}\right\}$-gons, where $k \geq 0$ takes any value such that $2^{k+1}a < 11$. These features, described for $b = 11$, would be found with any odd number b. However, this tape has a special symmetry as a consequence of its *odd* period; namely, if it is "flipped" about the

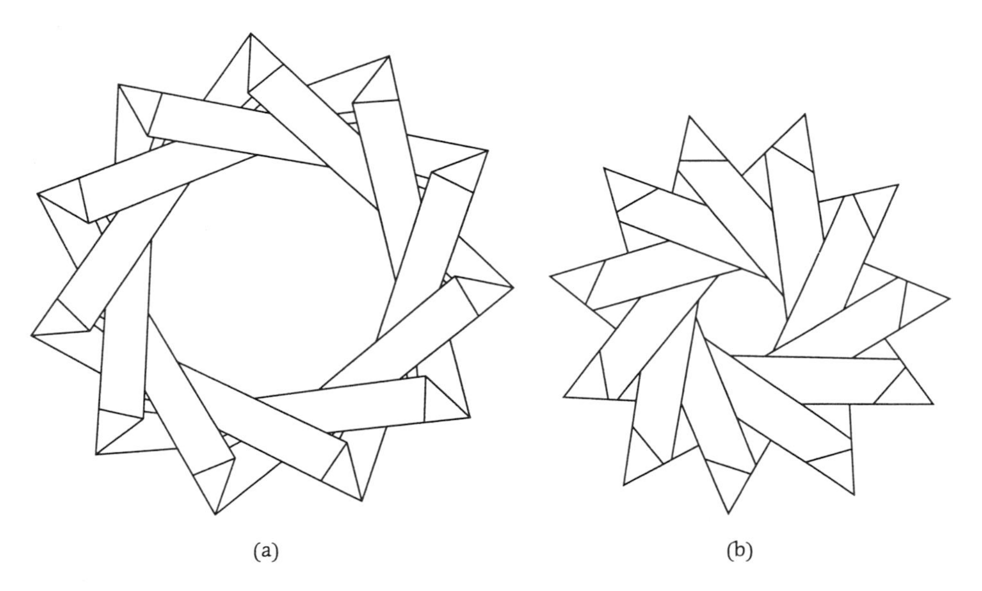

(a) (b)

FIGURE 11

horizontal line halfway between its parallel edges, the result is a *translate* of the original tape. As a practical matter this special symmetry of the tape means that we can use either the top edge or the bottom edge of the tape to construct our polygons. On tapes with an *even* period, the top edge and the bottom edge of the tape are not translates of each other (under the horizontal flip), which simply means that care must be taken in choosing the edge of the tape used to construct a specific polygon. Figures 11(a) and (b) show the completed $\{\frac{11}{3}\}$-, $\{\frac{11}{4}\}$-gons, respectively.

Now, to set the scene for the number theory of Section 6, let us look at the patterns in the *arithmetic* of the computations when $a = 3$ and $b = 11$. Referring to Figure 10(b) we observe that[7]

The smallest angle to the right of A_n where	is of the form $\frac{a}{11}\pi$ where	and the number of bisections at the *next* vertex
$n = 0$	$a = 3$	$= 3$
$n = 1$	$a = 1$	$= 1$
$n = 2$	$a = 5$	$= 1$
$n = 3$	$a = 3$	$= 3$
$n = 4$	$a = 1$	$= 1$
$n = 5$	$a = 5$	$= 1$

We could write this in shorthand form as follows:

$$(b =)\ 11 \left| \begin{matrix} (a =)\ 3 & 1 & 5 \\ 3 & 1 & 1 \end{matrix} \right| . \tag{18}$$

Observe that, had we started with the putative angle of $\frac{\pi}{11}$, then the *symbol* (18) would have taken the form

$$(b =)\ 11 \left| \begin{matrix} (a =)\ 1 & 5 & 3 \\ 1 & 1 & 3 \end{matrix} \right| . \tag{18$'$}$$

In fact, it should be clear that we can *start anywhere* (with $a = 1, 3,$ or 5) and the resulting symbol, analogous to (18), will be obtained

[7]Notice that, referring to Figure 10(b), to obtain an angle of $\frac{3\pi}{11}$ at $A_0, A_6, A_{12}, \ldots$, the folding instructions would more precisely be $U^3D^1U^1D^3U^1D^1 \ldots$. But we don't have to worry about this distinction.

by cyclic permutation of the matrix component of the symbol, placing our choice of a in the first position along the top row.

In general, suppose we wish to fold a $\left\{\frac{b}{a}\right\}$-gon, with b, a odd and $a < \frac{b}{2}$. Then we may construct a symbol[8] as follows. Let us write

$$b \left| \begin{matrix} a_1 & a_2 & \cdot & \cdot & \cdot & a_r \\ \\ k_1 & k_2 & \cdot & \cdot & \cdot & k_r \end{matrix} \right| \tag{19}$$

where b, a_i $(a_1 = a)$ are odd, $a_i < \frac{b}{2}$, and

$$b - a_i = 2^k a_{i+1}, \quad i = 1, 2, \ldots, r, \quad a_{r+1} = a_1. \tag{20}$$

We emphasize, as we will prove at the beginning of Section 6, that, given any two odd numbers a and b, with $a < \frac{b}{2}$, there is always a completely determined unique symbol (19) with $a_1 = a$. At this stage, we do not assume that $\gcd(b, a) = 1$, but we have assumed that the list $a_1, a_2, \ldots, a_r$ is without repeats. Indeed, if $\gcd(b, a) = 1$ we say that the symbol (19) is ***reduced***, and if there are no repeats among the a_i's we say that the symbol (19) is ***contracted***. (It is, of course, theoretically possible to consider symbols (19) in which repetitions among the a_i are allowed.) We regard (19) as encoding the general folding procedure to which we have referred.

• • • **BREAK**

If we wish to fold a 17-gon we may start with $b = 17$, $a = 1$ and construct the symbol

$$(b =)\, 17 \left| \begin{matrix} (a =)\, 1 \\ 4 \end{matrix} \right|$$

which tells us that folding U^4D^4 will produce tape (usually denoted (4, 4)-tape, though it would be correct to call it simply (4)-tape) that can be used to construct a FAT 17-gon. In fact, this tape can also be used to construct FAT

$$\left\{\frac{17}{2}\right\}\text{-},\quad \left\{\frac{17}{4}\right\}\text{-},\quad \text{and} \quad \left\{\frac{17}{8}\right\}\text{-gons.}$$

[8] More exactly, a 2-symbol. Later on, we introduce a more general t-symbol, $t \geq 2$.

However, if we wish to fold a $\{\frac{17}{3}\}$-gon we start with $b = 17$, $a = 3$ and construct the symbol

$$(b =)\ 17 \left| \begin{array}{rrr} (a =)\ 3 & 7 & 5 \\ 1 & 1 & 2 \end{array} \right|$$

which tells us to fold $U^1 D^1 U^2 D^1 U^1 D^2$—or, more simply, to use the (1, 1, 2)-folding procedure—to produce (1, 1, 2)-tape from which we can fold the FAT $\{\frac{17}{3}\}$-gon. Again, we get more than we initially sought, since we can also use the (1, 1, 2)-tape to construct

$$\text{FAT} \left\{\frac{17}{6}\right\}\text{-},\quad \left\{\frac{17}{7}\right\}\text{-},\quad \text{and} \left\{\frac{17}{5}\right\}\text{-gons.}$$

We can combine all the possible symbols for $b = 17$ into one *complete* symbol, adopting the notation

$$17 \left| \begin{array}{c|ccc} 1 & 3 & 7 & 5 \\ & & & \\ 4 & 1 & 1 & 2 \end{array} \right| .$$

Check your understanding of the complete symbol for $b = 93$ by doing the calculation to fill in the blanks below:

$$93 \left| \begin{array}{cccc|cccc|cccccc} 1 & 23 & 35 & 29 & 5 & 11 & 41 & 13 & 7 & 43 & 25 & 17 & 19 & 37 \\ & & & & & & & & & & & & & \\ 2 & 1 & 1 & 6 & 3 & ? & ? & 4 & 1 & ? & ? & ? & ? & ? \end{array} \right| .$$

(Notice that no multiples of 3 or 31 appear in the top row. Why not?)

You may wish to calculate other complete symbols and study them, along with the complete symbols for 17 and 93 above, to see if you notice any consistent patterns or features in them.

Suppose we allow unreduced symbols. Calculate the symbol with $b = 51, a = 9$. What do you notice? Why is it pointless to allow unreduced symbols?

4.6 THE QUASI-ORDER THEOREM

The first claim we have to substantiate, made in the previous section, is that, given positive odd integers b, a with $a < \frac{b}{2}$, there is always a unique contracted symbol

$$b\begin{vmatrix} a_1 & a_2 & \cdot & \cdot & \cdot & a_r \\ \\ k_1 & k_2 & \cdot & \cdot & \cdot & k_r \end{vmatrix}, \quad a_1 = a, \quad a_i \neq a_j \quad \text{if } i \neq j, \tag{21}$$

where each a_i is odd, $a_i < \frac{b}{2}$, and

$$b - a_i = 2^{k_i} a_{i+1}, \quad i = 1, 2, \ldots, r, \; a_{r+1} = a_1. \tag{22}$$

We argue as follows. We fix b and let S be the set of positive odd numbers $a < \frac{b}{2}$. Given $a \in S$, define a' by the rule

$$b - a = 2^k a', \quad k \textit{ maximal}; \tag{23}$$

that is, we take as many factors of 2 as we can out of $b - a$. Notice that $k \geq 1$, since $b - a$ is certainly even. We claim that $a' \in S$. First, a' is obviously odd. Second, $2a' \leq 2^k a' = b - a < b$, so $a' < \frac{b}{2}$. Thus (23) describes a function $\psi: S \to S$, such that $\psi(a) = a'$. We will show that ψ is a *permutation* of the finite set S; it is sufficient to show that ψ maps S *onto* itself; and, to show this, it is certainly enough to exhibit a function $\varphi: S \to S$ such that $\psi\varphi(a') = a'$. We define φ as follows: given $a' \in S$, let k be *minimal* such that $2^k a' > \frac{b}{2}$ and set $\varphi(a') = a$, where

$$a = b - 2^k a'. \tag{24}$$

Notice that $k \geq 1$, since $a' < \frac{b}{2}$, so that a is odd; that $a < \frac{b}{2}$, since $2^k a' > \frac{b}{2}$; and that $b > 2^k a'$, since $2^{k-1} a' < \frac{b}{2}$, so that a is positive. Thus $a \in S$; and comparison of (23), (24) shows that, as claimed, $\psi\varphi(a') = a'$. Thus ψ is a permutation and φ is the inverse permutation.

The permutation ψ has one more important property. We write $\psi(a) = a'$, as above, and claim that

$$\gcd(b, a) = \gcd(b, a'). \tag{25}$$

For it is clear from (23) that if $d \mid b$ and $d \mid a'$, then $d \mid b$ and $d \mid a$. Conversely, if $d \mid b$ and $d \mid a$, d is *odd* and $d \mid 2^k a'$ so $d \mid b$ and $d \mid a'$. Thus if a_1 in (21) is coprime to b, so are $a_2, a_3, \ldots, a_r$, and we may, if we wish, confine ψ and φ to the subset S_0 of S consisting of those $a \in S$ which are coprime to b; that is, we may confine ourselves to reduced symbols.

We now use the fact that, given *any* permutation ψ of a finite set S_0 and any $a \in S_0$, then a must generate a ***cycle***, in the sense that, if we iterate ψ, getting

$$a,\ \psi(a),\ \psi^2(a),\ \psi^3(a),\ \ldots$$

$$(\text{here } \psi^2(a) = \psi(\psi(a)), \text{ etc., and we may write } \psi^0(a) \text{ for } a),$$

we must eventually repeat, that is, we will find $m > 0$ such that

$$a,\ \psi(a),\ \ldots,\ \psi^{m-1}(a)$$

are all different but $\psi^m(a) = a$. In case this is not clear to you, we give the easy proof. Certainly, since S_0 is finite, we must eventually repeat in the weaker sense that we find $s \geq 0$, $m > 0$ such that $\psi^s(a) = \psi^{s+m}(a)$. Suppose this is the *first* time we get a repeat. We claim that $s = 0$; for, if not, we have

$$\psi(\psi^{s-1}(a)) = \psi(\psi^{s-1+m}(a)).$$

But ψ is one-to-one, so $\psi^{s-1}(a) = \psi^{s-1+m}(a)$ and the given repeat wasn't our first repeat. Thus $s = 0$ and $a = \psi^m(a)$. Of course, with regard to (21), $m = r + 1$. This completes the proof of our claim.

So we have a universal algorithm for folding a $\left\{\frac{b}{a}\right\}$-gon, where a, b are coprime odd integers with $a < \frac{b}{2}$. But, from the number-theoretic point of view, it turns out that we have much more. For, reverting to (21), let

$$k = \sum_{i=1}^{r} k_i \quad (\text{we may call this the } \textbf{\textit{fold-total}}).$$

Then we prove what we call the ***Quasi-Order Theorem***.[9]

Theorem 16 *Suppose that* (21) *is not only contracted but also reduced. Then the quasi-order of* $2 \bmod b$ *is* k. *That is,* k *is the smallest positive integer such that*

$$2^k \equiv \pm 1 \bmod b.$$

In fact, $2^k \equiv (-1)^r \bmod b$.

Proof The proof is really a triumph of technique! First, we find it convenient to think in terms of the φ-function rather than

[9]You may like to know that this result is very new; we first published it in 1985 (see [4, 6]).

the ψ-function. Thus we work ***backward*** in constructing our symbol (21). Also we will find it convenient to repeat the initial number a_1. Precisely, we write our ***modified symbol*** as

$$b\begin{pmatrix} c_1 & & c_2 & & c_3 & & \cdot & & \cdot & & \cdot & & c_r & & c_1 \\ & \ell_1 & & \ell_2 & & \ell_3 & & \cdot & & \cdot & & \cdot & & \ell_r & \end{pmatrix}, \tag{26}$$

where[10] (compare (22) and (24)) ℓ_i is minimal such that $2^{\ell_i}c_i > \frac{b}{2}$ and

$$b - c_{i+1} = 2^{\ell_i}c_i, \quad i = 1, 2, \ldots, r \quad (c_{r+1} = c_1). \tag{27}$$

We set $\ell = \sum_{i=1}^{r} \ell_i$, and our first task will be to prove that

$$2^\ell \equiv (-1)^r \bmod b. \tag{28}$$

To this end, consider the $(\ell + 1)$ numbers, all less than $\frac{b}{2}$,

$$c_1, 2c_1, \ldots, 2^{\ell_1-1}c_1, c_2, 2c_2, \ldots, 2^{\ell_2-1}c_2, c_3, \ldots, c_r, \ldots, 2^{\ell_r-1}c_r, c_1. \tag{29}$$

In this sequence there are r places where we *switch* from c_i to c_{i+1}. If we rewrite the sequence (29) as

$$n_1, n_2, n_3, \ldots, n_{\ell+1},$$

then

$$n_{j+1} \begin{cases} = 2n_j & \text{if there is no switch,} \\ \equiv -2n_j \bmod b & \text{at a switch, by (27).} \end{cases} \tag{30}$$

Since there are r switches, we conclude from (30) that

$$n_{\ell+1} \equiv (-1)^r 2^\ell n_1 \bmod b. \tag{31}$$

But $n_{\ell+1} = n_1 = c_1$, and c_1 is coprime to b. Thus (31) implies that

$$2^\ell \equiv (-1)^r \bmod b,$$

which is (28).

To show that ℓ is the quasi-order of 2 mod b, we must show that, for every positive $m < \ell$, the congruence

$$2^m \equiv \pm 1 \bmod b \tag{32}$$

[10] In fact, if we compare (21) and (26), $k_i = \ell_{r+1-i}$, $a_i = c_{r+2-i}$.

is *false*. Now (32) implies, in the light of (30), that $n_{m+1} \equiv \pm c_1 \bmod b$, with $m + 1 < \ell + 1$. We first show that

$$n_{m+1} \equiv c_1 \bmod b$$

is impossible. Now, as we have remarked, it follows from the definition of the φ-function that all the numbers n_j in the sequence (29) satisfy $n_j < \frac{b}{2}$. Thus

$$n_{m+1} \equiv c_1 \bmod b \quad \text{implies} \quad n_{m+1} = c_1.$$

But either n_{m+1} is even or it is some c_i different from c_1. Thus, since the symbol (26) is contracted and, of course, c_1 is odd, $n_{m+1} = c_1$ is impossible.

Finally, we show that $n_{m+1} \equiv -c_1 \bmod b$ is impossible. For $n_{m+1} + c_1$ is a positive integer less than b, so it is not divisible by b. This completes the proof of the theorem. □

Theorem 16 is striking in that, given b, we compute k starting with *any* a which is odd, less than $\frac{b}{2}$, and coprime to b. Of course, the choice $a = 1$ is always available and is the one to make when we seek folding instructions for producing a regular convex b-gon.

• • • **BREAK**

1. Why is it that, given a *finite* set S, a function $\psi: S \to S$ is a permutation if and only if it maps S onto itself.
2. Is it true that ψ is a permutation if and only if it is one-to-one?

In Section 3 we emphasized the importance of generalization and, almost throughout, replaced the base 2 by an arbitrary base $t \geq 2$. Such a generalization is perfectly possible here and leads to a generalization of Theorem 16. However, the algorithm for executing the ψ-function is more complicated, so we will not go into details here; the ambitious reader may consult [6]. We will, though, just show you what happens if $t = 3$. We form our 3-symbol starting with a and b coprime to 3, a coprime to b, and $a < \frac{b}{2}$ (notice this: we do *not* require $a < \frac{b}{3}$). The (modified) ψ-function now allows us to consider $b - a$ or $b + a$. Exactly one of these is divisible by 3, and that is the one we take. We adjoin a third row to our symbol; in this row we write 1 if we took $b - a$ and 0 if we took $b + a$. The second row records the number of times we took the factor 3 out of $b - a$ or $b + a$ to get a'. Here's an example. We have (with $t = 3$, $b = 19$, and $a_1 = 1$) the 3-symbol

$$19 \left| \begin{array}{cccccc} 1 & 2 & 7 & 4 & 5 & 8 \\ 2 & 1 & 1 & 1 & 1 & 3 \\ 1 & 0 & 1 & 1 & 0 & 0 \end{array} \right| .$$

This symbol records the following calculations:

$$19 - 1 = 3^2 \cdot 2,$$

$$19 + 2 = 3^1 \cdot 7,$$

$$19 - 7 = 3^1 \cdot 4,$$

$$19 - 4 = 3^1 \cdot 5,$$

$$19 + 5 = 3^1 \cdot 8,$$

$$19 + 8 = 3^3 \cdot 1.$$

We will be content simply to state the general Quasi-Order Theorem (see [6]) for details). Thus, given a base t, we select positive integers b, a such that $a < \frac{b}{2}$, b is prime to t, and $t \nmid a$. We may then form a t-symbol

$$b \left| \begin{array}{cccccc} a_1 & a_2 & \cdot & \cdot & \cdot & a_r \\ k_1 & k_2 & \cdot & \cdot & \cdot & k_r \\ \varepsilon_1 & \varepsilon_2 & \cdot & \cdot & \cdot & \varepsilon_r \end{array} \right|, \qquad \varepsilon_i = 0 \text{ or } 1.$$

Let $k = \sum_{i=1}^{r} k_i$, $\varepsilon = \sum_{i=1}^{r} \varepsilon_i$. Then the quasi-order of t mod b is k and, in fact, $t^k \equiv (-1)^\varepsilon \bmod b$.

Reverting to our example above, the conclusion is that the quasi-order of 3 mod 19 is 9 ($= 2 + 1 + 1 + 1 + 1 + 3$) and that, in fact,

$$3^9 \equiv -1 \bmod 19,$$

choosing the minus sign since there is an *odd* number of 1's in the third row.

• • • **BREAK**

1. Why were we right in saying that, in constructing a 3-symbol, exactly one of $b - a$, $b + a$ would be divisible by 3?
2. Try constructing some examples of complete 3-symbols yourself, say, with $b = 20$, $b = 25$, and $b = 37$. (Some 3-symbols may have more than one part, which is analogous to what was shown to be the case with the 2-symbol when $b = 17$ and when $b = 93$ in the last break in Section 4.)
3. Why didn't we need a third row when we constructed symbols with $t = 2$?

There are two rather remarkable examples of our symbol (21), that is, with $t = 2$. One is

$$23 \left| \begin{matrix} 1 & 11 & 3 & 5 & 9 & 7 \\ \\ 1 & 2 & 2 & 1 & 1 & 4 \end{matrix} \right|$$

telling us that

$$2^{11} \equiv 1 \bmod 23, \quad \text{or} \quad 23 \mid 2^{11} - 1. \tag{33}$$

This is remarkable because $2^{11} - 1$ is a ***Mersenne number***, that is, a number of the form $2^p - 1$, where p is prime. Abbé Mersenne hoped that all these numbers would be prime; but (33) shows that this is not so (of course, this was already known long before the invention of the symbol (21)).

The second example is even more remarkable; it is

$$641 \left| \begin{matrix} 1 & 5 & 159 & 241 & 25 & 77 & 141 & 125 & 129 \\ \\ 7 & 2 & 1 & 4 & 3 & 2 & 2 & 2 & 9 \end{matrix} \right| \tag{34}$$

telling us that

$$2^{32} \equiv -1 \bmod 641, \quad \text{or} \quad 641 \mid 2^{32} + 1. \tag{35}$$

This is striking because $2^{32} + 1$ is a ***Fermat number*** F_n, that is, a number of the form

$$2^{2^n} + 1.$$

The French mathematician, Pierre Fermat hoped that all those numbers would be prime; but (35) shows that this is not so (the factorizability of $2^{32}+1$ was first noticed by the Swiss mathematician Leonhard Euler).

Apparently, our symbol only gives one factor of $2^{11}-1$ or $2^{32}+1$, not the other. We will show that this is not so—we get both factors. The context for this part of the story is the following natural question, which was first raised in [8].

Given a folding procedure $(k_1,k_2,\ldots,k_r)$, *what polygons can we fold with it?*

The equivalent arithmetical (or algebraic) question is this:

Given $(k_1,k_2,\ldots,k_r)$, *find* b, $a_1,a_2,\ldots,a_r$, *so that*

$$b \left| \begin{matrix} a_1 & a_2 & \cdots & a_r \\ k_1 & k_2 & \cdots & k_r \end{matrix} \right| \tag{21}$$

is a reduced and contracted (2-) *symbol.*

Of course, for (21) to be contracted, it is necessary that $(k_1, k_2, \ldots, k_r)$ should not consist of repetitions of some (shorter) sequence $(k_1, k_2, \ldots, k_s)$. (This condition is also, in fact, sufficient.)

We now show how to find b, $a_1, a_2, \ldots, a_r$ in (21). In fact, we solve the simultaneous equations (22), that is, $b - a_i = 2^{k_i}a_{i+1}$, $i = 1, 2, \ldots, r$, $a_{r+1} = a_1$, for the "unknowns" $a_1, a_2, \ldots, a_r$, obtaining

$$Ba_i = bA_i, \quad i = 1, 2, \ldots, r, \tag{36}$$

where

$$B = 2^k - (-1)^r \tag{37}$$

and

$$A_i = 2^{k-k_{i-1}} - 2^{k-k_{i-1}-k_{i-2}} + 2^{k-k_{i-1}-k_{i-2}-k_{i-3}} - \cdots + (-1)^r 2^{k_i} - (-1)^r. \tag{38}$$

Let us explain the notation of (37), (38). First remember that $k = \sum_{i=1}^{r} k_i$. Then, in the expression for A_i, we interpret the subscripts on the k's to be "mod r"; thus, for example, if $i = 1$, then $k_0 = k_r$, $k_{-1} = k_{r-1}$, etc. (This makes good sense since the symbol (21) would repeat if allowed, so that we may think of it as written on a cylinder.)

Now B and A_i are determined by $(k_1, k_2, \ldots, k_r)$ and then (36) tells us that

$$\frac{b}{a_i} = \frac{B}{A_i}.$$

Thus, since b and a_i are to be coprime, we find them by reducing the fraction $\frac{B}{A_i}$. Notice that this works, that is, the b we get is independent of i, because $\gcd(B, A_i)$ is itself independent of i. The argument is just as for (25), being based on the fact that B is odd and, as you may prove from (37), (38),

$$B - A_i = 2^{k_i} A_{i+1}.$$

If we write d for $\gcd(B, A_i)$, then

$$B = db, \quad A_i = da_i. \tag{39}$$

Let us, as an example, show how to apply our algorithm to fill in the reduced symbol

$$b \left| \begin{array}{cccc} a_1 & a_2 & a_3 & a_4 \\ & & & \\ 1 & 2 & 2 & 3 \end{array} \right| \quad (k = 8). \tag{40}$$

From (37) we find $B = 2^8 - 1 = 255$; and from (38) we find

$$A_1 = 2^5 - 2^3 + 2^1 - 1 = 25,$$

$$A_2 = 2^7 - 2^4 + 2^2 - 1 = 115,$$

$$A_3 = 2^6 - 2^5 + 2^2 - 1 = 35,$$

$$A_4 = 2^6 - 2^4 + 2^3 - 1 = 55.$$

Now $\gcd(B, A_1) = 5$; hence, dividing by 5, we find (from (39))

$$b = 51, \quad a_1 = 5, \quad a_2 = 23, \quad a_3 = 7, \quad a_4 = 11.$$

• • • BREAK

1. Start with $b = 51$ and $a_1 = 5$ and construct the symbol (21) from scratch. You should see, with some satisfaction, that the bottom row of the symbol will be just as in (40)—and, of course, the a_i's will be as given above.

2. Construct the *complete* 2-symbol associated with $b = 43$ and with $b = 51$. Verify that, in these symbols and the symbols you constructed during the last break in Section 4, the fold-total k and the parity[11] of r depend only on b and not on a.

Let us make one remark before passing on to the complementary factors for our Mersenne number $2^{11} - 1$ and our Fermat number $2^{32} + 1$. As you saw in the exercise suggested in the break, once we have found b and a_1 it is easier to complete the symbol (40) just by applying the ψ-function (this would also give us a check on our calculation). Thus a more efficient algorithm is this:

(i) calculate B, A_1 from (37), (38);
(ii) calculate $d = \gcd(B, A_1)$;
(iii) determine b, a_1 from $B = db$, $A_1 = da_1$;
(iv) complete the symbol (21) using the ψ-function.

Now consider again the symbol

$$23 \left| \begin{array}{cccccc} 1 & 11 & 3 & 5 & 9 & 7 \\ \\ 1 & 2 & 2 & 1 & 1 & 4 \end{array} \right|.$$

Then (37) tells us that $B = 2^{11} - 1$ and so (36) tells us that

$$2^{11} - 1 = 23A_1.$$

Thus we will complete the factorization of $2^{11} - 1$ by calculating A_1. From (38) this yields

$$A_1 = 2^7 - 2^6 + 2^5 - 2^3 + 2^1 - 1 = 89,$$

so that

$$2^{11} - 1 = 23 \times 89. \tag{41}$$

Likewise, applying (37) and (36) to the symbol (34) tells us that $B = 2^{32} + 1$ and $2^{32} + 1 = 641A_1$, where, by (38),

$$A_1 = 2^{23} - 2^{21} + 2^{19} - 2^{17} + 2^{14} - 2^{10} + 2^9 - 2^7 + 1 = 6{,}700{,}417,$$

[11] "Parity" means "oddness or evenness."

so that

$$2^{32} + 1 = 641 \times 6{,}700{,}417. \tag{42}$$

It is striking—don't you agree?—that (41) and (42) have been obtained without ever expressing $2^{11} - 1$ or $2^{32} + 1$ in base 10—and that the fact that 641 is a factor of $2^{32} + 1$ was established without ever using a number bigger than 641.

• • • **BREAK**

It has recently been shown that the smallest prime factor of the Fermat number

$$F_6 \ (= 2^{2^6} + 1) \quad \text{is } 274177;$$

and that the smallest prime factor of the Fermat number

$$F_9 \ (= 2^{2^9} + 1) \quad \text{is } 2424833.$$

You may wish to *check* that this is so by constructing the symbol (21). You'll need no piece of technology more sophisticated than a hand calculator, though you could program a computer if you wanted to. We found (see [11]) that the symbol with $b = 274177$ that begins with $a_1 = 1$ has 19 entries and we have not found any symbol with $b = 274177$ having fewer entries. Further, with $b = 2424833$, the symbol beginning with $a_1 = 1$ has 237 entries, but the symbol beginning with $a_1 = 65537$ has only 213 entries. We don't know whether or not the symbols we have found verifying the factors of F_6 and F_9 are the shortest possible or not. Can any reader do better? (Notice that $65537 = F_4$!)

4.7 APPENDIX: A LITTLE SOLID GEOMETRY

The folded strips of paper that produce equilateral triangles and regular pentagons can be used to construct the models shown in Figure 12. In particular, the models shown in Figure 12(a) are all *braided* together with straight strips of equilateral triangles (from the D^1U^1-tape), squares (folded exactly), or the strips that produce regular convex pentagons (from the D^2U^2-tape). By contrast, the models in Figure 12(b) are all constructed by gluing together parts

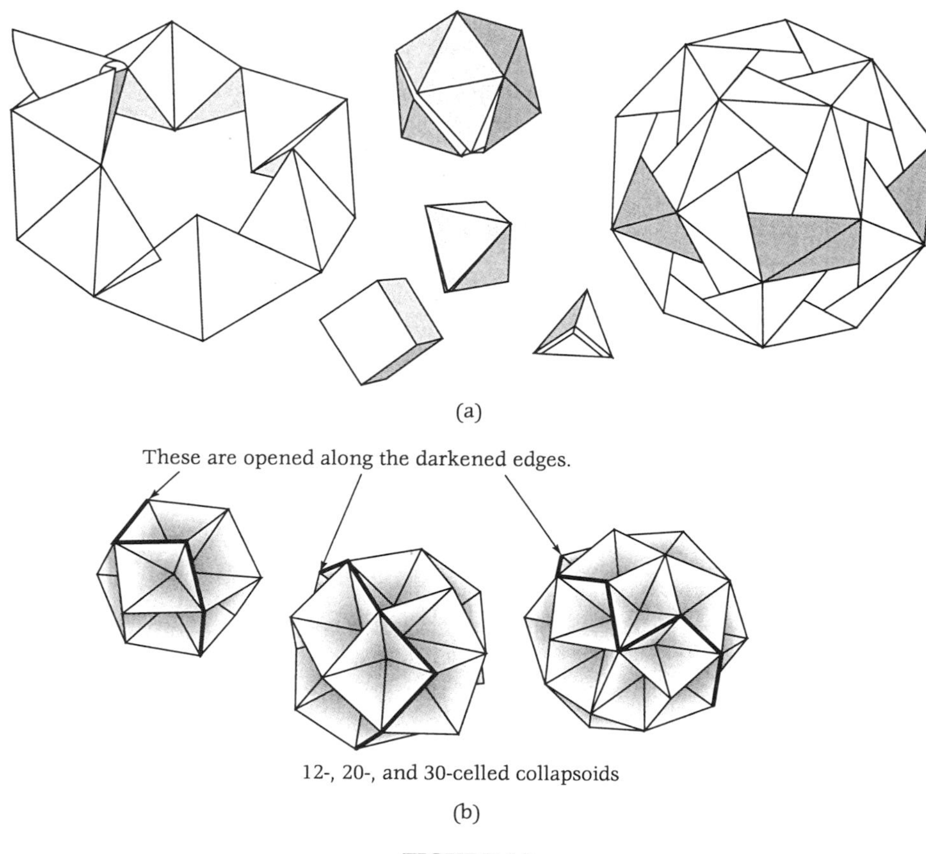

FIGURE 12

of the strips of equilateral triangles and they collapse in interesting ways (when a slit is made on the surface of the model along a sequence of edges connecting any pair of opposite vertices). Because of their inherent beauty and symmetry, you may wish to construct some of these models. In fact, we very strongly encourage you to do so. We have provided very detailed instructions on precisely how to carry out the construction of these and other models in [7].

Once you have the models in hand you will be in a good position to appreciate, or possibly to discover for yourself, Euler's famous formula connecting the number of vertices, edges, and faces of a polyhedron; you may also use these models to verify Descartes' formula (see [2]) concerning the total angular deficiency on a given closed surface. We have found that the mathematics connected with these, and similar, models is both accessible and interesting to mathematics students at all levels (again, see [7]). We show the models here to emphasize yet another way in which mathematics leads us in surprising directions—this time to very important ideas in topology.

In an actual presentation at Hunter College, City University of New York, by one of us (see [12]) the models in Figure 12(a), which include a ***rotating ring of tetrahedra*** and a ***Golden Dodecahedron***, were taken apart to show that they were constructed from straight strips; and the models in Figure 12(b) (called ***collapsoids***) were collapsed (after releasing triangular tabs attached to a "seam" running along edges connecting opposite vertices) to show that they could be folded up into $\frac{4}{6}$, $\frac{5}{6}$, and $\frac{6}{6}$ of a complete hexagon, respectively. Reproducing the effect of a real-life presentation on the printed page is impossible. However, a careful inspection of Figure 12(a) may reveal, at least in some cases, how those models could have been braided together from straight strips, and Figure 13 illustrates how the collapsoids of Figure 12(b) look at an intermediate stage as

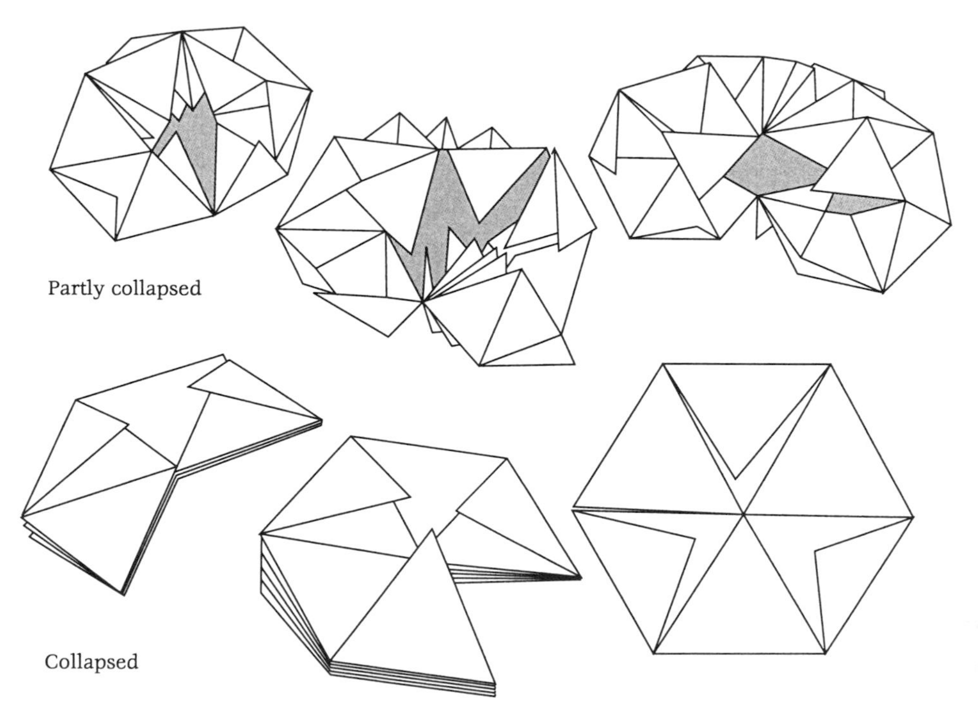

FIGURE 13

they collapse. So we hope you won't lose too much by not having been able to attend the Hunter College presentation!

• • • **FINAL BREAK**

Using gummed tape (or some alternative) try to construct a regular tetrahedron, a cube, and a regular octahedron. See Figure 12(a) for inspiration.

(**Harder**) Use D^2U^2-tape to construct the Golden Dodecahedron (see Figure 12(a) again, and for the same reason).

REFERENCES

1. Coxeter, H.S.M., *Regular Polytopes*, Macmillan Mathematics Paperbacks, New York, 1963.
2. Hilton, Peter, and Jean Pedersen, Descartes, Euler, Poincaré, Pólya and polyhedra, *L'Enseign. Math.*, **27**, 1981, 327–343.
3. Hilton, Peter, and Jean Pedersen, Approximating any regular polygon by folding paper; An interplay of geometry, analysis and number theory, *Math. Mag.*, **56**, 1983, 141–155.
4. Hilton, Peter, and Jean Pedersen, Folding regular star polygons and number theory, *Math. Intelligencer*, **7**, No. 1, 1985, 15–26.
5. Hilton, Peter, and Jean Pedersen, Certain algorithms in the practice of geometry and the theory of numbers, *Publ. Sec. Mat. Univ. Autonoma Barcelona*, **29**, No. 1, 1985, 31–64.
6. Hilton, Peter, and Jean Pedersen, *Geometry in Practice and Numbers in Theory*, Monographs of Undergraduate Mathematics, No. **16**, 1987, 37 pp. (Available from the Department of Mathematics, Guilford College, Greensboro, North Carolina 27410, USA)
7. Hilton, Peter, and Jean Pedersen, *Build Your Own Polyhedra*, Addison-Wesley, Menlo Park, California, 1987, (reprinted 1994), 175 pp.
8. Hilton, Peter, and Jean Pedersen, On the complementary factor in a new congruence algorithm, *Int. J. Math. and Math. Sci.*, **10**, No. 1, 1987, 113–123.

9. Hilton, Peter, and Jean Pedersen, On a generalization of folding numbers, *Southeast Asian Bulletin of Mathematics*, **12**, No. 1, 1988, 53–63.

10. Hilton, Peter, and Jean Pedersen, Folding regular polygons and number theory. *Mathematics in Education* (edited by Themistocles M. Rassias), University of La Verne, Le Verne, California, 1992, pp. 17–50.

11. Hilton, Peter, and Jean Pedersen, On factoring $2^k \pm 1$, *The Mathematics Educator*, **5**, No. 1, 1994, 29–32.

12. Hilton, Peter, and Jean Pedersen, Geometry: A gateway to understanding, *The College Mathematics Journal* **24**, No. 4, 1993, 298–317.

13. Wilde, Carroll O., *The Contraction Mapping Principle*, **UMAP Unit 326** (available from COMAP, Inc., 57 Bedford Street, Lexington, MA 02173-4428), 1978.

5 CHAPTER Quilts and Other Real-World Decorative Geometry

5.1 QUILTS

Quilts can be incredibly beautiful—and interesting too. Look at the four quilts that we've got on the next few pages (see Figures 1, 2, 3, and 4). Unfortunately this book is only in black and white but even so you can get some idea of the exquisiteness of these common household objects.

You might like to trace over them and add the color that is missing. The original (see [3]) shown in Figure 1(a) is simply red on a white background. In the central part the squarish shapes are alternately red and white, while the scrolls on the sides and the butterfly arabesques in the corners are red. The concentric circles of shapes give the feel of a spiral coming out of the center. This quilt, part piece-work and part appliqué, was made in New Jersey, USA, around 1910.

The quilts in Figures 1(b) and (c) were made in the 1980s by Mary Whitehead, a present-day prize-winning quilt-maker, from Menlo Park, California, USA. What is so remarkable about both of these quilts is the repeated use of the square motif. You might like to study the designs in each of these quilts to see if you can figure out how they were created, or to

(a)

(b) Tamsen Whitehead (a mathematician, left) with her mother, Mary Whitehead (right), the creator of this quilt.

(c) A second quilt created by Mary Whitehead.

FIGURE 1

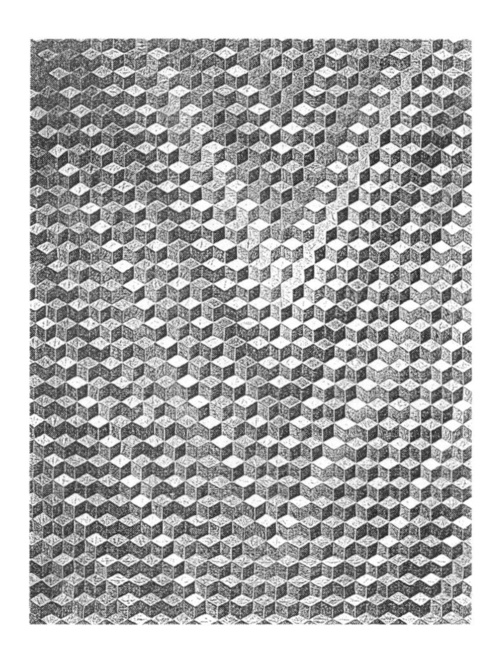

FIGURE 2

compare the differing symmetric features of the various components.

The rich colors of the quilt of Figure 2 make the original a much more marvellous work of art than any black and white reproduction could possibly show. Colored rhombi have been carefully sewn together so as to create a V-shaped pattern. If you think of three rhombi together making a box shape (in three dimensions), then the tops of the boxes making each of the large V's are the same color. The same is true for the corresponding sides of each of these boxes. This quilt is of Amish origin and was made in Ohio about 1935.

The quilt in Figure 3 also comes from Ohio but was made about 15 years earlier than the previous one. Its vibrant colors elevate the simple design of squares to a work of art. Each square had a sidelength of $\frac{3}{4}''$ ($1''$ means one inch, or approximately 2.54 cm), so you can calculate how big the complete article is. Does that present a problem?

Quilts go back quite a long way. The hexagonal pattern of Figure 4 was made around 1850 and is from New York State. Here

FIGURE 3

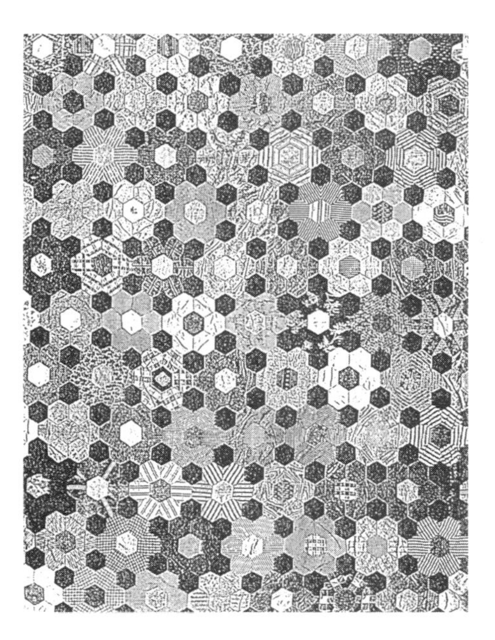

FIGURE 4

we have a quilt that is 37″ by 32″. So what is the sidelength of the individual regular hexagons that make up the pattern?

This last quilt is the only one we have shown so far which is made from pieces that were not cut from plain material. Even in the black and white photographs, you can see that striped, checked, and patterned fabrics have been used to make it.

But why make a quilt that is only 37″ by 32″? Surely the whole point of a quilt is that it is used as a warming cover over a bed? There can't be too many people who would fit into a bed that was 37″ by 32″!

From the eighteenth century in America, quilts became very popular, as a way of keeping warm in bed. However, they go back considerably further than that. The oldest known quilt in existence is a fifteenth-century quilt from Sicily which is decorated by scenes from the legend of Tristan and Isolde.

There are three layers to a quilt. The bottom layer is the ***backing***. Generally, this is fairly plain and made either from a single piece of material or from a few large strips. The ***batting***, as it is called, is the middle layer and usually consists of cotton, wool, or some synthetic fiber. On the top is the ***decorative layer***. We have shown examples of quilt tops in Figures 1, 2, 3, and 4. The top layer was often pieced together from scraps of fabric that had been saved for the purpose.

The construction of quilts has mainly been the work of women, though there are some notable exceptions. Many quilts were made by groups of women. In fact, at the stitching stage, when the three layers were put together, women often got together in what was known as a ***quilting bee***. At such gatherings, more than one quilt could be finished in a day.

We have been very careful in our choice of quilt examples so far. If you look at Figures 3 and 4, you'll notice that the whole of the quilts concerned are covered by square and hexagonal pieces, respectively. What other shapes could be used to cover a quilt top? We just want to repeat the same shape over and over again until the whole top layer is covered. Well, Figure 2 will give you a third example. The rhombus has been used with great effect to cover that particular quilt.

So we have three examples: square, regular hexagon, and rhombus. Are these the only shapes, all of the same size, that will cover a quilt? If there are more, are there only a finite number of shapes or are there an infinite number? Which shapes are practical as far as quilts are concerned? Which shapes by themselves have actually been used to form the top layer of a quilt?

• • • **BREAK**

Use some square paper to make several copies of any shape you like. Will that shape cover a quilt? Try some other shapes.

We know that squares and regular hexagons will cover our quilt. They are both regular polygons. In order to cut our search for quilt covering pieces down to a manageable size, let's first of all see if we can find all the ***regular*** polygonal shapes that will do the covering. (Naturally, this restriction eliminates the rhombus and many other polygons.) Further, let's be systematic. Perhaps the best way to do this is to start with polygons of 3 sides and work up.

• • • **BREAK**

So what is the usual name for a 3-sided regular polygon? Will that cover a quilt? Is this obvious from what we have done so far anyway?

Those last questions are fairly easy to answer, so let's go on to the 4-sided regular polygon. Of course, that's a square. And we know that works (see Figure 3). Which takes us on to the regular 5-gon or regular pentagon.

• • • **BREAK**

Will enough regular pentagons completely cover a quilt?

Just in case you were troubled by the 3-sided regular polygon, we should remind you that this is usually called an ***equilateral triangle***. To see that enough of these will cover a quilt, you only have to start with a regular hexagon and cut it up into six equilateral triangles (see Figure 5). Because regular hexagons will do the job (see Figure 4), then so will equilateral triangles. (Of course, we could also see this by cutting an appropriate rhombus in two.)

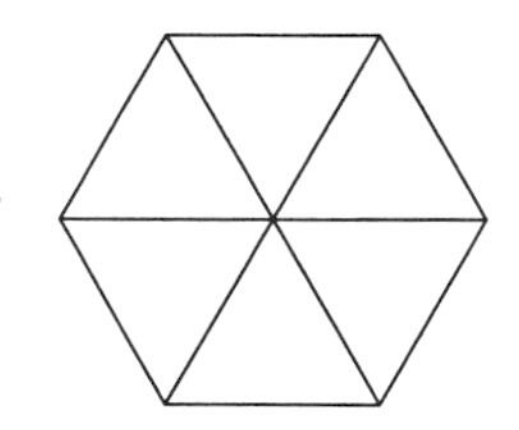

A hexagon made of six equilateral triangles.

FIGURE 5

But we suspect that you probably had a lot more trouble with the regular pentagon. There really does seem to be a difficulty in getting regular pentagons to fit together. Is it that they are just naturally awkward or are their angles all wrong?

Have a look at Figure 6. If regular pentagons are going to fit together, then a certain number of them will have to fit at a vertex. (You'd better make sure you believe that.) If, say, three of them do fit together, then three times the interior angle of a regular pentagon has to equal 2π.

• • • BREAK

But that would mean that the interior angles of a regular pentagon are $\frac{2}{3}\pi$? Is that the case?

While you're thinking about that one, we'll move up to 6. So, it's regular hexagon time. Figure 4 will have already convinced you that these will cover a quilt. That puts polygons with 7 sides in the frame.

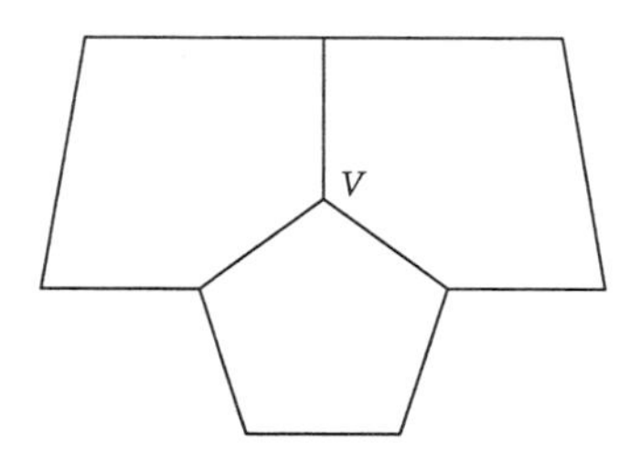

Can three regular pentagons meet at a single vertex *V*?

FIGURE 6

• • • **BREAK**

Can we put together a group of regular heptagons at a single vertex?

We think you've probably been having some difficulty with regular pentagons and heptagons. One or two of you, we can see, have convinced yourselves that it can't be done. The reason for that may have something to do with the fact that 5 and 7 are odd, so let's move on to regular octagons.

• • • **BREAK**

Maybe we can fit regular octagons together?

Well, we don't see too much success there. Indeed, if we were going to form a conjecture on the basis of what you've done so far, we'd have to say that regular n-gons will only cover a quilt if $n = 3, 4$, or 6. Nothing else seems to fit. But proving this conjecture doesn't look too easy. We might get a proof by finding the size of the interior angle of a regular n-gon and showing that only when $n = 3, 4$, or 6, is 2π an integral multiple of this angle.

Let's try that for the pentagon. What's the magnitude of the interior angle of a regular pentagon? You should be able to work that out from what we did in Chapter 4. But if you look at Figure 7, you'll see another way to do it. There we've divided the pentagon up into five triangles. The sum of all the angles of all these triangles is $5 \times \pi$. Now the angles in the triangles which are at the center, C, of the pentagon must add up to 2π. So the sum of the interior angles is $5\pi - 2\pi = 3\pi$. Any one interior angle must therefore be $\frac{3}{5}\pi$, because all of these angles are equal. Clearly $\frac{3}{5}\pi$ doesn't go into 2π exactly. So regular pentagons won't cover a quilt!

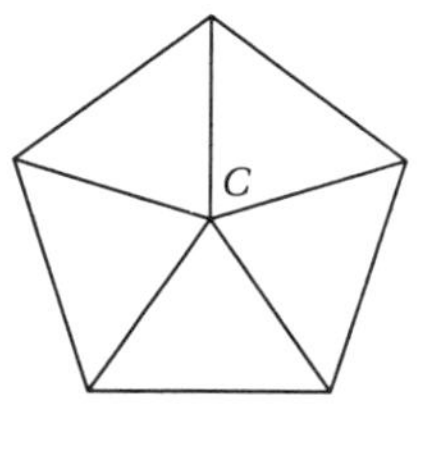

FIGURE 7

• • • **BREAK**

Does this proof work for the regular heptagon, regular octagon, or regular nonagon? Try it out.

How did it go? We think it's probably not worth doing these special cases. We're sufficiently confident now to try to take on the regular n-gon. We first find the size of an interior angle of a regular polygon using the method that we used above for the pentagon.

In Figure 8, we illustrate a regular n-gon split up into n triangles. The angles of all the triangles add up to $n\pi$. Again the angles of the triangles at the center, C, of the n-gon add up to 2π. Hence the sum of the exterior angles of the regular n-gon is $n\pi - 2\pi$.[1] But there are n interior angles in this figure. So each interior angle α, is given by

$$\alpha = \frac{n\pi - 2\pi}{n} = \pi - \frac{2\pi}{n}.$$

Suppose we've got a group of n-gons which meet at a vertex. Then α must divide 2π. (Further, $\frac{2\pi}{\alpha}$ is the number of regular n-gons which meet at that vertex.) But

$$\frac{2\pi}{\alpha} = \frac{2\pi}{\pi - \frac{2\pi}{n}} = \frac{2n\pi}{n\pi - 2\pi} = \frac{2n}{n-2}.$$

• • • **BREAK**

The whole question then boils down to, for what $n \geq 3$ does $n - 2$ divide $2n$? How are we going to crack that one?

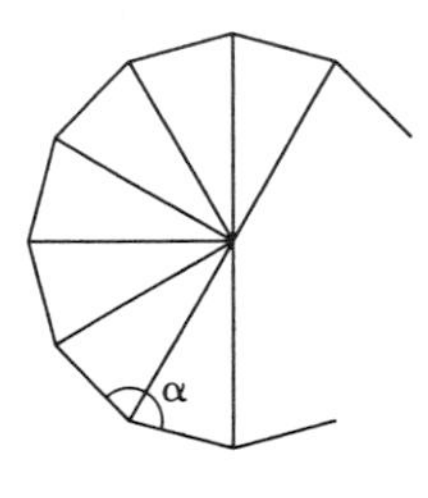

FIGURE 8

[1]This was shown by another method at the end of Section 1 of Chapter 4.

At moments like this it's always worth checking that we haven't made any algebraic or other errors. If we are on target, $n - 2$ will divide $2n$, at least for $n = 3$, 4, or 6. This is because we know that equilaterial triangles, squares, and regular hexagons will cover a quilt. So let's check.

$$n = 3: \qquad \frac{2 \times 3}{3 - 2} = \frac{6}{1} = 6;$$

$$n = 4: \qquad \frac{2 \times 4}{4 - 2} = \frac{8}{2} = 4;$$

$$n = 6: \qquad \frac{2 \times 6}{6 - 2} = \frac{12}{4} = 3.$$

Things seem to be working out. You will have already noticed that 6 equilateral triangles, 4 squares and 3 regular hexagons meet at a vertex.

So when is $\frac{2n}{n-2}$ an integer? And an integer greater than or equal to two, because it is the number of regular n-gons which meet at a point. First notice that a little algebra gives

$$\frac{2n}{n-2} = \frac{2n-4}{n-2} + \frac{4}{n-2} = 2 + \frac{4}{n-2}.$$

And when does $n - 2$ divide 4? Surely when $n - 2 = -4, -2, -1, 1, 2$, or 4. The negative values of $n - 2$ aren't any use to us. But $n - 2 = 1, 2$, or 4, gives $\frac{2n}{n-2} = 6, 4$, or 3—all integral values bigger than one. So the only values of n that make any sense in this context are 3, 4, or 6.

The algebraic method we've just used is often useful if you want to find integer values of quotients like $\frac{an}{bn+c}$, where a, b, c are integers, b divides a, and n is a variable integer. If you divide through by $bn + c$, then the original quotient is an integer when $bn + c$ divides the appropriate constant.

So you, er, we've done it!

• • • **BREAK**

Hang on a minute. What is it exactly that we have done?

Just to make sure that we know where we are, let's put down our conjecture formally.

Conjecture *Regular n-gons will only cover a quilt if* $n = 3, 4,$ *or* 6.

Now this is meant to say two things. First, it says that if $n = 3$, 4, or 6, then regular n-gons will cover a quilt. Second, it says that if $n \neq 3$, 4, or 6 then regular n-gons will not cover a quilt.

We can tidy up the second part by getting rid of the double negative. Rephrased then, it says that if a regular n-gon covers a quilt, then $n = 3$, 4, or 6.

This enables us to rephrase the whole conjecture as follows.

Conjecture (stated precisely) *A regular n-gon will cover a quilt if and only if* $n = 3, 4,$ *or* 6.

To prove this we have to show that: (1) if $n = 3$, 4, or 6, then regular n-gons cover a quilt; and (2) if regular n-gons cover a quilt, then $n = 3$, 4, or 6.

• • • **BREAK**

Have we proved both these things? Check back to see.

We think you should be happy now. The first part of the conjecture is covered by looking at Figures 3, 4, and 5. The second part follows from the $\frac{2n}{n-2}$ argument. There we started by assuming we had a covering. This led us to $\frac{2n}{n-2}$ being an integer. We then showed that n had to be 3, 4, or 6, and that clinched the deal. Our conjecture is therefore a theorem.

But how does Figure 2 fit into the picture? There the quilt was covered by rhombi! What other shapes will cover the quilt? Can you find a quilt that has been covered totally by equilateral triangles?

5.2 VARIATIONS

Despite the fact that a great deal of variety can be obtained from the basic square and regular hexagon shapes by the use of colored and patterned fabric, as well as by using appliqué, they are still somewhat limiting. Not surprisingly, quilt-makers have gone way past these simple shapes, at the same time still basing their designs on them. Let's see what we mean.

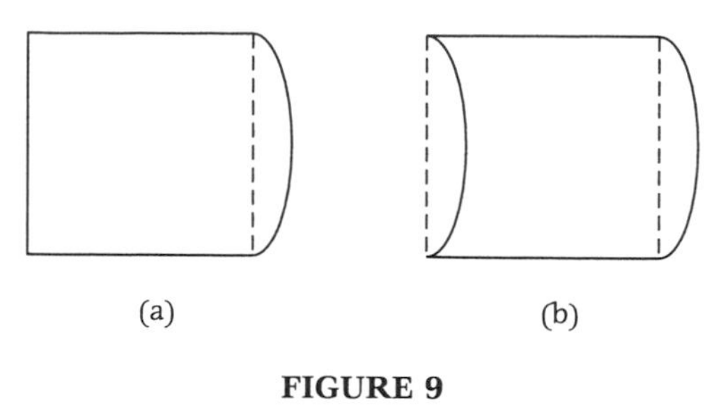

FIGURE 9

Starting with a square we could first add on a circular piece as shown in Figure 9(a). But then, to have all our quilt pieces the same, we would need to take the same circular piece out of the opposite side of the square, as in Figure 9(b). The shape obtained in Figure 9(b) certainly covers the quilt, although there are going to be some bumpy bits around the side.

Of course, there is no need to stick with circular bits on the vertical sides of the squares as we did in Figure 9. By adding circular sections both inside and outside the basic square, we can change things starting in Figure 9 and continuing to the sequence of Figure 10 (see also [5]).

This particular pattern is fairly common in quilting. It is the basis of what is known as the Double Wedding-Ring pattern. An example of a quilt with this pattern is shown in Figure 11 (it was made by Jean Pedersen's grandmother, Caroline Turpin in the 1930s). Once you've cottoned on to this idea, you can really let your imagination go.

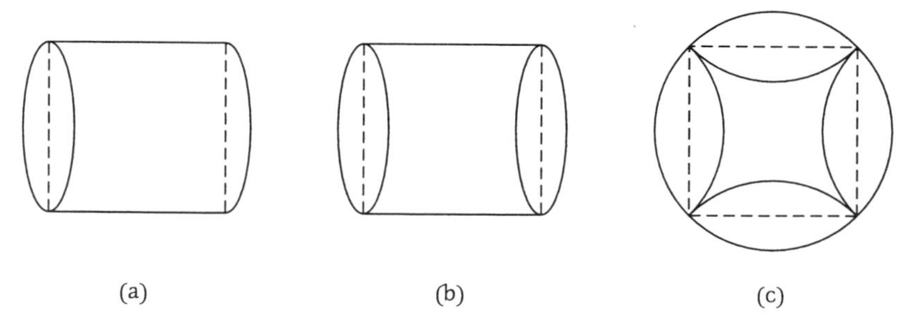

FIGURE 10

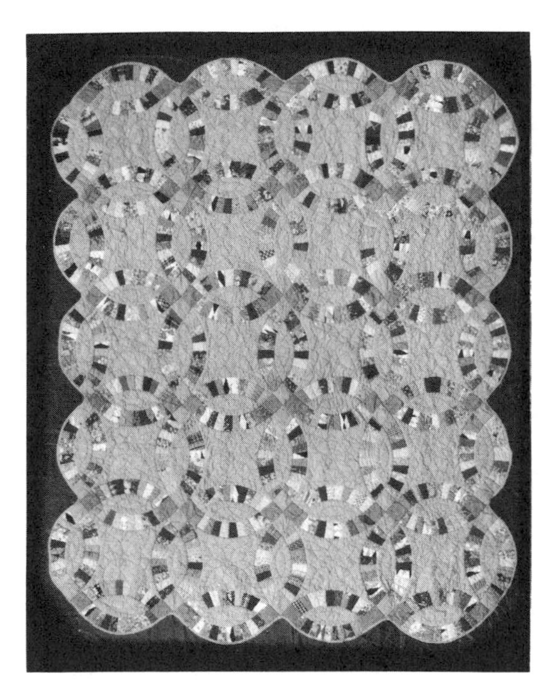

(Photo by Chris Pedersen)

FIGURE 11

• • • **BREAK**

What designs can you make based on the simple square grid layout? What can you do with the regular hexagonal grid?

But you don't have to be restricted to the basic idea shown in Figure 10. An alternative approach is shown in the appliquéd quilt of Figure 12. See what you can do with this method. Even using the basic square, a great deal of new and interesting designs can be produced. In fact, an infinite variety is possible.

The Dutch graphic artist M.C. Escher certainly let his hair down. He wasn't a quilt-maker but an engraver. His engravings have become so world-famous that many have tried to imitate his efforts. Figure 13 shows some very effective Escher-type figures that were created using the TesselMania® software program (a registered trademark and copyright owned by MECC in Minneapolis, Minnesota, USA). Each one uses a more complicated variation of the

FIGURE 12

square than we used in Figure 10. These illustrations look so intricate that it is sometimes hard to see the basic square on which the whole thing is based. Try to sort them out.

• • • BREAK

Place dots at the corners of the squares in the Escher-like prints. Can you see how he developed his shapes from the basic square?

Generally, quilt-makers don't go to the extremes that Escher did. We suspect this is because it is too complicated and time-consuming to cut round the fabric and even more difficult to sew the pieces together. It is also, presumably, difficult to make every shape exactly the same, although quilt-makers could and do use templates.

You can see then that it's not just quilt-makers who have got into the "covering" act. But then artists and quilt-makers don't have the field to themselves either. Next time you are in a tiled bathroom (or any tiled room for that matter) have a look at the shape of the tiles that have been used to cover the floor. You'll often find that

FIGURE 13

they are squares but sometimes you'll find equilateral triangles or even regular hexagons.

We've now given you three examples of common, everyday objects that are covered in some way by regular shapes. You might like to look around for some more.

Mathematicians call the activity that we've been engaged in ***tiling the plane***. A tiling of the plane is also called a ***tessellation***. (See [4].) So we know at least that we can tile the plane using equilateral triangles, squares, or regular hexagons.

• • • **BREAK**

What other shapes can we use to tile the plane? Can you think of a tessellation which isn't based on one of the three regular polygons we've discussed?

Suppose we go back to quilts and try covering them with a large number of congruent pieces. We know we can do this with three regular figures. Can we cover quilts with triangles, quadrilaterals, or hexagons that are not necessarily regular? Are there any pentagons that will do the job? Can we cover quilts with polygonal shapes of any given number of sides?

• • • **BREAK**

As before let's start with the smallest case—the triangle. Can you show that *any* triangle will tessellate?

Let's go on to quadrilaterals. It's not difficult to see that parallelograms can be used to tessellate. Clearly there are some non-parallelogram-type shapes that have been used in quilts. You can see one of these in Figure 14.

But will any quadrilateral work? The trick, of course, is to make the bits fit together at a vertex. The last time we tried this we had to find the formula for the sum of the interior angles of a regular n-gon. But that formula works for *any* polygon, whether it's regular or not (see Chapter 3). Hence we see that the interior angles of a quadrilateral add up to 2π. So there's a chance that we can fit congruent quadrilaterals together to tile the plane.

FIGURE 14

• • • **BREAK**

Will *any* congruent quadrilaterals tile the plane?

Perhaps the easiest way to start work on this problem is to cut four congruent quadrilaterals out of a piece of paper and try fitting them together without turning them over (see Figure 15). We've labeled the angles to make it easier to see how the quadrilaterals are oriented. Would Figure 15(b) convince you that the quadrilateral we chose will cover a quilt? Can you use this to show that *any* quadrilateral will tessellate the plane? Or is there a better way?

The first problem we need to grapple with is whether or not two sides of the same length have to be placed together. They certainly have been in Figure 15.

In Figure 16 we have tried another arrangement. A potential tessellation can only be continued at the point X, if the quadrilateral concerned has an angle equal to $\pi - D$. This will only happen for every special quadrilaterals. Cyclic quadrilaterals and trapezoids (also called trapezia) have this property. In general, however, there

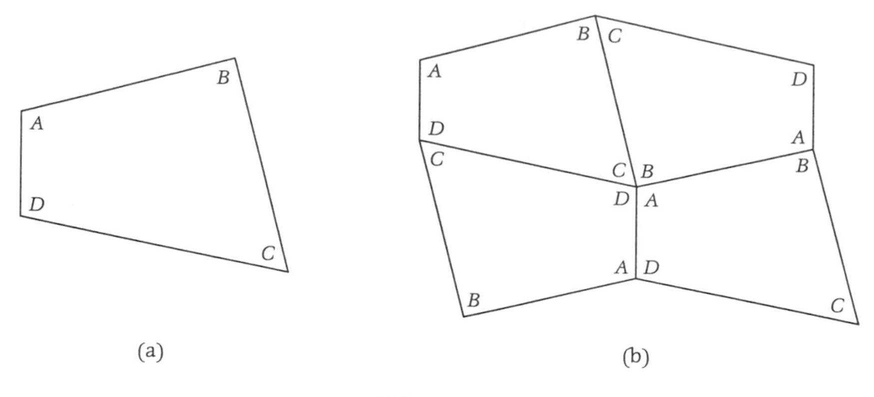

FIGURE 15

won't be an angle in the quadrilateral that is equal to π minus some other angle. In that case we *are* forced to put sides of equal length together in our tessellation. This means that it *may* be possible to make progress with the configuration of Figure 15.

• • • BREAK

Can the arrangement of Figure 15 be extended to a tiling of the plane?

Those of you who have already solved this problem will have to bear with us for a while. It may be that the argument that we are giving here is different from yours anyway. The first thing to do is to extend the configuration of Figure 15 to include several more quadrilaterals of the same type. In Figure 17, we've run quite a few more down the page, well, an infinite number really, although the page is too small to contain them. The net effect is to produce a

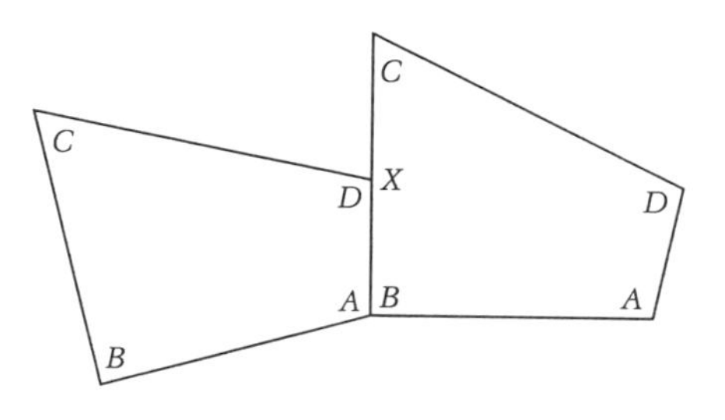

FIGURE 16

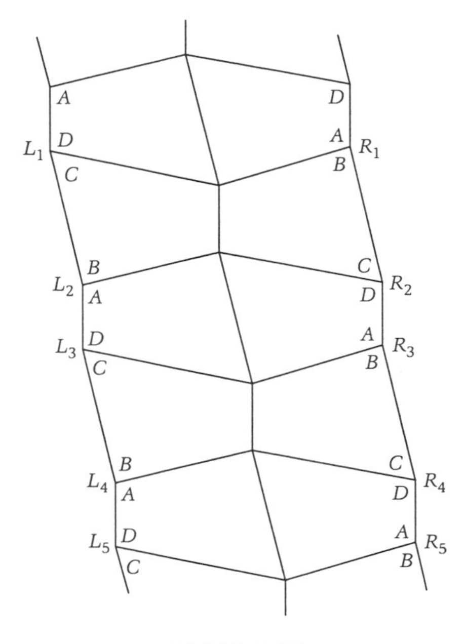

FIGURE 17

strip of quadrilaterals, two wide, down the page and off to infinity in both directions.

Have a good look at the boundaries of this strip. The same lengths are repeated on either side. There's the AD side, followed by the BC side, followed by the AD side, followed by the BC side, followed by the ....

And the angles that these lengths make with each other are the same too. To the left of vertex L_1, there is an angle of size $A + B$. To the right of vertex R_1, the angle's size is $2\pi - (A + B) = C + D$. This means that the angles to the left of L_1 and to the right of R_1 are equal. But the same thing can be said for the angles at L_2 and R_2, L_3 and R_3, and so on. This means another strip, exactly the same as that in Figure 17, will fit snugly up against the original strip. Keep going like this and eventually the whole plane will be covered. Any quadrilateral will indeed tile the plane.

Now any quadrilateral may tile the plane but we know the same isn't true for *any* pentagon—we have already seen that *regular* pentagons, for instance, don't tessellate. Some of you may already have played around with pentagons and have come up with something

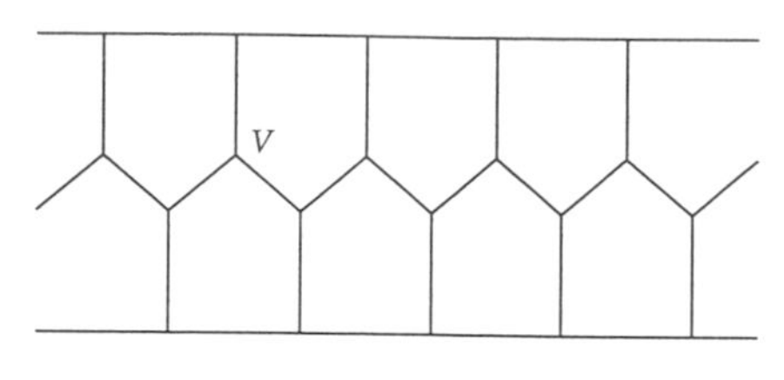

FIGURE 18

like Figure 18 which shows that *some* pentagons can tile the plane.

• • • **BREAK**

By suitably rearranging the shape of the pentagons at the vertices labeled V, you can extend the pentagons of Figure 18 to a tessellation by polygons of any odd number of sides. Can you extend it to *any* even number? Can you find any quilt, etching, floor tiling, or carpet that has a pattern on it like that shown in Figure 18?

It should now be clear that, given any number n bigger than two, there is a polygon with n sides that will tile the plane. However, there are polygons that won't tile the plane. The regular octagon is one of these. So how come you find regular octagons being used in floor coverings? How can that be if regular octagons don't tessellate? To give you some idea we've included a picture of part of a carpet (see Figure 19). This is really cheating, of course. The octagons by themselves won't cover the plane. They need to be accompanied by squares.

• • • **BREAK**

What other *pairs* of shapes together completely cover the plane?

5.3 ROUND AND ROUND

But there are other things that meet the eye in this quilt business. If you look at some quilts you'll see a certain amount of symmetry. For instance, in Figure 3, you can put a vertical line through the center of the quilt; then the quilt looks the same on either side. If

FIGURE 19

a mirror were placed on that vertical line, the reflection of the left side would look just like the right side, and vice versa.

There's a similar horizontal line of symmetry, as well as two diagonal lines through the corners of the quilt. What's more, if we rotated the whole quilt through $\frac{\pi}{2}$ clockwise, you wouldn't know that it had been moved. In fact, the same thing happens if you rotate through π, $\frac{3\pi}{2}$, or any multiple of $\frac{\pi}{2}$. So the quilt of Figure 3 has quite a lot of symmetry—at least eight types if we include "doing nothing."

One final thing, a symmetry followed by a symmetry is also a symmetry. Now you can easily check here that the effect of a reflection followed by a rotation is the same as that of some rotation followed by some reflection; and that the composite of two reflections can be achieved by a single rotation. So altogether there are no more than eight symmetries of this quilt, but only four of them can be achieved by actually moving the quilt. The rest can be seen by using the mirror method.

• • • BREAK

Check that the quilt in Figure 1 has precisely the same symmetries as the quilt in Figure 3. What symmetries can you find in the other quilts and in the Escher-like prints?

If we could delete the scrolly bits around the edge of the quilt of Figure 1, and just retain the circular part, we would suddenly find a whole lot more symmetry. There are now 52 lines of reflection passing through the center of the quilt and 52 rotational symmetries. As in the case above, we don't get any more symmetries by combining reflections and rotations. Hence, altogether, this (mutilated) quilt has 104 symmetries.

• • • **BREAK**

Figure 20 shows another quilt made by Caroline Turpin. How many symmetries has the star-shaped portion of this quilt?

Suppose for a moment that quilts weren't finite and that they extended off to infinity in all directions. Then some of the quilts which are laid out on a square grid, as opposed to those made of square pieces, have another type of symmetry. Instead of just lines of reflection and rotational symmetry, there is ***translational*** symmetry. We can move the whole infinite quilt onto itself if we slide it vertically or horizontally. The whole thing translates onto itself. If the translation cannot be broken down into smaller translations, then we call it a ***simple*** translation.

• • • **BREAK**

Imagine you are looking at a piece of infinite graph paper. You should then be able to see how this translation business works. How many translations are there? What simple translations are there? Can all translations be obtained from simple translations?

Have a look at the hexagonal tiling depicted in Figure 21. You should be able to see six rotational symmetries and six reflections about lines of symmetry. But if you slide or translate A to B, you'll shift the whole tessellation onto itself. If you did this simple translation with plane tiles when we had our backs turned, we would never know that you had done it. Nor would we know whether you had performed this type of symmetry or some other type.

But you could also translate the hexagons so that A went to C and this would still not be detectable. Likewise, moving A to coin-

(Photos by Chris Pedersen)

FIGURE 20

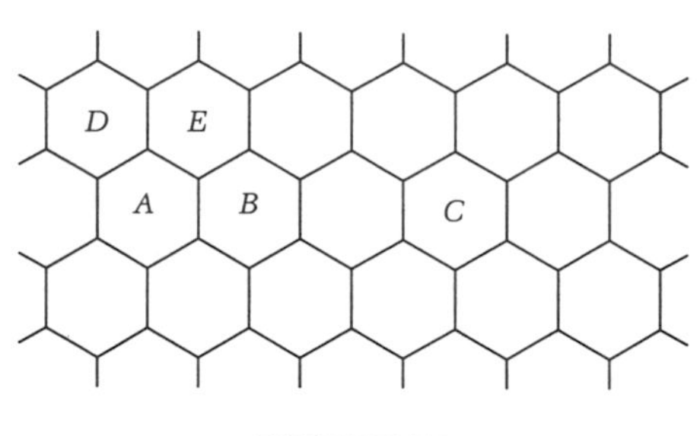

FIGURE 21

cide with *any* hexagon in that A, B, C row, we have undetectable moves—and an infinite number of them too.

Similarly, A could slide to D or E or any hexagon in the direction of A to D or A to E, and the hexagonal tiling would look unchanged.

Then again we could pick up the tiles and without rotating or reflecting them, we could put them down with A on top of any other hexagon at all. But any such *slide* symmetry would be the same as a certain number (maybe negative) of simple translations like A to B, followed by some number of simple translations of the form A to D.

• • • **BREAK**

Check out this last statement. Do you agree with it?

So all the symmetries of a regular hexagonal tiling can be found using all combinations, as many times as we like, of the six basic reflections (about the lines of symmetry), a rotation through $\frac{\pi}{3}$, a simple translation from A to B, and a simple translation from A to D. This tiling has an infinite number of symmetries!

Is there any end to this tiling business? Are there an infinite number of tessellations or is there some restriction on what can be done? Are there any limits on the kinds of symmetries that we can have in a tessellation or are the possibilities endless?

5.4 UP THE WALL

We now take up the story by looking at a wall. But before we look at wallpaper itself, we'll warm up on wallpaper friezes. You probably know that sometimes wallpapering is finished off by a strip, called a ***frieze***, which is usually pasted horizontally, and is often there to hide the rough edges of the wallpaper itself. In all wallpaper friezes, a basic region, called by mathematicians a ***generating region*** or a ***fundamental region***, is repeated endlessly in some symmetric way. We give two examples of friezes in Figures 22(a) and (b). They are both based on the fundamental region of Figure 22(c).

One of us has got a frieze on the wall of our downstairs toilet. It's a fairly basic pattern. The same asymmetric fundamental region is repeated endlessly (or at least from one end of the wall to the other). A rough approximation to that frieze is shown in Figure 23, along with its fundamental region.

Actually, this frieze is about the simplest pattern you can have. (The only way to make it simpler would be to have no fundamental region at all—just a plain single colored strip. Since manufacturers rarely make such plain friezes, we'll ignore them here.) Our toilet frieze just takes the fundamental region and moves it on along the strip in either direction for ever. By suitably translating the fundamental region, we get the whole frieze.

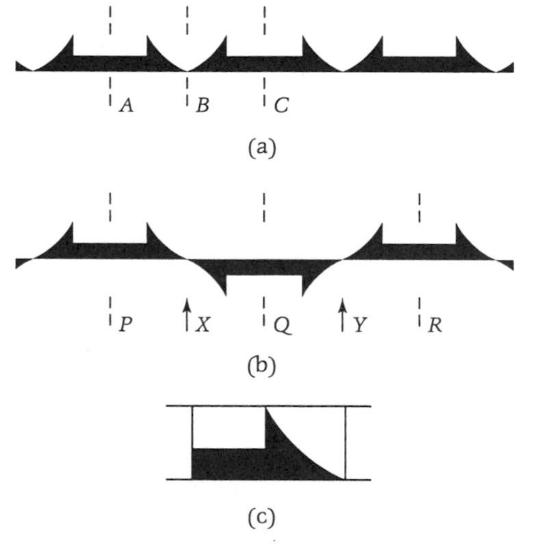

FIGURE 22

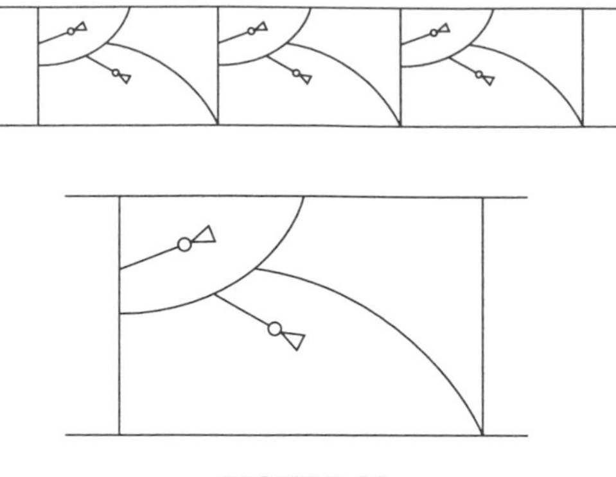

FIGURE 23

• • • **BREAK**

What do we have to do with the fundamental region of Figure 22(c) to construct the friezes of Figures 22(a) and (b)?

Frieze construction is fundamentally about two things—generating regions and symmetry. Surprisingly, there isn't a great deal that you can do with friezes once these two things are settled. It turns out that once you have decided on your generating region (where you can let your artistic temperament run riot), there is a limit to what you can do with it. There are only so many symmetries, one-to-one onto mappings or ways of moving the whole frieze so that it fits onto itself, that are available to use. As we said above, our toilet frieze is quite simple in this regard. The only symmetries it has are a simple movement along itself from one fundamental region to the next, a simple translation, or an arbitrary number of repetitions of a simple translation in either direction.

So what other kinds of symmetries can we expect? Apart from translations, there might be ***reflections*** about axes of symmetry in the plane of the frieze or ***rotations*** about axes perpendicular to the plane of the frieze. Assuming that the frieze is considered to be horizontal, the axes of any reflection can only be vertical or horizontal. Rotations are pretty limited too. They can only be about an axis through the middle of the frieze and through an angle of magnitude π.

These are the basic symmetries. Of course, you can have combinations of the above. The only combination that appears to be worth singling out, though, is the combination of a simple translation and a horizontal rotation. This is called a ***glide reflection***. In Figure 24, we show the effect of a glide reflection on the fundamental region of Figure 22(c).

• • • **BREAK**

Consider the effect of all possible combinations of simple translations, reflections, and rotations.

Now let's look into the friezes of Figure 22 in some detail. Clearly both of them have translational symmetry, though they don't have a simple translation which moves one fundamental region to the next. (In fact, every frieze has to have translational symmetry.) So let's concentrate on Figure 22(a) to see what other symmetries it has. There is a line of symmetry (marked by the dotted lines at A), but there are also parallel lines of symmetry through B, C, as well as through an infinite number of other similar points. There are, however, no rotations or glide reflections.

Moving along to Figure 22(b), it should now be clear that there are lines of symmetry at P, Q, R and an infinite number of other points. Rotations are also present and can be found about the points X, Y as well as an infinite number of others. On the other hand, there is no reflection about a horizontal line of symmetry. In case you thought there was a glide reflection, recall that such a symmetry only moves the fundamental region along by one. What you might have thought a glide reflection here, would have to move the end of the arrow at P along to Q, a distance of two units. All other symmetries of Figure 22(a) are simply combinations of the symmetries that we have already noted.

• • • **BREAK**

We said above that there are only a limited number of types of friezes available, once the fundamental region has been

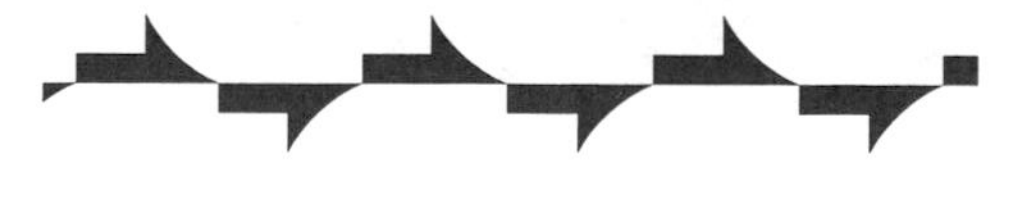

FIGURE 24

determined. Call such a type a ***pattern***. Try to find out how many such patterns there are (note that you have already seen three in Figures 22, 23, and 24). Can you find a systematic way to decide which pattern a particular frieze is?

In Figure 25 we have given four friezes based on a different fundamental region from those in Figure 22. How do these fit into your classification? Which, if any, have the same pattern as one of the earlier friezes?

There are, in fact, systematic ways to decide the pattern of any given frieze. The flow chart in Figure 26 is one such way. It's based on an idea due to Doris Schattschneider in [9].

By studying the flow chart in Figure 26 you should be able to see that there are only seven different frieze patterns. These seven patterns have been illustrated in Figures 22, 23, 24, 25, and 27. Precisely, I is shown in Figure 25(c), II in Figure 22(b) and 25(b), III in Figure 22(a) and 25(a), IV in Figure 25(d), V in Figure 24, VI in Figure 27, and VII in Figure 23.

Despite the limitation to seven patterns based on symmetry, it's amazing how much variety in friezes designers can get. The article [6] by McLeay gives some interesting examples of ironwork friezes. In [7] Pedersen extends the ideas to weaving. You might

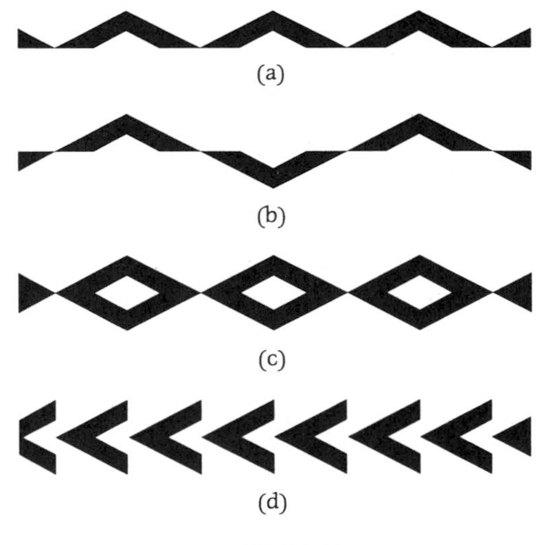

FIGURE 25

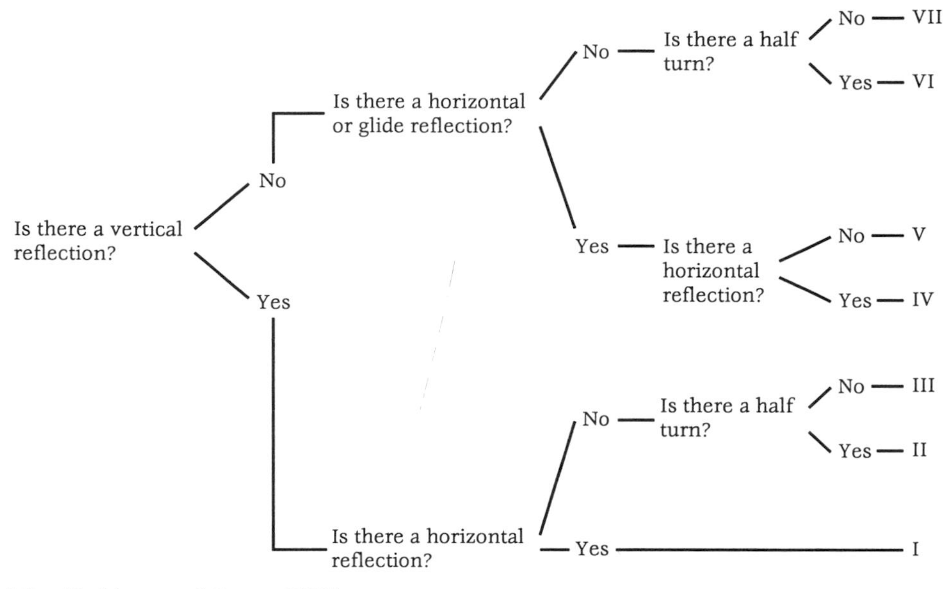

FIGURE 26

FIGURE 27

like to experiment for yourself and use graph paper or some other aid, to come up with some ideas and applications of your own.

As an aid to remembering the seven basic patterns, Budden in [1], suggests the mnemonic of Figure 28 (this is based on Figure 26.168 of his book). The key word to remember is SHEAR because each letter of the word "shear" gives one of the seven patterns. The other two follow by knowing "footsteps" and the "sine curve."

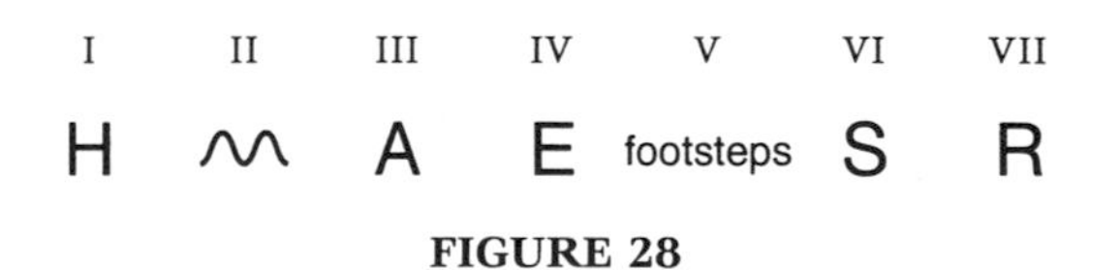

FIGURE 28

It should not be too difficult to see now that the friezes of Figure 22(b) and Figure 25(b) have the same pattern. They are described by the "sine curve" in the mnemonic. You might like to look out for friezes around your home or your friends' homes and see how they should be classified.

Naturally, what we've done here does not actually *prove* that there are only seven frieze patterns. All we have done is to show seven patterns that we know to exist and give some rationale for their existence. If you want to look up a proof that there are only seven, you'll find the book [2] by Coxeter useful.

That just about covers friezes, so let's move on to the larger problem of wallpaper patterns. As you would probably expect, there are wallpaper patterns that are not frieze patterns. But exactly how many are there? It's not even clear at the start whether there are only finitely many.

To get a handle on this problem, we need to be sure first of all what it is we are trying to describe. What is a wallpaper pattern? Well, wallpaper patterns also go under the name of ***periodic*** or ***repeating*** patterns, covering the whole plane. The essential idea is the same as for friezes. A finite region, the generating region, produces the whole pattern when acted on by the symmetries of the whole, and these symmetries include two simple translations which are at an angle to each other.

So we will first need to decide what symmetries are at work in the plane that can take our generating region around the whole pattern. Having compiled a basic list of these, we may be able to produce a flow chart which will do the job for wallpaper patterns that Figure 26 did for friezes.

• • • BREAK

Try to describe a program, like the one for friezes, which will determine both the number of possible wallpaper patterns and the pattern of a given wallpaper design.

Perhaps the first thing to note is that the two translations produce a set of lattice points on which to build the pattern. If we take an arbitrary point in the pattern and act on it by successive simple translations of each type, a lattice is formed. This is exactly what we suggested you do in the Escher picture of Figure 13. So how many types of lattice are there?

First of all, we can always think of the angle α between the two simple translations as being less than or equal to $\frac{\pi}{2}$. You may need to twist the plane around a bit to make sure of this, or take a translation backward rather than forward. To see the extra symmetries that are possible, we vary the angle α. Now if $\alpha = \frac{\pi}{2}$, then the lattice formed can be a square or a rectangle depending on the relative amount of displacement each translation provides. At slightly less than perpendicular, we get a lattice whose basic cell is a parallelogram. If the two translations move the pattern the same distance, then we get a rhombus, but the symmetry we get from the rhomboid lattice can be more easily obtained directly from the rectangular lattice. A special case of the rhomboid lattice occurs when $\alpha = \frac{\pi}{3}$; in this case, the cell can be thought of as being made up of two equilateral triangles. We show all of these situations in Figure 29.

Having constructed the lattices behind the patterns, we now have to worry about rotations and reflections. Earlier in this chapter we convinced you that the only conceivable rotations in the plane were through angles of magnitude $\frac{1}{3}\pi$, $\frac{1}{2}\pi$, $\frac{2}{3}\pi$, and π. As far as reflections go, there seem only to be glide reflections and

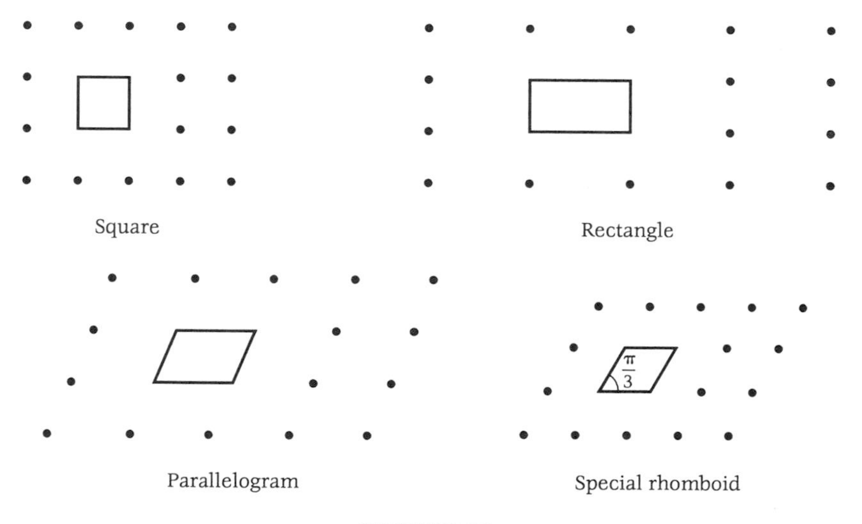

FIGURE 29

mirror images about axes parallel to the two translations—just as we had with friezes.

• • • **BREAK**

How many wallpaper patterns can you come up with armed with simple translations, rotations of magnitude $\frac{1}{3}\pi$, $\frac{1}{2}\pi$, $\frac{2}{3}\pi$, and π, reflections (mirror images), and glide reflections? Seven? More than seven? Finite or infinite?

What we are asking you to do, of course, is nontrivial. It's going to take a great deal of care to find the right number of patterns but it is well worth the effort, especially if you get the answer or come very close to it. Use an arrow similar to the one in Figure 22(c) as your fundamental region. Just to get you started we'll do an example.

Suppose that, for a start, we restrict ourselves to looking at the patterns we get when we only have the translational symmetries. Then, depending on which basic cell we have, we get one of the wallpaper patterns of Figure 30. We could, of course, consider each of the patterns of Figure 30 as different. However, they all have a sameness about them which will be enhanced when we look at some more examples. What's more, each of them is a special case of Figure 30(c). After all, a square, for example, is just a special case of a parallelogram. For these reasons we'll think of the patterns in Figure 30 as all being the same. That has got one pattern out of the way.

Now it is not too difficult to put in reflections. We've done this in Figure 31, using the lattice whose basic cell is a rectangle. Do we get a different pattern if we use some other basic cell here? If not, why did we bother talking about basic cells in the first place? Are there certain symmetries that only apply to particular basic cells? (See Figure 32.) Does the pattern in Figure 31 look like a translation of a frieze? Will that idea help produce more patterns?

You're on your own from here. See what progress you can make toward listing distinct wallpaper patterns. To help you along the way, we've put down 10 possible patterns in Figure 32. Did you get that many or more? Did you get some that are not in Figure 32?

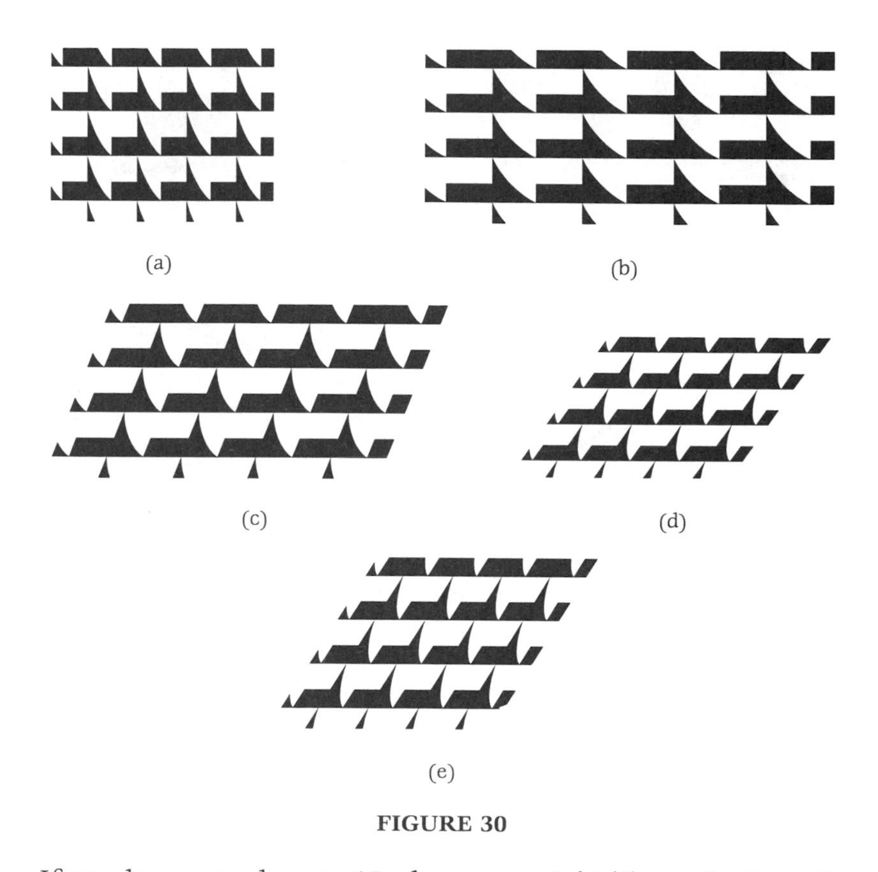

FIGURE 30

If you have got close to 15, then you might like to devise a flow chart for wallpaper patterns modified on the one in Figure 26. The one in Figure 33 is again based on work by Schattschneider [9] and is derived from Washburn and Crowe [10].

It's now plain for all to see that there are actually 17 wallpaper patterns. How close were you? Of course, the patterns of Figures 30 and 31 were included in Figure 32. Again we haven't actually *proved* that there are only 17 patterns. A start on the proof, though, is made in [2]. A complete proof was first given by Federov in 1891 (see

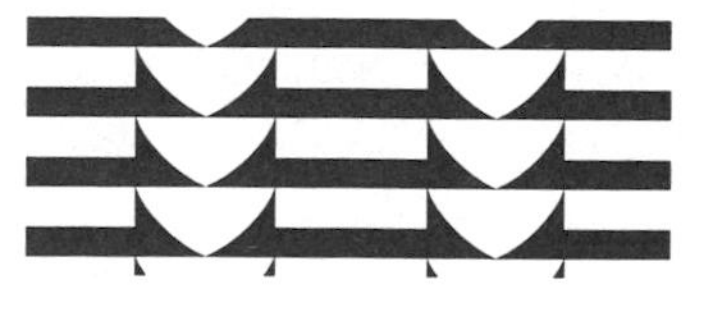

FIGURE 31

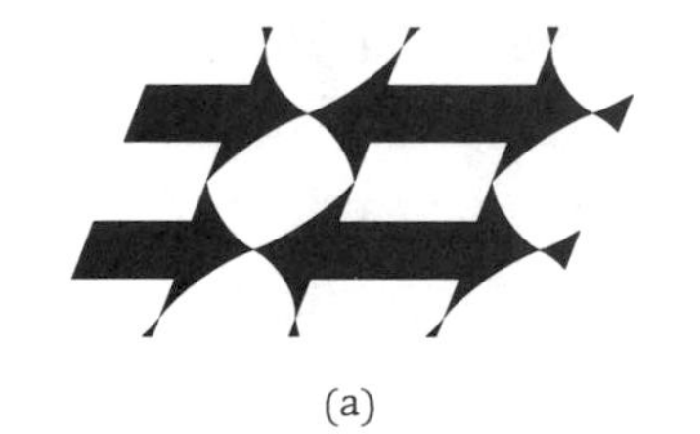

(a)

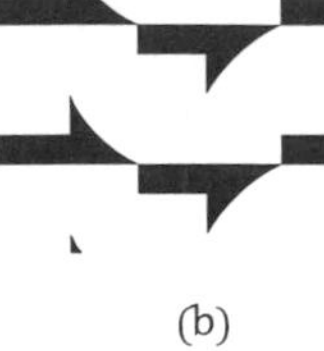

(b)

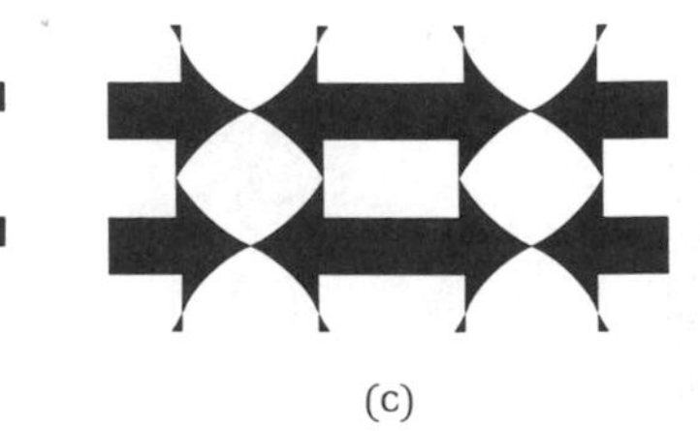

(c)

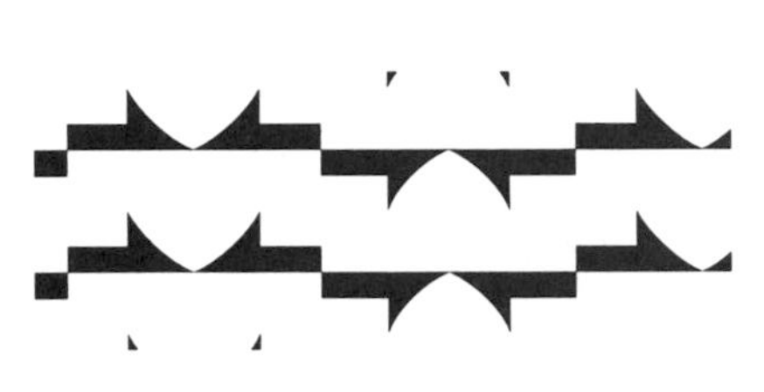

(d)

(e)

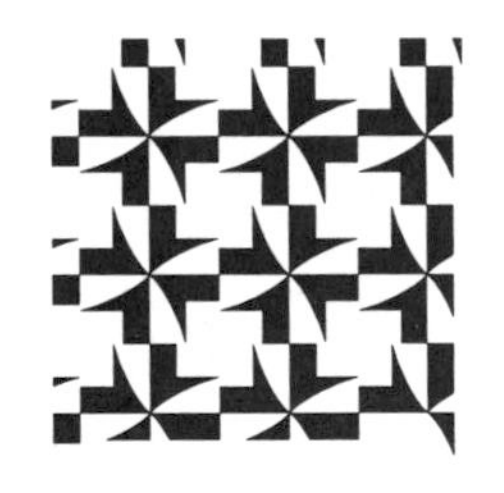

(f)

(g)

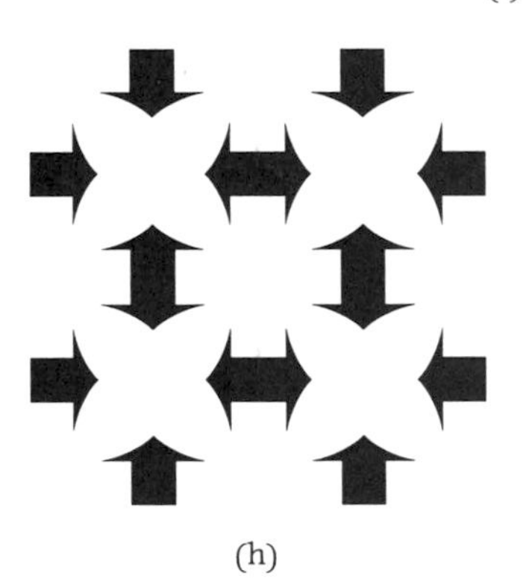

(h)

(i)

(j)

FIGURE 32

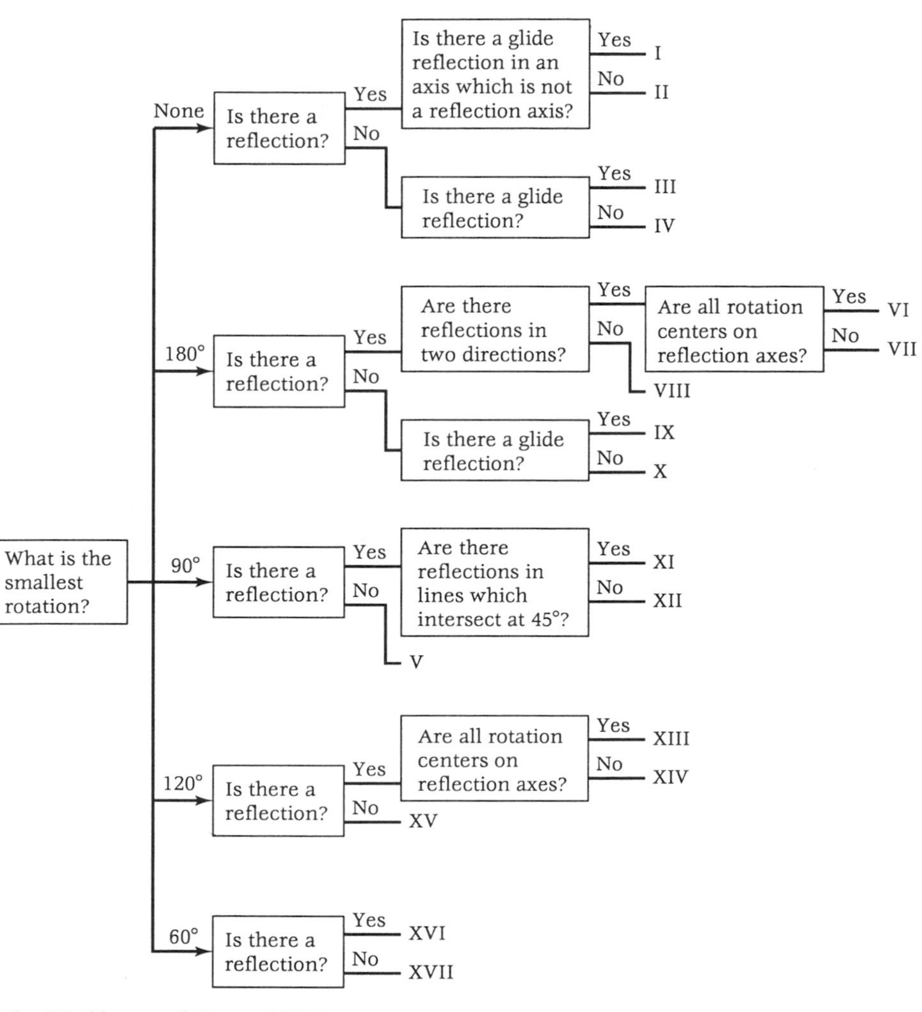

(after Washburn and Crowe, 1988.)

FIGURE 33

[4, pp. 55–56]), and Pólya rediscovered the proof in 1924 (see [8]). Of course, the 17 patterns were known implicitly to the ancient Egyptian craftsmen.

We now expect you to rush out and inspect wallpaper, carpets, quilts, and any other plane designs that happen to be hanging around, to see which of the 17 allowable patterns they contain. To

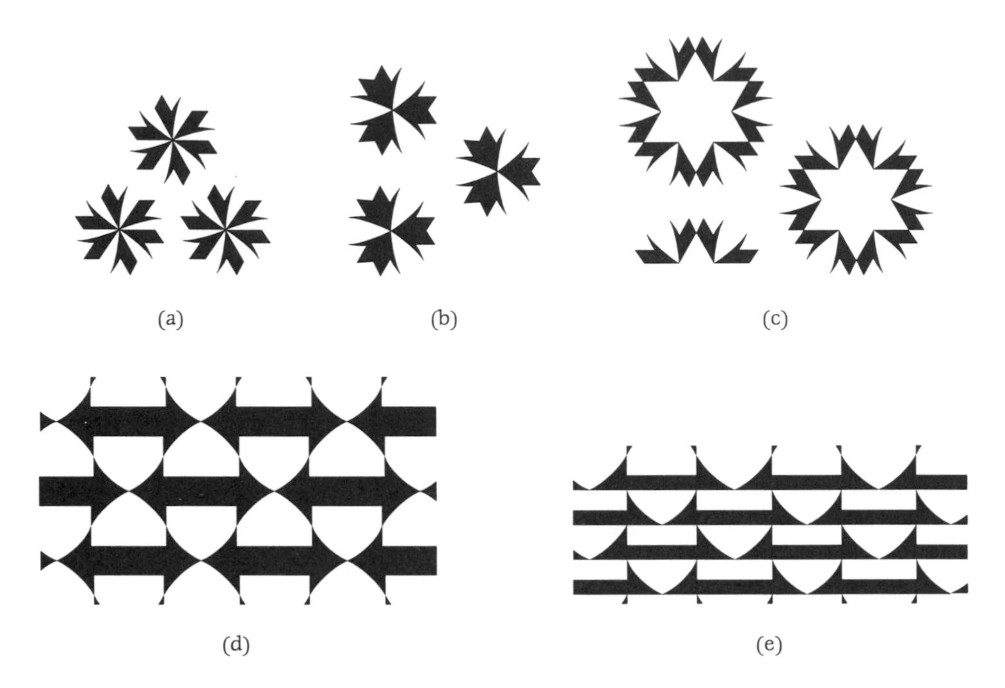

FIGURE 34

help you in Figure 33, we've given examples of six more patterns. The complete 27 are then to be found in Figures 31, 32, and 34.

• • • FINAL BREAK

Here are a few problems for you to try out your new skills on.

1. Find all *integral* solutions of $\frac{1}{x} + \frac{1}{y} = d$, where $d = 1, 2, 3$, or $\frac{1}{2}$.
2. A formula for the sum of the interior angles of a regular polygon was given in Section 1. Show that the same formula holds for nonregular polygons. Does the proof work for nonconvex polygons? Is the formula true for nonconvex polygons? Could you suggest a proof?
3. Give pairs of shapes, some which together do, and some which don't cover the plane.
4. Maybe we were a bit hasty in our conclusions that any quadrilateral will tile the plane. After all we only did it for a particular quadrilateral. Perhaps the proof fails for

other quadrilaterals. Show that the proof we gave is valid for any convex quadrilateral. Then check out nonconvex quadrilaterals.

5. What kind of spiral can you see in the quilt of Figure 1?
6. Show that every frieze has translational symmetry.
7. Use the flow chart of Figure 33 to identify the patterns of Figures 31, 32, and 34.
8. **A Baby Bib.** The pattern below is for a bib and will fit a child from about 8 months to 3 years of age. The scale is shown on the edge of the grid. (1 inch is about 2.5 cm.)

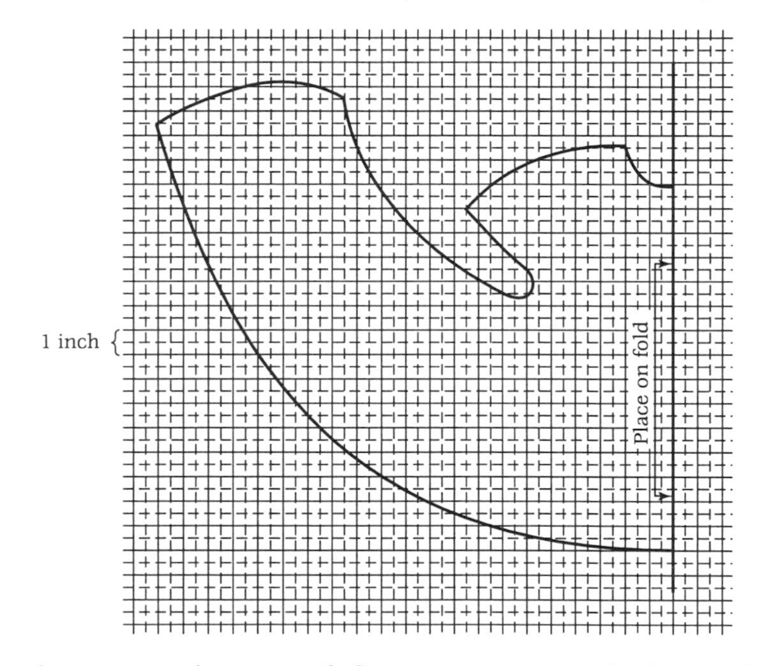

Basic Instructions (Feel free to create variations of your own.)
To construct the bib you will need a reasonably heavy-weight bath towel (take the pattern along to ensure that you buy the right size), a package of bias tape, and, of course, some thread. Begin by folding the towel in half and placing the pattern on it with the straight edge along the fold line. Cut out the bib and unfold it. It should look like the picture

below (without the letters, of course!).

Now lift the curved edge that is labeled AB in the illustration and pin it to the edge shown as $A'B'$ so that the corner labeled A goes on top of A' and the corner labeled B goes on top of B'. Then lift the curved edge that is labeled CD and pin it to the edge shown as $C'D'$ so that the corner labeled C goes on top of C' and the corner labeled D goes on top of D'. Sew each of these edges together using a $\frac{5}{8}$ inch seam allowance. Flatten the seams and zigzag the protruding part of the seam to the bib with matching thread (if you use matching thread the seam will be almost invisible). Finally, bind the unfinished edge with bias tape. You can start anywhere and you will find that there is only one edge. Here is a picture of Kirsten and Katrina Pedersen, Jean Pedersen's granddaughters; Katrina is modeling the bib.

(Photo by Chris Pedersen)

You may notice that the bib has two surfaces (for breakfast and lunch!) separated by the binding along the single edge. You should experiment trying it on a life-sized doll before putting it on a child. You'll need to know how it works to use it, but we can say from experience that most children can't get out of the bib without help.

9. **A Möbius Muff**
   Materials Required

   A jacket zipper of length x (about 10 inches works well).

   A jacket zipper of length $2x$.

   Thread.

Assembly Instructions

Part I. Take the shorter jacket zipper and lay it out as shown below (you don't need to be concerned about the "right" side being up, or whether the zipper tab is on the left or the right). Lift the end shown with the labels A, B and twist it (just a half turn) placing A on D and B on C. Sew it in this position.

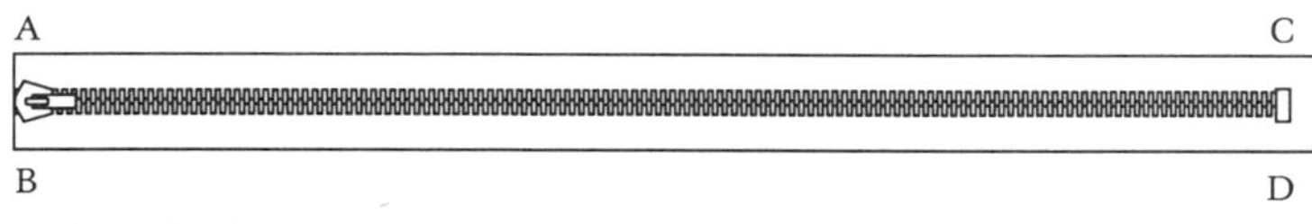

It should look like this:

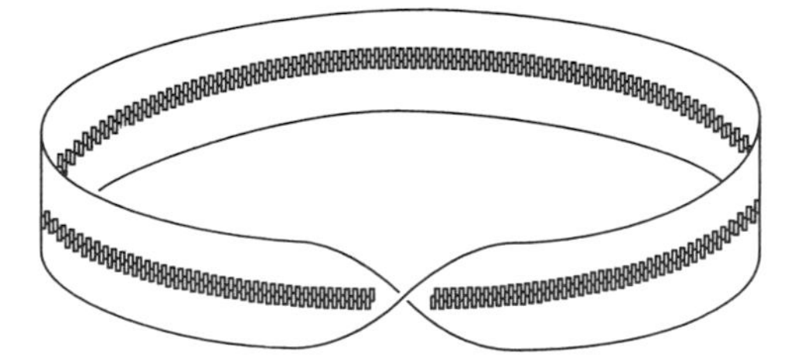

Part II. Unzip the Möbius zipper (to make it easier to handle at the sewing machine at the next stage). Then sew the longer zipper to the edge of the unzipped configuration that does not involve the teeth (or coil) of the zipper. A sewing machine with a zigzag stitch is most helpful for this. Complete the model by sewing each portion of the longer zipper to itself where the ends meet. Zip the first (shorter)

zipper together and the result should look something like this, a nonorientable muff!

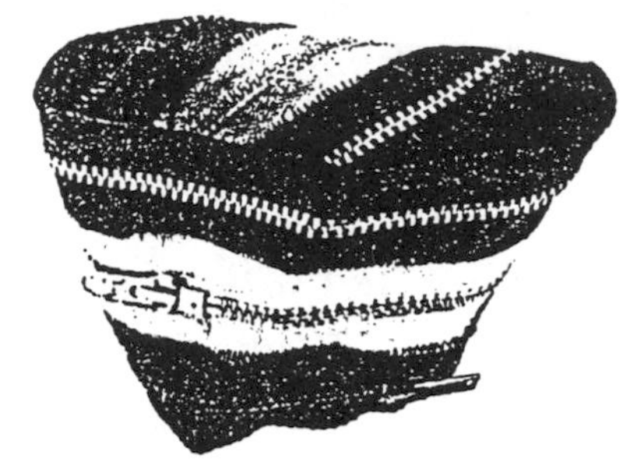

Now unzip everything and reassemble it. If the two zippers used to construct the model are of different colors the model, once unzipped, is fairly easy to reconstruct. You will find that it is substantially more difficult to reassemble this puzzle if the zippers are both of the same color.

REFERENCES

1. Budden, Francis *The Fascination of Groups*, Cambridge University Press, London, 1972.
2. Coxeter, H.S.M. *Introduction to Geometry*, Wiley, New York, 1961.
3. Duke, Dennis, and Deborah Handing, editors, *America's Glorious Quilts*, Park Lane, New York, 1987.
4. Grünbaum, Branko, and Geoffrey Shepherd, *Tilings and Patterns*, W.H. Freeman, San Francisco, 1987.
5. Holton, Derek, *Let's Solve Some Math Problems*, Canadian Mathematics Competitions, University of Waterloo, Waterloo, Ontario, 1993.
6. McLeay, Heather, A closer look at the cast ironwork of Australia, *Math. Intelligencer*, **16**, No. 4, 1994, 61–65.
7. Pedersen, Jean, Geometry: The unity of theory and practice, *Math. Intelligencer*, **5**(4), 1983, 37–49.
8. Pólya, George, Uber die Analogie der Kristallsymmetrie in der Ebene, *Zeits. Krystallog. Mineral*, **60**, 1924, 278–282.

9. Schattschneider, Doris, The plane symmetry groups: Their recognition and notation, *Amer. Math. Monthly*, **85**(6), 1978, 439–450.

10. Washburn, Dorothy K., and Donald W. Crowe, *Symmetries of Culture: Theories and Practice of Plane Pattern Analysis*, University of Washington Press, Seattle, 1988.

ANSWERS FOR FINAL BREAK

1. Notice first that, since x, y are integers, $\frac{1}{x} \leq 1$, $\frac{1}{y} \leq 1$. Also, $x = 0$ (or $y = 0$) cannot be a solution. If $d = 1$, then $(x-1)(y-1) = 1$. Hence $x - 1 = +1$ or -1, so $x = 0$ or 2. But clearly $x \neq 0$. So $x = 2$ and $y = 2$.
   If $d = 2$, then $x = 1$, $y = 1$.
   If $d = 3$, then there are no solutions.
   If $d = \frac{1}{2}$, then $x - 2$ divides 4. Hence, by symmetry, $x = -2$, $y = 1$, or $x = 1$, $y = -2$; $x = 3$, $y = 6$, or $x = 6$, $y = 3$; $x = 4$, $y = 4$, are the solutions.
2. For nonregular convex polygons, simply join any vertex of the polygon to all other vertices. This makes $n - 2$ triangles. The interior angles of the polygon are just the angles in the triangles.
   For nonconvex polygons divide the interior of the polygon into triangles by joining the appropriate vertices. You should again get $n - 2$ triangles. Of course, if you can find *one* pair of vertices such that the segment joining them lies in the interior of the polygon, you can then argue by induction on n.
3. Obviously, there are a lot which don't. It's all in the angles. Are there only a finite number of pairs that do tile the plane though?
4. The same argument does apply.
5. Equiangular? You'll need to take some careful measurements to be sure.
6. Since there is to be a pattern, that is something repeated, there has to be.
7.

I	II	III	IV	V
34(e)	31	32(b)	30	32(c)
VI	VII	VIII	IX	X
34(d)	32(d)	32(e)	32(a)	32(h)
XI	XII	XIII	XIV	XV
32(g)	32(f)	34(b)	32(j)	32(i)
XVI	XVII			
34(c)	34(a)			

CHAPTER 6 Pascal, Euler, Triangles, Windmills, . . .

6.1 INTRODUCTION: A CHANCE TO EXPERIMENT

In earlier chapters we have shown you how beginners may be encouraged to "make mathematics" themselves. Here we distinguish between "doing mathematics," which usually means "doing somebody else's mathematics," and "making mathematics," by which we understand "making one's own mathematics," that is, noticing mathematical patterns, making conjectures, constructing one's own proofs—or at least fully grasping somebody else's—and considering the possibility of formulating valuable generalizations.

We believe that no better laboratory for this type of experimental mathematical activity exists than the ***Pascal Triangle***.[1] This is the triangle, usually designed as an equilateral triangle, but for some purposes realigned as a right-angled triangle (or ***right-justified***, to use the rather strange official phrase), in which the entries in the row labeled n consist of the binomial coefficients $\binom{n}{r}$, $r = 0, 1, 2, \ldots, n$; here $\binom{n}{r}$ is the number of ways of selecting r objects from a set of n objects. For the Pascal Triangle is rich in pattern, and the numerical patterns are linked with the obvious geometrical patterns inherent in an equilateral triangle. Thus

[1] Or *Pascal's Triangle*—both expressions are in common use.

the serious reader may be able to discover patterns, conjecture theorems, and try to prove them.

We do not know how we can fully convey this experimental flavor in a book, since, in a book, the authors are doing all the talking! However, we do want you to understand that we are, in a sense, reporting on a prolonged, successful (and successfully repeated) experiment. Section 2 discusses different approaches to the definition of the binomial coefficients, each with its role to play in the further development of the subject. In Section 3 we consider the question of how to extend the definition of $\binom{n}{r}$ in such a way as to allow n, r to be *any* integers (negative as well as positive). This section may be thought of as a prototype for the art of generalization. We discuss what properties we would wish to preserve in making the generalization, together with the inescapable questions—is it worthwhile trying to generalize, and what benefits do we hope to derive from doing so?

In Section 4 we notice some very remarkable arithmetic-geometric patterns in the Pascal Triangle and in its extension to what we call the Pascal Hexagon (or Pascal Windmill). So far as we know, our first pattern here is quite new, but the second—the Generalized Star of David Theorem—is, as its name implies, a generalization of a known result. However, we claim that neither the original, restricted statement of the Star of David Theorem, nor the original proof, gave any clue as to how it might be successfully generalized—an entirely new approach was needed. It's worth mentioning that we (together with a fourth colleague) recently found another pattern of the same kind in the Pascal Triangle; we published this as a paper on baked beans and spaghetti! See reference [1] at the end of this chapter.

Sections 2, 3, and 4 constitute Part I of this chapter, which we have divided into two parts because it is rather long.

In Sections 5 and 6, which constitute Part II, we are, in a sense, even truer to the actual history of the evolution of the ideas described. The first significant observation was made by our student, Allison Fong, in one of our elementary university classes. After we had discussed her remarkable result and certain closely related results with the members of her class, it was clear that we were on the threshold of something big, but not something to which elementary students would be able to make a contribution, or which they would even properly appreciate. (We do, however, expect the readers of this book to be able to appreciate these ideas!)

So Sections 5 and 6 describe the consequent mathematical research which two of us undertook, taking a colleague at Eastern Michigan University, Tim Carroll, into collaboration. Thus it is bound, in some of its details, to be heavy going for you, the reader. What we principally want you to understand, however, is how the whole story evolved, starting with an observation made by an elementary student of mathematics, and growing into a powerful theoretical structure, at first involving another classical triangle of numbers, the Eulerian Triangle, and then requiring the discovery of new numbers, closely related to the Eulerian numbers, which are featured in the last section. The full story is to be found in [3].

Finally, we obtain, by the use of these new numbers, new results about binomial coefficients in which these new numbers do not figure at all! This is, for us, one of the sublime mysteries of mathematics, part of its magic. Elementary-seeming results emerge from a piece of very indirect reasoning involving concepts that do not appear at all in the statements of those results! Thus a collection of results about familiar mathematical objects is obtained; but the thread which links those results and which provides the ideas for their proofs, is made of very different, unfamiliar material which cannot, apparently, be replaced by something better known. Where but in mathematics do such intriguing complexities of human thought arise—and prevail?

I. PASCAL SETS THE SCENE

6.2 THE BINOMIAL THEOREM

We start by looking at two conceptual interpretations of the symbol $\binom{n}{r}$, where n, r are integers with $0 \leq r \leq n$, and at an arithmetical identity (1) which provides a rule for calculating $\binom{n}{r}$. In fact, (1) may be used to define and calculate $\binom{n}{r}$ for *any* real number n and any nonnegative integer r.[2] In this section we will also show how

[2]However, we will largely restrict our attention in this chapter to integer values of n.

these various interpretations of $\binom{n}{r}$, along with some elementary algebra, enable us to produce a series expansion for $(1+x)^n$ when n is *any* integer—and, in consequence, in Section 2 we are able to extend the well-known Pascal Triangle to obtain what we call the ***Pascal Hexagon*** or, if we want to emphasize the nonzero regions, the ***Pascal Windmill***—with the nonzero regions being referred to as ***blades***.

Let us begin. First, observe that ***combinatorially*** $\binom{n}{r}$ may be thought of as the *number of selections of r objects from n objects*, provided $r \geq 1$. This combinatorial interpretation of $\binom{n}{r}$ explains why we often read this symbol as "n choose r."

The number of such possible selections is seen to be

$$\binom{n}{r} = \frac{n(n-1)(n-2)\cdots(n-r+1)}{r!}. \tag{1}$$

We further adopt the convention

$$\binom{n}{0} = 1.$$

Now (1) may be rewritten rather neatly by multiplying the top and bottom by $(n-r)!$, obtaining the identity

$$\binom{n}{r} = \frac{n!}{r!\,(n-r)!}. \tag{2}$$

Note that (2) is valid for $r = 0$ or $r = n$, provided that we interpret $0!$ in the standard way as 1. Either (1) or (2) may be used to actually calculate $\binom{n}{r}$, so we call them ***arithmetical*** interpretations of $\binom{n}{r}$.

Third, we observe that ***algebraically*** $\binom{n}{r}$ may be thought of as the *coefficient of* x^r *in the binomial expansion of* $(1+x)^n$. This interpretation explains why the symbol $\binom{n}{r}$ is commonly referred to as a ***binomial coefficient***. We usually use the term "binomial coefficient" when speaking about a general $\binom{n}{r}$, but, when speaking about a particular $\binom{n}{r}$, it is helpful to say "n choose r," because without further explanation "binomial coefficient" doesn't tell us the values of n and r, and the recommended phrase reminds us of the precise combinatorial meaning.

In order to show that the algebraic interpretation of $\binom{n}{r}$ is equivalent to the combinatorial interpretation, we first look at the expansion of $(1+x)^n$ for some small values of n. In our display we write $1x^n$ for x^n simply to emphasize the symmetry in the array of coefficients.

Writing the expansions of $(1+x)^n$, for $n = 0, 1, \ldots, 5$, in an orderly fashion we see that

$$\begin{aligned}
(1+x)^0 &= 1 \\
(1+x)^1 &= 1 + 1x \\
(1+x)^2 &= 1 + 2x + 1x^2 \\
(1+x)^3 &= 1 + 3x + 3x^2 + 1x^3 \\
(1+x)^4 &= 1 + 4x + 6x^2 + 4x^3 + 1x^4 \\
(1+x)^5 &= 1 + 5x + 10x^2 + 10x^3 + 5x^4 + 1x^5
\end{aligned} \tag{3}$$

It is now evident from the array in (3) that an expansion of $(1+x)^n$ will be an expression of the form

$$c_0x^0 + c_1x^1 + c_2x^2 + \cdots + c_nx^n, \tag{4}$$

and our object is to find the coefficients $c_0, c_1, c_2, \ldots, c_n$. It should immediately be clear that, for all values of n, $c_0 = c_n = 1$; in fact, we could have deduced this without doing any expanding at all. But how do we obtain the remaining coefficients? First, we think of the product of n factors

$$\underbrace{(1+x)(1+x)\cdots(1+x)}_{n \text{ times}}. \tag{5}$$

Then we observe that, as we multiply up, we get a whole lot of terms of the form x^r where r can take any value from 0 to n; and the expansion of $(1+x)^n$ is obtained by adding up all these terms. So let us fix r and ask how often x^r occurs when we multiply out (5). The answer is this: we obtain x^r by choosing x in r of the factors $(1+x)$ in (5)—and choosing 1 in the remaining $(n-r)$ factors. In how many ways can we select the r factors $(1+x)$ in which we are going to choose x? We have n factors altogether, so we can select r factors in $\binom{n}{r}$ ways. Thus we have established that the coefficient of x^r in the expansion of $(1+x)^n$ is $\binom{n}{r}$. We have, in fact, shown that we may rewrite (4) as

$$(1+x)^n = 1 + \binom{n}{1}x + \binom{n}{2}x^2 + \cdots + \binom{n}{r}x^r + \cdots + x^n, \tag{6}$$

or, using the summation notation (used in Chapter 3 and discussed in Chapter 9), we could write this as

$$(1+x)^n = \sum_{r=0}^{n} \binom{n}{r} x^r. \qquad (6')$$

The expansion (6) or (6′) is the well-known ***binomial formula***, sometimes called the ***binomial expansion***. This powerful formula is actually very easy to use since all that is required is to determine the values of the binomial coefficients $\binom{n}{r}$. In general, we may always use (1) or (2) to calculate $\binom{n}{r}$, but, as we will soon show, there is an especially easy way to determine $\binom{n}{r}$ if n and r are reasonably small nonnegative numbers.

• • • **BREAK**

Use (6) and the fact that

$$(a+b)^n = \left[a\left(1+\frac{b}{a}\right)\right]^n = a^n\left(1+\frac{b}{a}\right)^n$$

to show that

$$(a+b)^n = a^n + \binom{n}{1}a^{n-1}b + \binom{n}{2}a^{n-2}b^2 + \cdots + \binom{n}{r}a^{n-r}b^r + \cdots + b^n.$$

This formula is also referred to as the ***binomial formula***. It explains the use of the word "***bi***nomial" because there are *two* indeterminates, a and b, in the expression $(a+b)^n$.

The algebraic interpretation of $\binom{n}{r}$ suggests that, if $n \geq 0$, we may extend the domain of definition to allow *any* integers r by setting

$$\binom{n}{r} = 0, \quad r < 0 \quad \text{or} \quad r > n. \qquad (7)$$

This we will do; but this extension is really part of the subject matter of Section 3.

One immediate consequence of (7) is that we can then rewrite (6′) even more simply as

$$(1+x)^n = \sum \binom{n}{r} x^r, \tag{6''}$$

understanding the absence of limits above and below the '$\sum$' to mean that we may sum over $-\infty < r < \infty$. However, in the light of (7), the terms in the sum will only be nonzero for $0 \le r \le n$.

Each of the three interpretations (combinatorial, arithmetical, and algebraic) of the symbol $\binom{n}{r}$ is important and useful, but their usefulness will depend on the context, as we will see. For example, we now derive some important identities involving binomial coefficients, and, for each identity, we will use the interpretation, or interpretations, we think most suitable for our purposes.

Really, the complete notation for the binomial coefficient $\binom{n}{r}$ is $\binom{n}{r\ s}$, where $r+s=n$; this, indeed, fully conforms with its interpretation as a coefficient in the expansion of $(a+b)^n$. This complete notation would not usually be worth using, but it *is* the notation which generalizes to ***trinomial*** coefficients obtained by expanding $(a+b+c)^n$. Thus

$$(a+b+c)^n = \sum \binom{n}{r\ s\ t} a^r b^s c^t,$$

where $r+s+t=n$, and $\binom{n}{r\ s\ t} = \frac{n!}{r!\ s!\ t!}$.

We now place the binomial coefficients in the famous Pascal Triangle, that is, we arrange them in a triangular array (see Figure 1) as suggested by the binomial expansions (3).

In the form (2) the binomial coefficient is immediately seen to satisfy what we call the ***Symmetry Identity***,

$$\binom{n}{r} = \binom{n}{n-r} = \binom{n}{s}, \quad \text{for } 0 \le r \le n, \text{ with } r+s=n. \tag{8}$$

Notice that the Symmetry Identity really holds for *all* integer values of r, by virtue of (7).

The entries lying along lines passing through the Pascal Triangle (Figure 1(a)) in directions which are parallel to the sides of the equilateral triangle underlying the grid have either n, r, or s constant as indicated in Figure 1(b). Thus we see that n has the same relationship to r as it has to s and that r and s play entirely

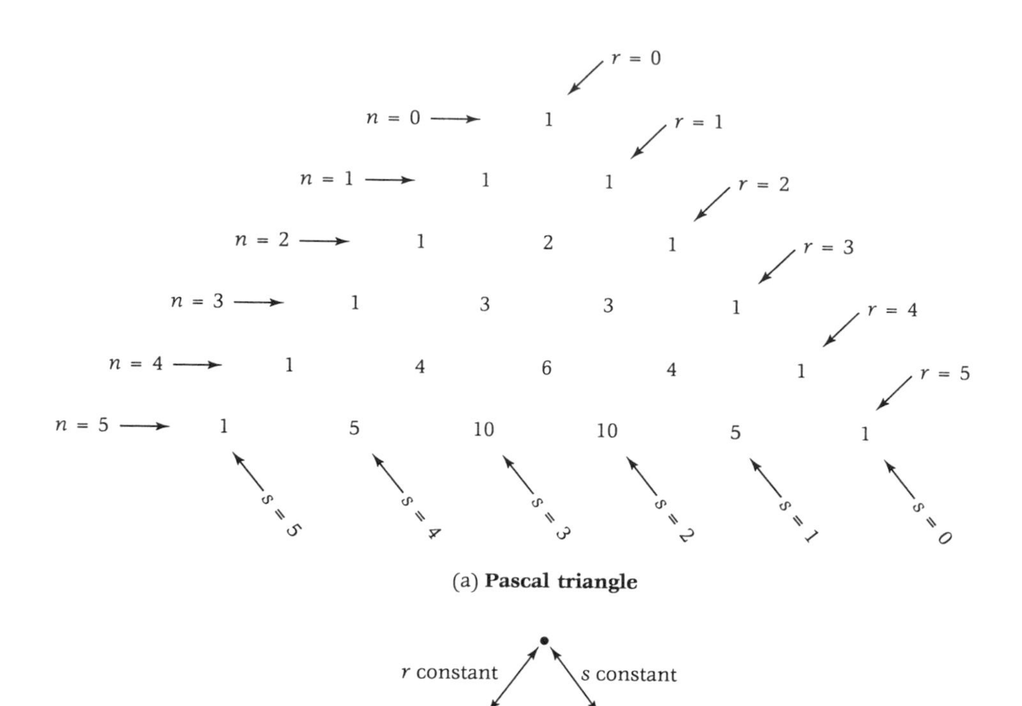

(a) **Pascal triangle**

(b) **Directions in which n, r, and s are constant.**

FIGURE 1

symmetric roles. One may express this symmetry by writing (8) in the form

$$\binom{n}{r\ s} = \binom{n}{s\ r}. \tag{8'}$$

It should now be crystal clear that the coefficient of $a^r b^s$ in the binomial expansion of $(a+b)^n$ is $\binom{n}{r\ s}$—good notation! (There are comments in Chapter 9 on good notation and on symmetry which are relevant to this paragraph.)

The Symmetry Identity is not only important in itself, but also has important practical consequences when we want to compute $\binom{n}{r}$. Using (1) to compute $\binom{12}{10}$ and $\binom{12}{2}$ should suffice to illustrate this practical aspect of (8).

In fact, (8) also has a simple combinatorial explanation. For example, if we are asked to select a group of 10 students from a class of 12 students to constitute a committee, we may instead simply select a group of 2 students from the class of 12 students who will *not* be members of the committee. The crucial observation in this conceptual proof is that each selection of a committee of r students from n students corresponds to a selection of a non-committee of $n - r$ students from n students, so that (8) is clearly true. Plainly, the algebraic interpretation also leads directly to a proof of (8).

Notice that our combinatorial proof of (8) does not use (1) or (2) per se. That is, the proof is *conceptual* and not *computational*—and thus it serves as an example of the power of combinatorial thinking. Such conceptual proofs are not only easier, since they do not involve us in complicated calculations, but they help to provide understanding as to *why* a given relationship is true. They do not merely compel belief.

There is another identity that plays a key role in the theory. Suppose we wish to select r students from a pool of $(n + 1)$ students. Let one of the students be designated by the letter S. Then we can divide our selections into two distinct classes; the class of those selections which include S, and the class of those selections which omit S. How many selections include S? Well, the remaining $(r - 1)$ students are to be selected from n students, so there are $\binom{n}{r-1}$ selections which include S. How many selections omit S? In this case, all r selected students must be drawn from the remaining n students, so there are $\binom{n}{r}$ selections which omit S. Thus we have proved the identity—known as the ***Pascal Identity***

$$\binom{n}{r-1} + \binom{n}{r} = \binom{n+1}{r}, \qquad \text{where } 1 \le r \le n. \tag{9}$$

What Pascal's Identity says from a geometric point of view is that any three adjacent entries in Pascal's Triangle located like this

$$\begin{array}{ccc} \bullet A & & \bullet B \\ & \bullet C & \end{array}$$

satisfy the relationship $A + B = C$. This is what makes it so easy to determine more rows of Pascal's Triangle. If one uses the symmetry identity (8) as well, the work involved in computing the entries in Pascal's Triangle is cut nearly in half (why not exactly half?)

Once again we see the benefit of conceptual reasoning. The only objection, from our point of view, is that the validity of our proof of (9) is restricted to integer values of n such that $n \geq r$, and we have, in (1), the potential for a definition of $\binom{n}{r}$ for any real number n. Thus there is some point in trying to prove (9) starting from (1). This we would call an ***arithmetic proof***. The arithmetic proof may be displayed as follows: we suppose $r \geq 1$, so that

$$\binom{n}{r-1} + \binom{n}{r}$$

$$= \underbrace{\frac{n(n-1)\cdots(n-r+2)}{1\cdot 2\cdots(r-1)}} + \frac{n(n-1)\cdots(n-r+1)}{1\cdot 2\cdots r}$$

multiply top and bottom by r to give both terms on the RHS the same denominator

observe that $(n-r+2)$ fits here

$$n(n-1)\cdots(n-r+2)r \quad . \quad n(n-1)\cdots(n-r+1)$$

Now BE OPTIMISTIC! Realizing that what we want is $\frac{(n+1)n(n-1)\cdots(n-r+2)}{1\cdot 2\cdots(r-1)r}$, factor $\frac{n(n-1)\cdots(n-r+2)}{1\cdot 2\cdots(r-1)r}$ from each term and hope that the sum of what remains is the factor $(n+1)$

$$= \frac{\qquad}{1\cdot 2\cdots(r-1)r}[r + (n-r+1)]$$

$= (n+1)$, our optimism is vindicated! We move $(n+1)$ to the left on the top of the fraction where it "fits" better

$$= \frac{(n+1)n(n-1)\cdots(n-r+2)}{1\cdot 2\cdots r}$$

$$= \binom{n+1}{r}.$$

Now we call your attention to several features of this proof. First, it shows that (9) is, indeed, true for *any real number n* and *any positive integer r*. Second, it enables us to use one of our principles (see Chapter 9 for more details), namely, BE OPTIMISTIC! But it does not give us any insight as to *why* (9) is true—and this is unsatisfying. We really prefer a proof that serves better as an *explanation*.

The algebraic interpretation is the only one we have not yet looked at. So let's give that a chance. Recall that, with n again a nonnegative integer, $\binom{n}{r}$ is the coefficient of x^r in the expansion of $(1+x)^n$. Now it is clear that

$$(1+x)(1+x)^n = (1+x)^{n+1}. \tag{10}$$

So we may ask for the coefficient of x^r, $r \geq 1$, on each side of equation (10)—for, whatever it is, it must be the *same* on both sides of the equation! On the right-hand side the coefficient of x^r is, by its very definition, $\binom{n+1}{r}$. On the left-hand side the coefficient of x^r will come from adding two terms; that is, you get x^r either

(i) by multiplying 1 by x^r, or

(ii) by multiplying x by x^{r-1}.

Hence the coefficient of x^r on the left-hand side is $\binom{n}{r} + \binom{n}{r-1}$, and we thereby obtain (9) again.

This last proof is conceptually better than the arithmetic one because it tells us something about *why* (9) is a valid identity. It also suggests further adventures. For example, (10) holds for any real number n—wouldn't it be marvellous if (6″) also held for any real number n? Then we'd have a beautiful algebraic proof of (9), valid in remarkable generality. (This idea is pursued in Section 3). But even more is true, for this approach also suggests that we may discover other interesting identities concerning binomial coefficients by considering the general algebraic identity

$$(1+x)^a(1+x)^b = (1+x)^{a+b}.$$

We remark that we now have available a very nice alternative proof that the combinatorial and algebraic definitions of $\binom{n}{r}$ coincide. For we have seen that, with either definition,

(i) $\binom{n}{0} = 1$;

(ii) $\binom{0}{r} = 0$, $r > 0$;

(iii) $\binom{n}{r-1} + \binom{n}{r} = \binom{n+1}{r}$, $r \geq 1$.

But these three identities determine $\binom{n}{r}$ for n and r non-negative integers. Thus we have a new proof that the combinatorial and algebraic definitions coincide which still makes no use whatsoever of the arithmetic identity (1), or (2).

• • • **BREAK**

1. Show that the Pascal Identity holds for $n \geq 0$ and any integer r.
2. (i) Use the Pascal Identity to prove that

$$\binom{n}{0} + \binom{n+1}{1} + \binom{n+2}{2} + \cdots + \binom{n+k}{k} = \binom{n+k+1}{k}.$$

(ii) Take a particular case of the result above, locate these coefficients in the Pascal Triangle, and enclose them, and no others, in a simple loop. You should see why this result is sometimes called the ***Christmas Stocking Theorem*** in the United States, and the ***Hockey Stick Theorem*** in New Zealand.
(iii) Use the Symmetry Identity to write another form of the Christmas Stocking Theorem (where the stocking hangs on the right-hand side of the Pascal Triangle).

6.3 THE PASCAL TRIANGLE AND WINDMILL

In this section we appeal to some notions of mathematical analysis to give meaning to the entries in the northeast and northwest blades of the Pascal Windmill (see Figure 2), and to justify the three zero regions. Thus we extend the meaning of $\binom{n}{r}$ over the full range of integers and show how the entries in Figure 2 with n or r negative have an important significance analogous to the algebraic interpretation of the entries in the ordinary Pascal Triangle. In the course of the discussion we show precisely how the individual entries for $\binom{n}{r}$ in Pascal's Windmill may be computed directly—that

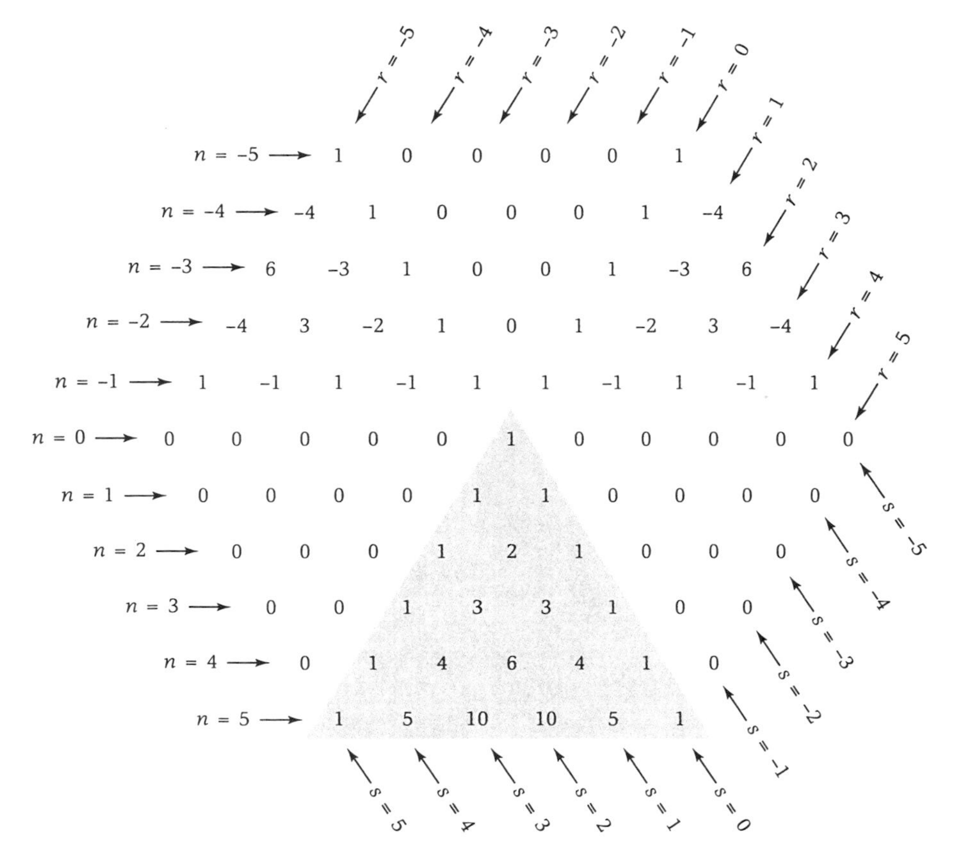

FIGURE 2 The Pascal Windmill (or **Hexagon**).

is, without having to construct any part of the hexagonal array of numbers. We will also draw explicit attention to symmetry properties (that the reader may discover, at this point, by simply studying Figure 2) which, as a practical matter, make it very easy to write out large sections of the hexagon in any direction.

First we point out that, by means of (7), we have already provided the two zero regions completing the lower half of the hexagon.

Now we turn to the upper half, that is, the part where n is a negative integer. We begin with $n = -1$ and ask, does it make sense to talk about an expansion of $(1 + x)^{-1}$? To answer the question we first look at the identity

$$\frac{1 - x^{r+1}}{1 - x} = 1 + x + x^2 + \cdots + x^r. \tag{11}$$

• • • **BREAK**

Show that

$$1 - x^{r+1} = (1 - x)(1 + x + x^2 + \cdots + x^r).$$

For which x's does this result imply that (11) is true?

Now it is certainly true that if $x = 1$ there is trouble in (11) so, for the moment, let us assume that $|x| < 1$. Notice that, by putting $x = 0$ in (11), we immediately see why the constant term on the right-hand side is 1.

Next observe that, if $|x| < 1$, and we let r become very large (or, if we want to sound more sophisticated, we let $r \to \infty$), then x^{r+1} would become arbitrarily small and (11) would imply that

$$\frac{1}{1-x} = 1 + x + x^2 + \cdots + x^r + \cdots, \quad \text{with } |x| < 1, \tag{12}$$

where the right-hand side has an *infinite* number of terms. We call the RHS of (12) the ***power series*** for $(1 - x)^{-1}$, valid if $|x| < 1$. It is a power series because it takes the form $\sum_{r=0}^{\infty} a_r x^r$; it is valid because it converges, that is, there is a function $f(x)$—in our case $\frac{1}{1-x}$—such that $\sum_{r=0}^{m} a_r x^r$ may be made as near to $f(x)$ as we want by taking m sufficiently large.

If we replace x by $-x$ in (12) we obtain

$$\frac{1}{1+x} = (1+x)^{-1} = 1 - x + x^2 - \cdots + (-1)^r x^r + \cdots, \quad \text{with } |x| < 1. \tag{13}$$

Of course neither (12) nor (13) makes any sense if $|x| > 1$. We will see later how to get a series for $\frac{1}{1+x}$ or $\frac{1}{1-x}$ in this case.

• • • **BREAK**

Experiment by choosing a value of x between -1 and $+1$ and, using your calculators, determine how many terms on the right of (13) will be required to approximate the function on the left to some specified degree of accuracy, say 0.0001. This should give you a feel for the meaning of convergence and for how fast the series converges for various values of x in the range of definition; this exercise is good preparation for the study of infinite series.

Notice that, if we adhere to our algebraic definition of the binomial coefficient, then (13) gives us the value of $\binom{-1}{r}$ for any $r \geq 0$, namely, $\binom{-1}{r} = (-1)^r$. Wonder of wonders, this agrees with the values we would obtain from (1) if we substituted $n = -1$. This strongly suggests that we have some chance of finding a series expansion of $(1+x)^N$ for all negative integers N and that we might reasonably hope that the value of the coefficients could always be obtained by means of (1). That is, we hope that, for any positive integer n, there is a power series expansion, for $|x| < 1$,

$$(1+x)^{-n} = \sum_{r=0}^{\infty} \binom{-n}{r} x^r,$$

and that the unknown coefficient $\binom{-n}{r}$ will turn out to be

$$\frac{-n(-n-1)(-n-2)\cdots(-n-r+1)}{r!}.$$

Now all the factors in the numerator of the last expression are negative numbers, so we can factor (-1) from each factor to obtain

$$\begin{aligned}
&\frac{-n(-n-1)(-n-2)\cdots(-n-r+1)}{r!} \\
&= (-1)^r \frac{(n+r-1)(n+r-2)\cdots(n)}{r!} \\
&= (-1)^r \binom{n+r-1}{r}.
\end{aligned}$$

Thus it is our hope that this last expression will, indeed, be the right value for $\binom{-n}{r}$. (Notice how optimistic we are!)

Our conjecture then is that, if $|x| < 1$ and $n > 0$, then

$$(1+x)^{-n} = \sum_{r=0}^{\infty} (-1)^r \binom{n+r-1}{r} x^r. \tag{14}$$

How might we try to prove this? Those who know enough differential calculus will find a quick proof of (14) when it is repeated a few pages later as (18). For others brave enough to try to understand a difficult argument, we offer you a proof now which does not require the calculus. To help you to understand the argument, we go over it again afterward in a *particular but not special case.*

★ First, we know that a convergent power series in x exists for $(1+x)^{-1}$ provided $|x| < 1$, namely, $1 - x + x^2 - x^3 + x^4 - \cdots$; and, from elementary algebra, we know that

$$(1+x)^{-(n+1)} = (1+x)^{-n}(1+x)^{-1};$$

thus, by induction on n, we know there *is* a convergent series for $(1+x)^{-n}$, provided $|x| < 1$. For, given a series for $(1+x)^{-n}$, we merely multiply that series by the series (13)—just as we multiply polynomials—to get a series for $(1+x)^{-(n+1)}$. So all that remains is to prove that the coefficient of x^r in the power series for $(1+x)^{-n}$ is $(-1)^r\binom{n+r-1}{r}$.

Stating it more formally, suppose we denote the coefficient of x^r in the expansion of $(1+x)^{-n}$ by $a_r^{(n)}$. Setting $x = 0$, we immediately infer that $a_0^{(n)} = 1$. We also know that $a_r^{(1)} = (-1)^r$. Then we wish to prove the following:

Theorem 1 $a_r^{(n)} = (-1)^r\binom{n+r-1}{r}$.

Proof We know this holds for $r = 0$. Thus we only have to prove it for $r \geq 1$. We are going to argue by induction on n, since it certainly holds for $n = 1$. Now

$$(1+x)^{-(n+1)} = \sum a_r^{(n+1)} x^r \quad \text{and} \quad (1+x)^{-n} = \sum a_r^{(n)} x^r,$$

so, basing ourselves on the obvious identity,

$$(1+x)^{-n} = (1+x)(1+x)^{-(n+1)},$$

and using the same algebraic argument as that for the Pascal Identity (9) we infer that

$$a_r^{(n+1)} + a_{r-1}^{(n+1)} = a_r^{(n)}, \quad r \geq 1. \tag{15}$$

Now, as we have said, we also know that the assertion of the theorem holds for $n = 1$; this is just (13). If we assume that, for some n,

$$a_r^{(n)} = (-1)^r\binom{n+r-1}{r},$$

then

$$a_r^{(n+1)} + a_{r-1}^{(n+1)} = (-1)^r\binom{n+r-1}{r} \quad \text{and} \quad a_0^{(n+1)} = 1.$$

Let's take stock. Our inductive hypothesis is that

$$a_r^{(n)} = (-1)^r \binom{n+r-1}{r};$$

and we know that $a_0^{(n+1)} = 1$. With this data we want to show that (15) implies that

$$a_r^{(n+1)} = (-1)^r \binom{n+r}{r}. \tag{16}$$

How should we do this? Well—induction seems to be a good idea, so let's try induction *with respect to* r, since (16) certainly holds for $r = 0$. Thus we may further assume that $r \geq 1$ and that

$$a_{r-1}^{(n+1)} = (-1)^r \binom{n+r-1}{r-1}.$$

Then (15) tells us that

$$\begin{aligned} a_r^{(n+1)} &= a_r^{(n)} - a_{r-1}^{(n+1)} \\ &= (-1)^r \binom{n+r-1}{r} - (-1)^r \binom{n+r-1}{r-1} \\ &= (-1)^r \left[\binom{n+r-1}{r} + \binom{n+r-1}{r-1} \right] \\ &= (-1)^r \binom{n+r}{r}, \end{aligned}$$

by the Pascal Identity, and we have the result we hoped for.

This completes our proof, so that we now know that, as we hoped,

$$a_r^{(n)} = \binom{-n}{r} = (-1)^r \binom{n+r-1}{r}. \qquad \square$$

If you had any trouble in following this ***double induction*** on n and r it may be well to look at a PARTICULAR BUT NOT SPECIAL CASE (see more on this in Chapter 9). Thus, for example, suppose that some student doesn't believe Theorem 1. For obvious reasons, we will call him Thomas. The dialogue might go as follows. We ask Thomas to tell us for what value of n he thinks the theorem fails for the first time. He replies "$n = 15$." We then ask, with $n = 15$, for what value of r does it first fail? He replies "$r = 18$." In other words, Thomas is asserting that he has checked and discovered that the theorem fails *for the first time* when you reach the coefficient of x^{18} in the expansion of $(1 + x)^{-15}$. He admits that Theorem 1 gives the right coefficient for x^r in the expansion of $(1 + x)^{-n}$ for $n = 1, 2, \ldots, 14$ and all $r \geq 0$, but, for $n = 15$, he believes that it is only true for $r = 0, 1, 2, \ldots, 17$ and then it fails. Specifically, he claims that $a_{18}^{(15)} \neq (-1)^{18}\binom{32}{18}$.

How do we argue that doubting Thomas must be wrong? Well, we know, from our algebraic argument for the Pascal Identity, that, for all n, and $r \geq 1$,

$$a_r^{(n+1)} + a_{r-1}^{(n+1)} = a_r^{(n)}.$$

So if we let $n = 14$ and $r = 18$ we have

$$a_{18}^{(15)} + a_{17}^{(15)} = a_{18}^{(14)}.$$

or, solving for $a_{18}^{(15)}$, we obtain

$$a_{18}^{(15)} = a_{18}^{(14)} - a_{17}^{(15)}. \tag{17}$$

However, Thomas has already agreed that each of the terms on the right-hand side of (17) is given by our theorem. Thus we know that

$$\begin{aligned} a_{18}^{(15)} &= (-1)^{18}\left[\binom{31}{18} + \binom{31}{17}\right] \\ &= (-1)^{18}\binom{32}{18}, \quad \text{by the Pascal identity,} \end{aligned}$$

and we have shown that Thomas was wrong—$a_{18}^{(15)}$ *is* what it's supposed to be! Of course, this kind of argument will apply wherever Thomas has his first doubts. ⋆

We repeat that our proof of the series expansion for $(1 + x)^{-n}$ was specially adapted to the needs of those of you who have not taken calculus. The usual (and much shorter) proof that

$$(1 + x)^{-n} = \sum_{r=0}^{\infty}(-1)^r\binom{n + r - 1}{r}x^r \quad \text{if} \quad |x| < 1, \tag{18}$$

suitable for those familiar with the basic facts of the differential calculus, is via the general argument for the existence of a convergent Taylor–Maclaurin series. Specifically, the differential calculus tells us that

$$f(x) = \sum_{r=0}^{\infty} \frac{f^{(r)}(0)}{r!} x^r$$

within any interval surrounding $x = 0$ in which $f(x)$ is analytic (see [6, 7, 10]). Here $f^{(r)}(x)$ is the rth derivative of the function $f(x)$.

By way of example, we now display $(1 + x)^N$ for[3] $N = -1, -2, -3$. We have arranged the terms so that the reader may verify that the Pascal Identity is valid (as we have already shown that it should be) for these examples. In this display the coefficients are located in the precise order in which they appear in the northeast blade of Figure 2, on the rows $N = -1$, $N = -2$, $N = -3$, respectively.

$$\begin{array}{lccccccccccc}
\bullet & & & & & \bullet & & \bullet & & \bullet & & \bullet \\
\bullet & & & & \bullet & & \bullet & & \bullet & & \bullet & \\
(1+x)^{-3} = & & & +1x^0 & & -3x^1 & & +6x^2 & & -10x^3 & & +15x^4 \\
(1+x)^{-2} = & & +1x^0 & & -2x^1 & & +3x^2 & & -4x^3 & & +5x^4 & \\
(1+x)^{-1} = & +1x^0 & & -1x^1 & & +1x^2 & & -1x^3 & & +1x^4 & & -x^5
\end{array}$$

We may now rewrite the Pascal Identity in its extended form

$$\binom{N}{r-1} + \binom{N}{r} = \binom{N+1}{r}, \quad \text{for } r \geq 1 \text{ and } N \text{ any integer.} \quad (19)$$

However, we see that (7) is actually *untrue* in the extended domain of the binomial coefficient, that is, if we no longer require $n \geq 0$; for example, $\binom{-1}{1} = -1$. On the other hand, we have proved, by establishing (18), the

Second Symmetry Identity

$$\binom{-n}{r} = (-1)^r \binom{n+r-1}{r}, \quad \textit{for } r \geq 0,\ n \geq 0. \qquad (8'')$$

We may, of course, regard (8″) as constituting a *definition* of $\binom{-n}{r}$, for $n > 0$, $r \geq 0$.

[3] We adopt the notation of "upper case N" in order to preserve, for the time being, the exclusive use of "lowercase n" for a non-negative integer.

• • • **BREAK**

Show that $\binom{-N}{r} = (-1)^r\binom{N+r-1}{r}$, for $r \geq 0$ and *any* integer N.

So far, in attaching a meaning to $\binom{N}{r}$, with N negative, r has always been a nonnegative integer; but might there not be a situation in which we would like to attach a meaning to the symbol $\binom{N}{R}$ with R negative? To see that, in fact, there is such a situation, observe that, if $|x| > 1$, then $|\frac{1}{x}| < 1$; and

$$(1+x)^N = x^N\left(1+\frac{1}{x}\right)^N. \tag{20}$$

Now, if $N \geq 0$, either side of (20) gives the usual binomial expansion of $(1+x)^N$. This leads us to the rule, already formulated in (7),

$$\binom{N}{R} = 0, \quad N \geq 0,\ R < 0;$$

for no *negative* powers of x appear in this expansion.

We are thus led finally to the case $N < 0$, $R < 0$; we write $N = -n$, so that $n > 0$. Assuming $|x| > 1$ for convergence, we obtain, from the RHS of (20),

$$(1+x)^{-n} = x^{-n}\sum_{s=0}^{\infty}\binom{-n}{s}x^{-s}, \quad n > 0. \tag{21}$$

Now $\binom{-n}{-r}$, $n > 0$, $r > 0$, should represent the coefficient of x^{-r} on the right-hand side of (21). This produces the definition

$$\binom{-n}{-r} = \left\{\begin{matrix} 0, & r < n, \\ \binom{-n}{r-n}, & r \geq n, \end{matrix}\right\} \quad n > 0,\ r > 0. \tag{22}$$

Once we adopt definition (22), our planned extension of the domain of the binomial coefficients is complete.

Notice that we may express (22) by setting, for n, r positive,

$$\binom{-n}{-r} = 0 \quad \text{if } r < n,$$

and then requiring that the first Symmetry Identity (8) holds universally. That is,

$$\binom{N}{R} = \binom{N}{N-R}$$

for *all* integers N, R.

Before closing this section let us look at the cases $N = -1, -2, -3$. These are displayed next, in an unconventional, but suggestive, arrangement that incorporates (20) and places the coefficients in precisely the same locations (relative to each other) as those they occupy in the northwest blade of the Pascal Windmill (shown in Figure 2), along the rows $N = -1$, $N = -2$, $N = -3$, respectively. Remember that we are assuming $|x| > 1$.

$$\begin{array}{ccccccccccc}
\bullet & & \bullet & & \bullet & & \bullet & & & = & \bullet \\
-10x^{-6} & & +6x^{-5} & & -3x^{-4} & & +1x^{-3} & & & = & (1+x)^{-3} \\
 & -4x^{-5} & & +3x^{-4} & & -2x^{-3} & & +1x^{-2} & & = & (1+x)^{-2} \\
+1x^{-5} & & -1x^{-4} & & +1x^{-3} & & -1x^{-2} & & +1x^{-1} & = & (1+x)^{-1}
\end{array}$$

The reader should now be able to fill in additional rows to expand the array of values in the northwest blade of the hexagon of Figure 2. We have already pointed out that an easy way to extend the southern blade is to use the Pascal Identity and the fact that $\binom{N}{N} = \binom{N}{0} = 1$. In the other two blades, however, a little study of the existing values in Figure 2 should enable you to reduce the labor involved in expanding the array in these blades. First, notice the pattern of the signs in the northeast and northwest blades. Next, think of what you would have if you ignored the signs and rotated those blades in either a clockwise, or anti-clockwise, direction through 120°.

Now that the extension to the Pascal Hexagon is complete we revert to the usual notation $\binom{n}{r}$, and allow n and r to represent *any* integer.

We should point out that the Pascal Identity holds *everywhere* in the Pascal Hexagon, *except* at the exact center where it is clear that $1 + 1 \neq 1$. However, this is not so surprising—nor should it cause us any undue distress, as we will explain.

In our analytical interpretation, the nonzero entries in the Pascal Windmill may be viewed as two overlapping arrays of numbers. In the first array (lying on, and below, the $\binom{n}{0}$ row of 1's running from the southwest to the northeast corner of the windmill) the entries $\binom{n}{r}$, with $r \geq 0$, correspond to the coefficients of x^r in the Taylor expansion of $(1+x)^n$, valid when $|x| < 1$. Similarly, in the second array (lying on, and below, the $\binom{n}{n}$ row of 1's running from the northwest to the southeast corner of the windmill) the entries $\binom{n}{r}$, with $r \leq n$, correspond to the coefficients of x^r in the Laurent expansion (21) of $(1+x)^n$, valid when $|x| > 1$. Notice that when $0 \leq r \leq n$ these two arrays overlap to form the fundamental region (the Pascal Triangle) and, as expected, in this case $\binom{n}{r}$ is the coefficient of x^r in the *finite* expansion of $(1+x)^n$, valid for *all* values of x. Where the Pascal Identity breaks down is the *unique* place where we would be (foolishly!) attempting to make both the *Taylor* expansion (18) and the *Laurent* expansion, based on (20), valid. So the one failure is really a success—it alerts us to a subtlety we might otherwise have missed.

• • • BREAK

1. We encouraged you earlier to think of the *binomial* theorem in its original sense of

$$(a+b)^n = a^n + \binom{n}{1}a^{n-1}b + \cdots + \binom{n}{r}a^{n-r}b^r + \cdots + b^n,$$

for n a positive integer. Now let us consider a negative integer N instead of n. There are two cases to consider

(i) $|a| > |b|$; and

(ii) $|a| < |b|$.

What happens in these two cases? Why can we ignore $|a| = |b|$?

2. Try rephrasing the description above of the nonzero regions of the Pascal Windmill using n, r, and s, noting the gain in symmetry.

★ Before we go on to discuss some of the properties of the entries in the Pascal Windmill we would like to call the reader's attention to the fact that there are many other triangular arrays, apart from the Pascal Triangle, which may also be extended to hexagonal arrays (see [9 and 11]).

One such array consists of the ***harmonic coefficients*** (due to Leibniz), denoted by $\left[{n \atop r}\right]$, which may be defined as

$$\left[{n \atop r}\right] = \frac{1}{(n+1)\binom{n}{r}}, \qquad 0 \leq r \leq n,$$

$$\text{where } \binom{n}{r} \text{ is the ordinary binomial coefficient.} \tag{23}$$

A second array consists of the ***q-analogues of the binomial coefficients***, which we denote here by $\binom{n}{r}_q$, defined for $0 \leq r \leq n$ as $\binom{n}{0}_q = 1$ and

$$\binom{n}{r}_q = \frac{(q^n - 1)(q^{n-1} - 1)\cdots(q^{n-r+1} - 1)}{(q^r - 1)(q^{r-1} - 1)\cdots(1 - 1)}, \quad r \geq 1, \tag{24}$$

so that $\binom{n}{r}_q$ is the q-analogue[4] of $\binom{n}{r}$.

• • • BREAK

1. Use what you know about the entries in the Pascal Triangle to write out the first ten rows of the Harmonic Triangle. Save this array—we will point out some of its surprising properties in the next section.
2. Write out the first few q-analogues and see (i) that they are always polynomials in q (which is somewhat surprising); and (ii) that when $q = 1$ the polynomial $\binom{n}{r}_q$ assumes the value of the corresponding binomial coefficient $\binom{n}{r}$. The proof of (i) uses the q-analogue of the Pascal Identity,

[4] Precisely, $\binom{n}{r}_q$ is a polynomial in q with integer coefficients and

$$\lim_{q \to 1} \binom{n}{r}_q = \binom{n}{r}.$$

In fact, $\binom{n}{r}_q$ is often called a *Gaussian polynomial.* A study of Gaussian polynomials, along the lines of [3], is to be found in [10]. The development of combinatorial interpretations of these q-analogues appears in [12]. The curious reader might also like to look at [4], where Gauss first introduced them.

namely (there are actually two versions, equally valid!)

$$q^r\binom{n}{r}_q + \binom{n}{r-1}_q = \binom{n+1}{r}_q, \quad 1 \le r \le n,$$

or

$$\binom{n}{r}_q + q^{n+1-r}\binom{n}{r-1}_q = \binom{n+1}{r}_q, \quad 1 \le r \le n;$$

while the proof of (ii) is an easy application, for those who know some differential calculus, of the famous l'Hôpital's Rule, which may be found in any calculus textbook.

A third array consists of the ***q-analogues of the harmonic coefficients*** (see [10]), which we denote here by $\begin{bmatrix} n \\ r \end{bmatrix}_q$, defined for $0 \le r \le n$ as

$$\begin{bmatrix} n \\ 0 \end{bmatrix}_q = \frac{q-1}{q^{n+1}-1}$$

and

$$\begin{bmatrix} n \\ r \end{bmatrix}_q = \frac{(q^r-1)(q^{r-1}-1)\cdots(q-1)}{(q^n-1)(q^{n-1}-1)\cdots(q^{n-r+1}-1)} \cdot \frac{q-1}{q^{n+1}-1}, \quad r \ge 1, \tag{25}$$

so that

$$\begin{bmatrix} n \\ r \end{bmatrix}_q = \frac{\begin{bmatrix} n \\ 0 \end{bmatrix}_q}{\binom{n}{r}_q}.$$

An important feature common to the binomial coefficients $\binom{n}{r}$, the harmonic coefficients $\begin{bmatrix} n \\ r \end{bmatrix}$, and the q-analogues of the binomial coefficients $\binom{n}{r}_q$, which accounts for their sharing some surprising *geometrical* properties, is that each of them is a ***separable function***. Let us be more specific. A function F of three variables n, r, s where $r + s = n$ is said to be ***separable*** if

$$F(n, r, s) = f(n)g(r)h(s). \tag{26}$$

We readily see from (8′) that the binomial coefficient, restricted to the Pascal Triangle, is a special case of (26), since the arithmetical definition of $\binom{n}{r}$ may be given as

$$\binom{n}{r} = \frac{n!}{r!\,(n-r)!} = \frac{n!}{r!\,s!}, \quad \text{where } 0 \le r \le n,\ 0! = 1,\ r+s = n.$$

Thus, in this case, f is the factorial function, while g and h are ★ both the reciprocals of the factorial function.

• • • BREAK

(Starred problems concern the starred material above.)

*1. What do you think is the q-analogue of the factorial function $n!$?

*2. (a) Show that the harmonic coefficient $\left[{n \atop r}\right]$ is a separable function and describe f, g, and h.
(b) Show that the q-analogue of the binomial coefficient $\binom{n}{r}_q$ is a separable function and describe f, g, and h.
(c) Show that the q-analogue of the harmonic coefficient $\left[{n \atop r}\right]_q$ is a separable function and describe f, g, and h.

Exercises 3 through 5 will set the scene for Section 6.4, so don't skip them.

3. Consider the vertices of a parallelogram in the Pascal Triangle as shown, when we choose, for example, $n = 10$, $\ell = 2$, $k = 4$ (and *you may choose* r). We box the dot next to $\binom{n}{r}$ and the dot next to the binomial coefficient opposite $\binom{n}{r}$.

$$\bullet \binom{n-\ell}{r} \qquad\qquad \boxed{\bullet} \binom{n-\ell}{r+k}$$

$$\boxed{\bullet} \binom{n}{r} \qquad\qquad \bullet \binom{n}{r+k}$$

(a) Using our suggested values of n, ℓ, k (and your r) calculate the value of the product of the binomial coefficients next to the boxed dots divided by the product of the other two binomial coefficients.

(b) Slide this parallelogram one unit (either way) in the direction in which s is constant and repeat Exercise (a). What do you notice?

(c) Try sliding the parallelogram any number of units (either way) you please in the direction in which s is constant and repeat Exercise (a). What do you notice? Note that, if you slide the parallelogram u units in the s direction, the binomial coefficients of the new parallelogram will then appear as follows:

$$\bullet \binom{n-\ell+u}{r+u} \qquad \boxed{\bullet} \binom{n-\ell+u}{r+k+u}$$

$$\boxed{\bullet} \binom{n+u}{r+u} \qquad \bullet \binom{n+u}{r+k+u}$$

4. Repeat Exercise 3(a), with the roles of ℓ and k reversed; that is, with $n = 10$, $\ell = 4$, $k = 2$. Does this result surprise you?

5. Observe that in Exercises 3 and 4 we took parallelograms with sides parallel to the directions in which n and r are constant—and that we were sliding the parallelogram in the direction in which s was constant. Suppose we exchanged the roles of r and s. Our parallelogram would then appear as follows:

$$\boxed{\bullet} \binom{n-\ell}{r-k-\ell} \qquad \bullet \binom{n-\ell}{r+\ell}$$

$$\bullet \binom{n}{r-k} \qquad \boxed{\bullet} \binom{n}{r}$$

Repeat Exercises 3 and 4 replacing "s" with "r"—you will be sliding the parallelogram in the direction in which r is constant. What do you notice?

6. Conjecture a similar phenomenon in which we slide parallelograms in the Pascal Triangle in the direction in which n is constant. Test your conjecture.

6.4 THE PASCAL FLOWER AND THE GENERALIZED STAR OF DAVID

In this section we exhibit some patterns[5] taken from [7, 8, 9], for which the proofs depend only on the fact that the binomial coefficient is a separable function. Hence these patterns will be equally valid for each of the three examples mentioned at the end of Section 3. We ourselves, and our students, have found the search for patterns in the Pascal Hexagon stimulating and rewarding—we hope you do too.

Notice that in Figure 3 there are three types of parallelogram. Each parallelogram has one vertex located at $\binom{n}{r}$ and sides of length k and ℓ. In particular,

- P_1 has sides parallel to the r and s directions (i.e., the directions of constant r and constant s);
- P_2 has sides parallel to the s and n directions; and
- P_3 has sides parallel to the n and r directions.

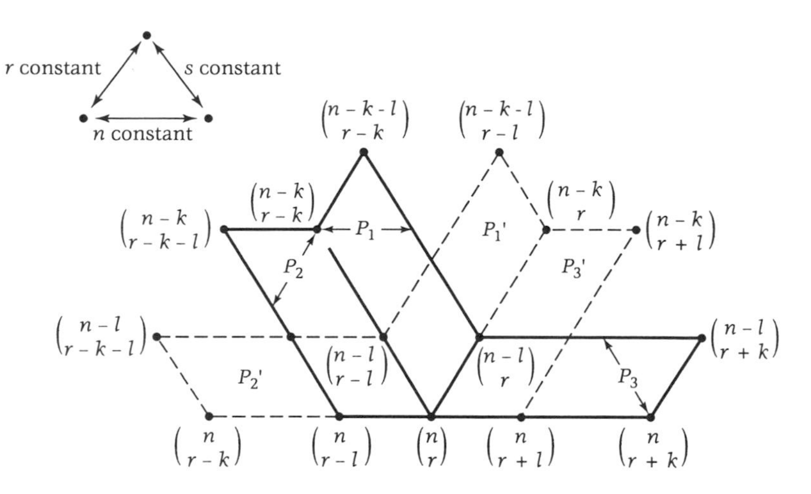

FIGURE 3 The Pascal Flower. The arrows show the directions in which the parallelograms are to be slid.

[5] By a *pattern* we mean a specific relationship that holds for the entries that lie at the vertices of a well-defined family of geometric figures within the Pascal Windmill. Thus the Pascal Identity is a pattern.

Each of these parallelograms P_i is shown, with its corresponding reflected[6] parallelogram P_i' indicated by the dashed lines. All of the parallelograms are *anchored* at $\binom{n}{r}$. By anchoring the parallelograms at $\binom{n}{r}$ we can give a geometric description of the relationship we wish to consider. In each case we define the ***cross-ratio*** (or ***weight***), W_i (or W_i'), of the corresponding parallelogram, P_i (or P_i'), to be the product of the binomial coefficient $\binom{n}{r}$ with the binomial coefficient at the vertex opposite $\binom{n}{r}$, divided by the product of the binomial coefficients at the other two vertices of the parallelogram. When all vertices are actually in the Pascal Triangle we can calculate, from definition (2), the following:

$$\begin{aligned}
W_1 &= \binom{n}{r}\binom{n-k-\ell}{r-k}\Big/\binom{n-k}{r-k}\binom{p-\ell}{r} \\
&= \frac{n!\,(n-k-\ell)!}{(n-k)!\,(n-\ell)!}, \\
W_2 &= \binom{n}{r}\binom{n-k}{r-k-\ell}\Big/\binom{n}{r-\ell}\binom{n-k}{r-k} \\
&= \frac{(r-k)!\,(r-\ell)!}{r!\,(r-k-\ell)!}, \\
W_3 &= \binom{n}{r}\binom{n-\ell}{r+k}\Big/\binom{n-\ell}{r}\binom{n}{r+k} \\
&= \frac{(n-r-k)!\,(n-r-\ell)!}{(n-r)!\,(n-r-k-\ell)!} = \frac{(s-k)!\,(s-\ell)!}{s!\,(s-k-\ell)!}.
\end{aligned} \tag{27}$$

We now notice several remarkable features. First, observe that, for given k and ℓ, although, a priori, W_i depends on n, r, s, subject to $r+s=n$, in fact,

$$\begin{aligned}
&W_1 \text{depends only on } n \text{ and is symmetric in } k \text{ and } \ell, \\
&W_2 \text{depends only on } r \text{ and is symmetric in } k \text{ and } \ell, \\
&W_3 \text{depends only on } s \text{ and is symmetric in } k \text{ and } \ell.
\end{aligned} \tag{28}$$

The symmetry part of (28) means that for each type of parallelogram the value of the weight remains fixed if k and ℓ are exchanged

[6] *Reflection* is the geometric equivalent of interchanging the roles of k and ℓ. For example, P_3' is obtained from P_3 by reflecting in the line joining $\binom{n}{r}$ to $\binom{n-\ell}{r+\ell}$.

with each other; that is, the weight is unaffected by reflecting that parallelogram across the line bisecting the angle between the sides emanating from $\binom{n}{r}$. Thus

$$\begin{aligned} W_1 &= W_1' \\ W_2 &= W_2' \\ W_3 &= W_3' \end{aligned} \tag{29}$$

But (28) also says that W_i depends only on n, r, s for $i = 1, 2, 3$, respectively. This means that if any of the parallelograms P_1, P_2, P_3 (or, by (29), P_1', P_2', P_3') is *slid* in the direction of constant n, r, s, respectively, *the weight will be unchanged*!

The properties embodied in the relations (28), (29) are called ***properties of the Pascal Flower*** (see Figure 3). They remain true in the other two blades of the Pascal Windmill, as you may verify, using (8′) and (22).

Each equality in (29) may be exploited to yield a generalized ***Star of David Theorem***.[7] Thus, for example, since $W_1 = W_1'$, we infer from (27) one aspect of this generalization, namely, that

$$\binom{n-\ell}{r-\ell}\binom{n-k}{r}\binom{n-k-\ell}{r-k} = \binom{n-k}{r-k}\binom{n-\ell}{r}\binom{n-k-\ell}{r-\ell}. \tag{30}$$

The geometric significance of (30) is seen in Figure 4. Each side of the equation is the product of the vertices attached to one of the two triangles making up the (slightly distorted) Star of David surrounding S_1. Two other identities, each similar to (30), can be read off Figure 4 for the Stars surrounding S_2 and S_3.

If n and r are not both nonnegative integers, and if the entire parallelogram P_i lies in one of the blades of the Pascal Windmill, then, as we have said, the dependency and symmetry statements given in (28) are still valid, so that a generalized Star of David Theorem also holds. Moreover, using geometrical (or, perhaps, agricultural) language, we may transport a parallelogram, in the appropriate

[7] We refer here to a generalization of the theorem to be found in [6]. The authors of [6] did not use the term "Star of David," which was coined by Gould in [5]. There is another Star of David Theorem conjectured in [5], which states that

$$\gcd\left(\binom{n-1}{r-1}, \binom{n}{r+1}, \binom{n+1}{r}\right) = \gcd\left(\binom{n-1}{r}, \binom{n+1}{r+1}, \binom{n}{r-1}\right).$$

This, however, does not seem to generalize beyond the hexagon formed by the points of the Pascal Triangle in the immediate vicinity of $\binom{n}{r}$.

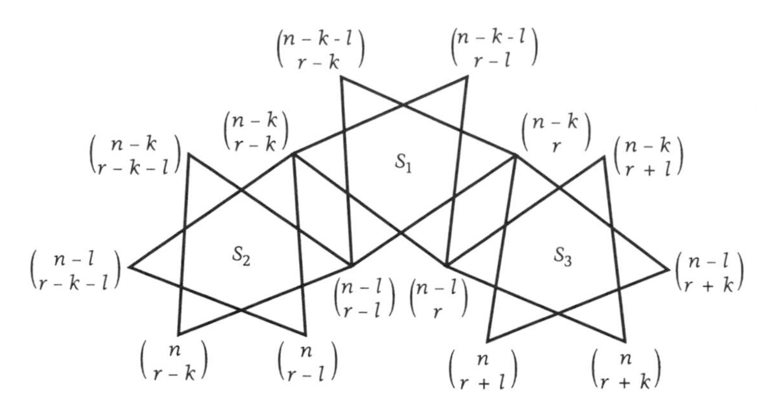

FIGURE 4 The Generalized Stars of David.

direction, from one blade of the windmill, across a zero zone, to another blade, and the weight will first become indeterminate and will then resume its original value.

There are two observations to be made about the calculation of the weights which led to the statements in (28), (29), and subsequently to the Star of David Theorems. The first observation is that, for a given parallelogram we could have used any vertex as anchor and obtained identical results; or, in other words, k or ℓ (or both) may be negative. The second, and perhaps more remarkable, observation is that (28), (29) and the Generalized Star of David Theorem were in no way dependent on $\binom{n}{r}$ being defined in terms of factorial functions of n, r, s. Indeed, confining ourselves once more to the Pascal Triangle, and referring to formula (2) for $\binom{n}{r}$, we point out:

> ***Any separable function of n, r, s, replacing*** $\frac{n!}{r!\,s!}$ ***would have given a similar result!***

★ For example, recall that the definition of the *Leibniz harmonic coefficient* $\begin{bmatrix} n \\ r \end{bmatrix}$ was given as

$$\begin{bmatrix} n \\ r \end{bmatrix} = \frac{1}{(n+1)\binom{n}{r}}, \quad 0 \le r \le n,$$

$$\text{where } \binom{n}{r} \text{ is the ordinary binomial coefficient.} \quad (22)$$

Thus $\left[{n \atop r}\right] = f(n)g(r)h(s)$, where $f(n) = \frac{1}{(n+1)!}$, $g(r) = r!$, $h(s) = s!$. Likewise $\binom{n}{r}_q = f(n)g(r)h(s)$, where

$$f(n) = (q^n - 1)(q^{n-1} - 1) \cdots (q - 1),$$

$$g(r) = \{(q^r - 1)(q^{r-1} - 1) \cdots (q - 1)\}^{-1},$$

$$h(s) = \{(q^s - 1)(q^{s-1} - 1) \cdots (q - 1)\}^{-1}.$$

Figure 5 shows a portion of the Pascalian harmonic triangle.

It is natural to ask which of the various properties that we have talked about for binomial coefficients hold for the harmonic coefficients. If we begin by testing formula (9) we quickly discover that the relationship is not precisely the same, although it is certainly similar. What is true is that all triples of adjacent entries A, B, C positioned like this

• C

• A • B

satisfy a relationship very similar to the Pascal Identity, namely, $A + B = C$. Geometrically, this means that whenever you add two adjacent elements in a row you obtain the number which is between them in the row *above* (rather than the row *below*, as was the case with the binomial coefficients). Or, more formally, the

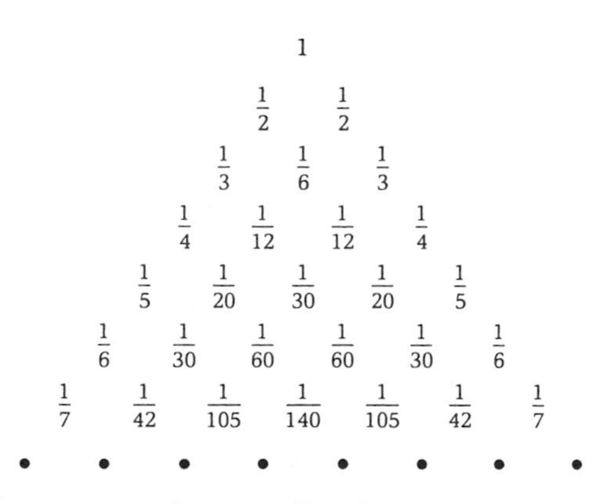

FIGURE 5 The Pascalian harmonic triangle.

Pascalian Identity for harmonic coefficients is

$$\begin{bmatrix} n \\ r \end{bmatrix} + \begin{bmatrix} n \\ r+1 \end{bmatrix} = \begin{bmatrix} n-1 \\ r \end{bmatrix}, \quad 0 \le r \le n-1. \tag{9'}$$

It is clear, from Figure 5, and from the definition, that the symmetry identity similar to equation (8) will also be valid. In fact, in [11], results analogous to those we obtained for $\binom{n}{r}$ concerning the weights of the parallelograms in the Pascal Flower of Figure 3 and the Star of David Theorem in Figure 4 are obtained as a direct result of the fact that $\left[\begin{smallmatrix} n \\ r \end{smallmatrix}\right]$ is a separable function. It is, indeed, instructive to try out some numerical examples of a Star of David Theorem by simply replacing the rounded brackets by square brackets in Figures 3 and 4, and equations (27) and (30).

Before we begin the next section we point out that if we have any separable function $F(n, r, s) = f(n)g(r)h(s)$, satisfying the symmetry condition

$$F(n, r, s) = F(n, s, r),$$

then we may choose g, h so that $g = h$. For $g(r)h(s) = g(s)h(r)$ for *all* r, s, so that $\frac{g(r)}{h(r)}$ is a constant c and, since we may replace $g(r)$ by $\lambda g(r)$ and $h(s)$ by $\lambda^{-1}h(s)$, we may simply choose $\lambda = c^{-1/2}$, so that $\frac{\lambda g(r)}{\lambda^{-1}h(r)} = 1$, or $\lambda g(r) = \lambda^{-1}h(r)$. ⋆

• • • BREAK

1. Referring to Figure 4, show that the stars S_1, S_2, S_3 are genuine Stars of David if and only if $k = 2\ell$.

2. Consider the equilateral triangle with vertices at

$$\binom{n}{r}, \binom{n-k-\ell}{r}, \binom{n-k-\ell}{r-k-\ell}$$

in the Pascal Hexagon. Show how, by cutting off appropriate small congruent equilateral triangles at each vertex, we obtain the star S_1 of Figure 4.

II. EULER TAKES THE STAGE

6.5 EULERIAN NUMBERS AND WEIGHTED SUMS

We begin this section by discussing a technique for discovering theorems about binomial coefficients and we look carefully at a discovery (and its consequences) that was actually made by a student of Jean Pedersen and Peter Hilton, Allison Fong (then a freshman at Santa Clara University). We spend the rest of this section developing a generalization of Fong's discovery—and this involves us with the classical ***Eulerian numbers*** and a generalization of them. The reader should regard this as a case study in mathematical method—the method for actually *doing* mathematics.

Look at Figure 6. If you can already conjecture and prove a theorem about the given equations shown, you can cover the next few

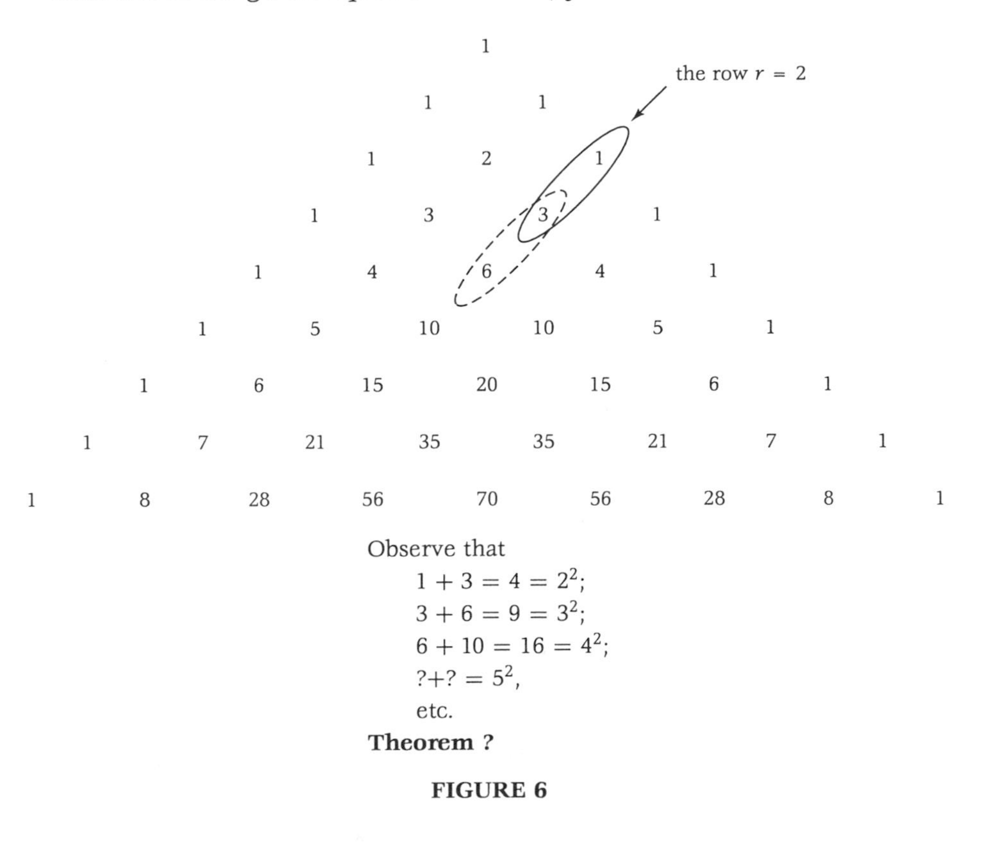

FIGURE 6

pages very quickly. However, don't skip this section entirely, because we'll soon be onto something new. Our presentation in the next few pages is based on our own actual presentation to students at a much more elementary level than yours. (But the pace will quicken!)

How do we find the statement for the Theorem in Figure 6? One way to proceed is first to convert each of the numbers on the left-hand side of the equations in Figure 6 into the equivalent binomial coefficient (by observing their location in the Pascal Triangle). Thus, we see that

$$1 + 3 = 2^2 \quad \text{is equivalent to} \quad \binom{2}{2} + \binom{3}{2} = 2^2,$$

$$3 + 6 = 3^2 \quad \text{is equivalent to} \quad \binom{3}{2} + \binom{4}{2} = 3^2,$$

$$6 + 10 = 4^2 \quad \text{is equivalent to} \quad \binom{4}{2} + \binom{5}{2} = 4^2.$$

Observing the pattern of these equations, and being optimistic, we hope that the next line should read

$$10 + 15 = 5^2 \quad \text{is equivalent to} \quad \binom{5}{2} + \binom{6}{2} = 5^2,$$

and it does!

At this point it is reasonable to hope that there is a theorem of the form

$$\binom{\ }{2} + \binom{\ }{2} = (\ )^2.$$

But how should we determine what variables to put in the three vacant positions? As a matter of fact, we may *choose any variable we like* to place in *one* of these three spots—and we can then figure out, in terms of that choice, by looking at the numerical examples, what should go in the other two positions. Remember, we're in charge (mathematics is created by human beings!). We can choose how we want to state our theorem. There is no unique *correct* choice—although there may be choices that some people will perceive as "prettier" or "easier to understand" than other choices. Let us demonstrate this last point.

Suppose we decide we would like our theorem to involve the variable n and to appear in the form

$$\binom{\ }{2} + \binom{\ }{2} = n^2.$$

All that remains, then, is to determine the top numbers in the binomial coefficients. It is clear from the pattern in our examples that the first binomial coefficient should have n, and the second should have $n + 1$, in the top. Thus we have arrived at a conjectured theorem.

Theorem A $\binom{n}{2} + \binom{n+1}{2} = n^2$.

A proof, appealing to (1), would appear as

$$\binom{n}{2} + \binom{n+1}{2} = \frac{n(n-1)}{2} + \frac{(n+1)n}{2} = \frac{n^2 - n + n^2 + n}{2} = n^2.$$

Notice that this proof tells us that Theorem A is true for *any real number n* (and this fact should then be appended to Theorem A). You might wish to verify that Theorem A is true for $n = -1, -2, -3, \ldots$ by sliding the oval shown in Figure 3 *up* the diagonal (along $r = 2$) into the northeast blade of the Pascal Windmill (Figure 2).

There are, of course, alternative formulations. Some might prefer that the variable n appear in the second term of the LHS. We would then obtain a theorem looking a little different from Theorem A. Thus

Theorem B $\binom{n-1}{2} + \binom{n}{2} = (n-1)^2$.

Theorem B is, of course, logically equivalent to Theorem A.

Which formulation do you prefer? Most people we've encountered, students and mathematicians alike, have preferred that of Theorem A—but we think it is very important to recognize that there is room for choice, and the choice is on aesthetic, and not on mathematical grounds.

We now turn to an account of the discovery, mentioned above, that was made in our class for liberal arts students. After being shown Theorem A (and discussing its variations), the students were asked to look for patterns themselves in Pascal's Triangle and try to write, and even perhaps prove, theorems of their own. Allison Fong returned to class the following day with the information shown in Figure 7. Her observation of these numerical facts led her to conjecture the following theorem, which was new to us.

Theorem C $\binom{n+2}{3} - \binom{n}{3} = n^2$.

A proof of Theorem C may also be achieved by using (1), so that Theorem C holds for any real number n. In fact, we will supply what we believe is an elegant and revealing "proof without words" later, of a much more general result; but what we want to emphasize here is that Theorem C is very much *like* Theorem A—and that the discovery of Theorem C seems to be a strong indication that these two theorems, taken together, might be part of a much larger picture.

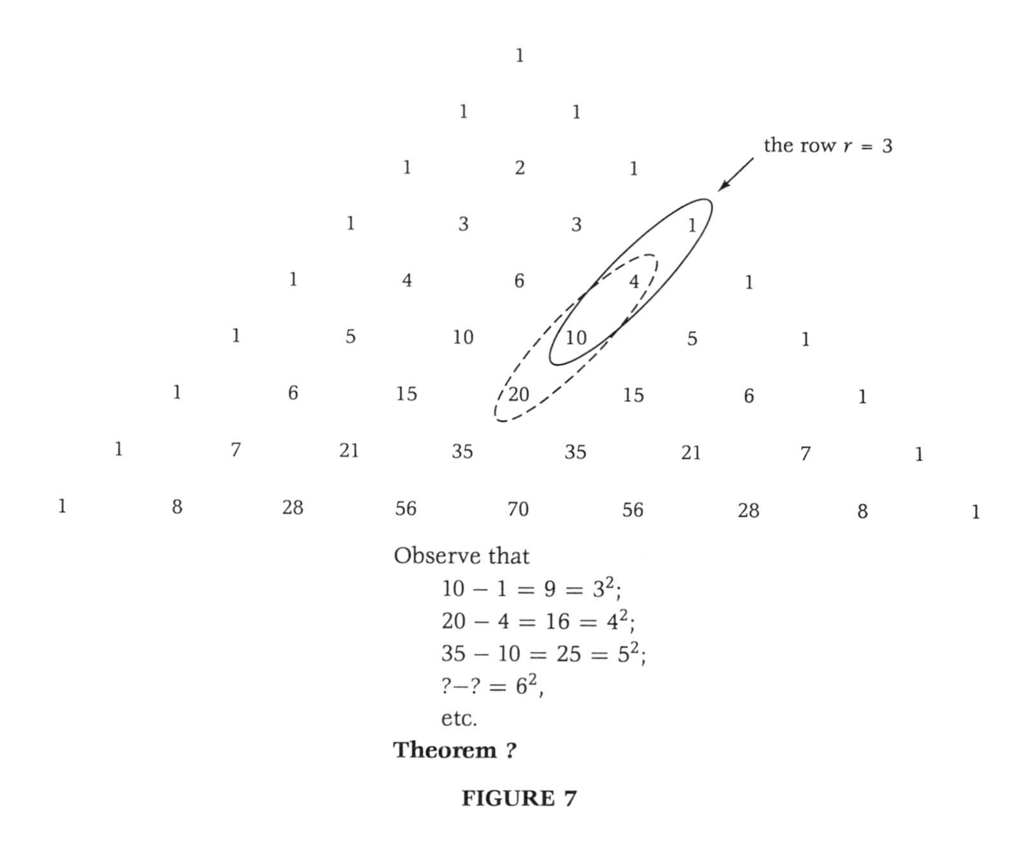

FIGURE 7

First, let us observe that the Pascal Identity provides a link between Theorems A and C. For

$$\binom{n+2}{3} - \binom{n}{3} = \left[\binom{n+2}{3} - \binom{n+1}{3}\right] + \left[\binom{n+1}{3} - \binom{n}{3}\right]$$
$$= \binom{n+1}{2} + \binom{n}{2}.$$

Thus we see, by virtue of the Pascal Identity, that Theorems A and C are equivalent—one is true because the other is true. This simple reasoning provides the key to much of the subsequent mathematical reasoning in this section.

So as to make the forms of Theorems A and C more consistent, with each other, we rewrite Theorem C to include the binomial coefficient $\binom{n+1}{3}$ (with coefficient 0), and rearrange the terms on the left-hand side to obtain the joint assertion

$$\left.\begin{aligned} \binom{n}{2} + \binom{n+1}{2} &= n^2, \quad \text{for all } n,\\ -\binom{n}{3} + 0\binom{n+1}{3} + \binom{n+2}{3} &= n^2, \quad \text{for all } n. \end{aligned}\right\}$$

The format of these two equations is suggestive, for we immediately see that, in the first case,

$$\text{we have two constants}^{8}\ S(2,0) = 1,\ S(2,1) = 1,$$
$$\text{such that } S(2,0)\binom{n}{2} + S(2,1)\binom{n+1}{2} = n^2,$$

while, in the second case,

$$\text{we have three constants } S(3,0) = -1,\ S(3,1) = 0,\ S(3,2) = 1,$$
$$\text{such that } S(3,0)\binom{n}{3} + S(3,1)\binom{n+1}{3} + S(3,2)\binom{n+2}{3} = n^2.$$

[8]We are planning ahead in seeking to use *appropriate notation* for the weights $S(-,-)$. We want to choose a symbol that will allow us to see emerging patterns. Thus, we choose a notation that incorporates both the value of the r-row from which we are selecting the binomial coefficients and the term to which the weight will be attached. Indexing the first weight with 0, rather than 1, makes the resulting formula much neater, and therefore easier to remember.

In fact, in each case, we have what is called a ***linear combination***, or ***weighted sum***, of binomial coefficients (in the first case lying along the row where r is fixed at 2, in the second case lying along the row where r is fixed at 3) whose value is *always* n^2. Notice that we can't do anything like this with the row $r = 1$; we must start with $r = 2$. This must surely be related to the fact that our linear combinations produce the *second* power of n.

We are now beginning to get glimpses of a larger picture. Thus we are tempted to ask the question:[9]

Question 1 For a fixed $r \geq 2$ can we always find constants (i.e., numbers independent of n) $S(r, i)$, $0 \leq i \leq r - 1$, such that

$$\sum_{i=0}^{r-1} S(r,i)\binom{n+i}{r} = n^2, \quad \text{for all } n? \tag{31}$$

However, there may well be an even larger picture. Must we stick with the square of n? Might we not ask:

Question 2 Can we do an analogous thing to obtain n^3, or n^4, or in general n^k, for any $k \geq 1$? In other words, given $k \geq 1$, can we always find constants (replacing the $S(r, i)$ above) such that they can be used to produce weighted sums of binomial coefficients, all lying in a row of Pascal's Triangle given by a fixed r, summing to n^k? By analogy with the example above we would expect this phenomenon to start in the row given by $r = k$.

To write the formula that would be analogous to (31) for a general k, we need to refine our notation even further, because the symbols $S(r, i)$ in (31) give no hint that the sum on the left will ultimately involve a *square* (of course, since we had so far only looked at weighted sums that resulted in the sum of n^2, our notation was adequate for Question 1). However, if we want to state a result for general values of k, then it would be better to

[9]This is a sure sign that we are really *doing* mathematics; for genuine mathematics very often proceeds from answer to question—and not, as most current textbooks would lead students to believe, merely from question to answer. See Chapter 9.

incorporate the chosen *power* into the weight notation. Remember *we* are in charge and we may CHOOSE OUR NOTATION. We will record the chosen power by inserting an appropriate subscript after S, so that we adopt the notation $S_k(r, i)$. Thus our original $S(r, i)$ becomes $S_2(r, i)$. Some might prefer the notation $S(k, r, i)$ to $S_k(r, i)$, but we prefer to distinguish sharply between the role of k and that of the other variables on which the weight depends.

In this new notation the first question would read:

Question 1 (Restated) For a fixed $r \geq 2$ can we always find weights

$$S_2(r, i), \quad 0 \leq i \leq r - 1,$$

such that

$$\sum_{i=0}^{r-1} S_2(r, i)\binom{n+i}{r} = n^2, \quad \text{for all } n? \tag{31'}$$

With this notation it is easy to state the generalization as follows; note that, as we have said, it is a good guess that, since, in Question 1, we took $r \geq 2$, we will need, in the generalization, to take $r \geq k$.

Question 2 (Restated) For a fixed $r \geq k$, can we always find weights $S_k(r, i)$, $0 \leq i \leq r - 1$, such that

$$\sum_{i=0}^{r-1} S_k(r, i)\binom{n+i}{r} = n^k, \quad \text{for all } n? \tag{32}$$

Of course, if we can answer the second question we will have an answer to the first, and some of you might now like to try to tackle that question directly. But we will proceed at a more leisurely pace, hoping to learn something from the case $k = 2$ that might help us with the more general situation.

So we ask about the validity of (31′) when $r = 4$, since it is already established for $r = 2, 3$. In other words, can numbers $S_2(4, i)$ be found such that

$$S_2(4,0)\binom{n}{4} + S_2(4,1)\binom{n+1}{4} + S_2(4,2)\binom{n+2}{4} + S_2(4,3)\binom{n+3}{4} = n^2? \quad (33)$$

One approach some people might try (we emphasize that this is certainly not the best, but the first approach seldom is) would be to use (1) to express the LHS of (33) as a polynomial in n, and then to equate coefficients of like powers of n on both sides of (33), to obtain simultaneous linear equations for the $S_2(4, i)$, and hence determine them. For the purposes of *this* calculation it would be less messy to CHOOSE A SIMPLER NOTATION for the coefficients. We choose the letters A, B, C, D for the coefficients of $\binom{n}{4}$, $\binom{n+1}{4}$, $\binom{n+2}{4}$, $\binom{n+3}{4}$, respectively. Writing out the computation we first obtain, using (1),

$$\begin{aligned} A\frac{n(n-1)(n-2)(n-3)}{4!} &+ B\frac{(n+1)(n-1)(n-2)}{4!} \\ &+ C\frac{(n+2)(n+1)(n)(n-1)}{4!} \\ &+ D\frac{(n+3)(n+2)(n+1)n}{4!} = n^2, \end{aligned}$$

and note that we may divide everything on both sides by n. Then, multiplying by 4! and expanding each of the terms on the left and REORGANIZING the terms according to the powers of n we have[10]

$$\begin{aligned} &(A+B+C+D)n^3 + (-6A-2B+2C+6D)n^2 \\ &\quad + (11A-B-C+11D)n + (-6A+2B-2C+6D) = 4!\ n. \end{aligned}$$

Equating the coefficients of like powers of n on both sides of this equation, we have the system of linear equations

$$\begin{aligned} A+B+C+D &= 0 \\ -6A-2B+2C+6D &= 0 \\ 11A-B-C+11D &= 4! \\ -6A+2B-2C+6D &= 0 \end{aligned}$$

[10]By looking at the general format for the numerator of each term, and by using the expanded form of the product $(n+a)(n+b)(n+c)$, you should be able to obtain the next step without *slogging it out*!

Solving this system of equations gives

$$(A, B, C, D) = (1, -1, -1, 1).$$

In a similar fashion one could find the weights in the next row, and so on. However, this approach plainly has no hope of providing a general answer to Question 1. Suppose that, somehow or other, you have calculated some of the $S_2(r, i)$. Figure 8 shows the Pascal Triangle with some of the $S_2(r, i)$, in parentheses, to the left of the binomial coefficient $\binom{n+i}{r}$, where n is the starting level down the r-row.

Figure 9 is simply Figure 8 with the binomial coefficients replaced by dots so that it is easier for you to see the relationship between the entries in the array of weights $S_2(r, i)$. Study Figure 9 to

$n = 0 \longrightarrow$ 1

$n = 1 \longrightarrow$ 1 1

$n = 2 \longrightarrow$ 1 2 (1)1

$n = 3 \longrightarrow$ 1 3 (1)3 (1)1

$n = 4 \longrightarrow$ 1 4 6 (0)4 (1)1

$n = 5 \longrightarrow$ 1 5 10 (1)10 (-1)5 (-1)1

$n = 6 \longrightarrow$ 1 6 15 20 (-1)15 (2)6 (1)1

1 7 21 35 (1)35 (0)21 (-3)7 (-1)1

1 8 28 56 70 (-2)56 (2)28 (4)8 (1)1

1 9 36 84 126 (1)126 (2)84 (-5)36 (-5)9 (-1)1

1 10 45 120 210 252 (-3)210 (0)120 (9)45 (6)10 (1)1

The array of parenthetical coefficients should be regarded as free to slide in the direction of the parenthetical arrows.

Weighted sums with sum n^2 (n = starting level).

FIGURE 8

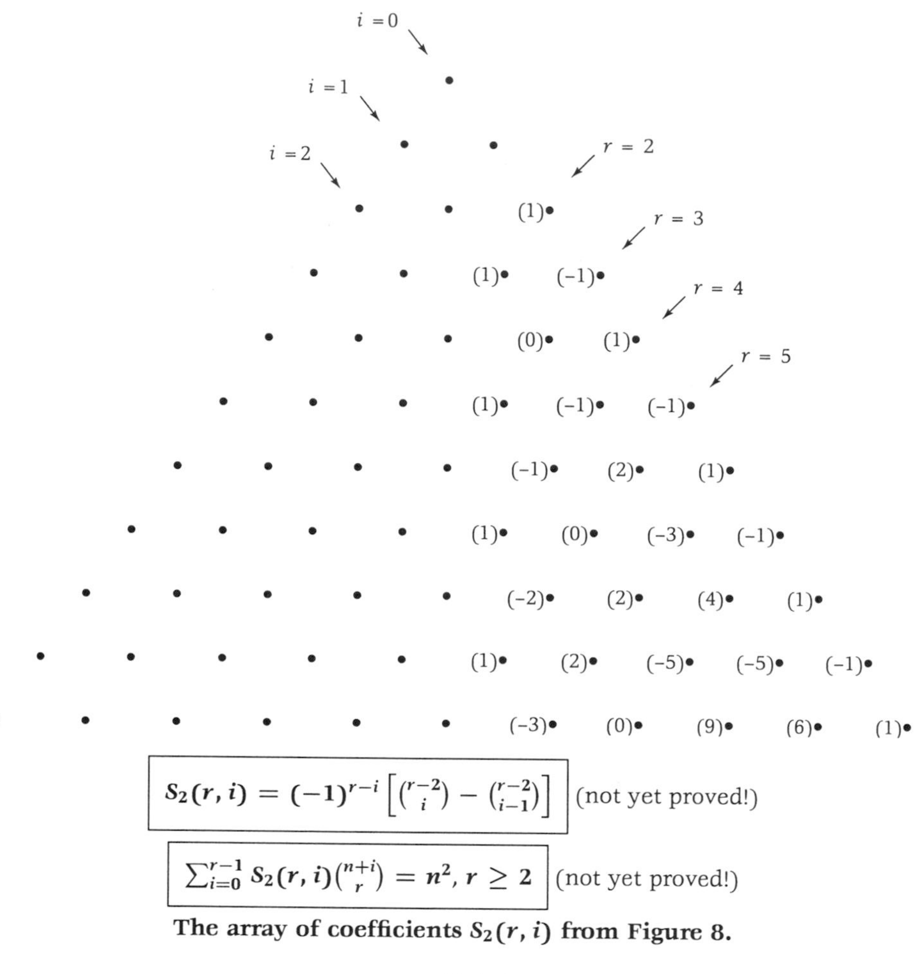

The array of coefficients $S_2(r, i)$ from Figure 8.

FIGURE 9

see if you can detect an *easy* way to extend the array of coefficients $S_2(r, i)$. The key observation is shown, geometrically, in Figure 10. We call this recurrence relationship the ***Transition Identity***, and we may express it very precisely as follows (with $r \geq 2$):

$$\left.\begin{aligned} S_2(r,i) &= 0, \quad i < 0 \text{ or } i \geq r, \\ S_2(r,i) &= S_2(r-1,i-1) - S_2(r-1,i), \quad r \geq 3. \end{aligned}\right\} \tag{34}$$

These identities, together with the already established initial values (initial with respect to r)

$$S_2(2,0) = S_2(2,1) = 1$$

entirely determine the weights $S_2(r,i)$. The explicit formula for $S_2(r,i)$ shown in Figure 9 will be proved in Section 6.

Now we will have the answer to Question 1 if we can prove the following:

Theorem 2 *If* $S_2(2,0) = S_2(2,1) = 1$, *and* (34) *holds, then*

$$\sum_{i=0}^{r-1} S_2(r,i)\binom{n+i}{r} = n^2, \quad r \geq 2. \tag{35}$$

Proof We proceed by an inductive argument. Suppose $r \geq 3$. Then

$$\sum_{i=0}^{r-1} S_2(r,i)\binom{n+i}{r}$$

$$= \sum_{i=0}^{r-1} [S_2(r-1,i-1) - S_2(r-1,i)]\binom{n+i}{r},$$

by the Transition Identity (34)

The expression on the right is organized according to the binomial coefficient $\binom{n+i}{r}$. We now REORGANIZE it instead according to the weight $S_2(r-1,i)$ to obtain[11]

$$= \sum_{i=0}^{r-2} S_2(r-1,i)\left[\binom{n+i+1}{r} - \binom{n+i}{r}\right],$$

(using $S_2(r-1,1) = S_2(r-1,r-1) = 0$)

$$= \sum_{i=0}^{r-2} S_2(r-1,i)\binom{n+i}{r-1}, \qquad \text{by the Pascal Identity.} \ \square$$

This completes the proof, since what we have shown is that formula (35) holds for r provided it held for $r-1$, and we know that it

[11] Some of you may find it useful to look at a particular but not special value of i (say, $i = 6$) to understand this step.

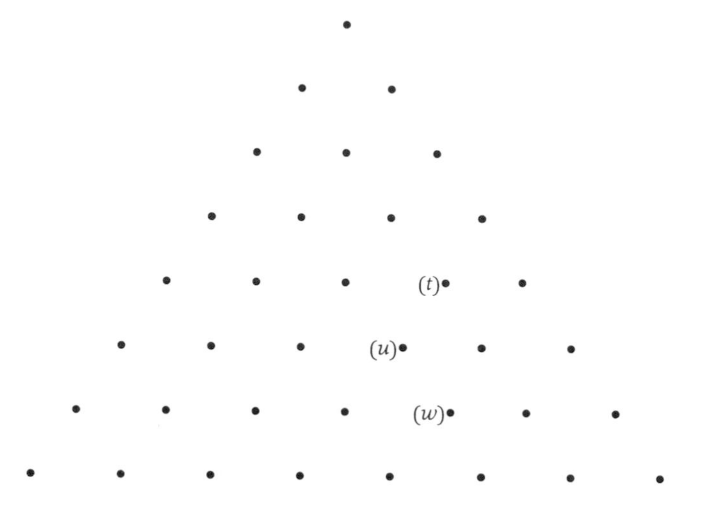

The Transition Identity, $t - u = w$,
where t, u, w stand for the parenthetical entries in Figure 9.

FIGURE 10

holds when $r = 2$. In this proof the reader should see just how, as claimed earlier, the truth of (35) for various values of r reflects the Pascal Identity. Indeed, one may say that the Transition Identity itself is a reflection of the Pascal Identity. Of course, the proof given renders the monstrous calculations given in the case $r = 4$ quite unnecessary—but this does not imply there was no value in doing them. Notice also that it was quite unnecessary to use the elaborate symbol $\sum_{i=0}^{r-1}$; the simple summation symbol $\sum$ would have sufficed—in fact, it would have been better at a key step in the proof of Theorem 2. Do you see where? Notice that formula (35) has been proved for all *real* numbers n.

We now turn to Question 2. It should be clear, from the proof we have just given in the case $k = 2$, that, if we now look at the case $k = 3$, all we need is to find first the initial values (with respect to r) of $S_3(r, i)$, that is, the values of $S_3(3, i)$ which will provide a weighted sum of n^3, and then use the Transition Identity to obtain $S_3(r, i)$ for all $r \geq 3$. Here we are fortunate, for there already exists (see [2]) a well-known array of numbers, called the ***Eulerian Numbers*** (because they were discovered by Euler), which, as we will see, do provide the initial values for *any* positive integer k. See Figure 11

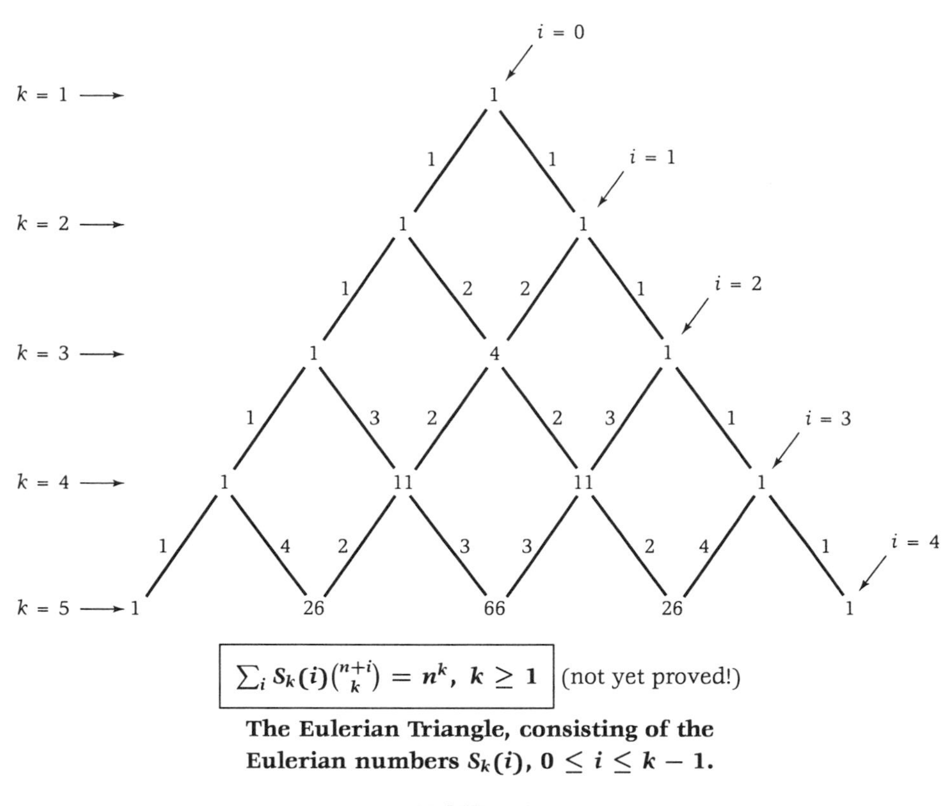

The Eulerian Triangle, consisting of the Eulerian numbers $S_k(i)$, $0 \leq i \leq k-1$.

FIGURE 11

for the ***Eulerian Triangle***[12] of values $S_k(i)$ with $0 \leq i \leq k-1$, for $k \leq 5$. These Eulerian numbers, it will turn out, do exactly what we want our numbers $S_k(k, i)$ to do, so we will be writing $S_k(i)$ instead of $S_k(k, i)$. Having $S_k(i)$, the Transition Identity gives us $S_k(r, i)$ for all $r \geq k$.

We may state the rules for generating the triangle of Figure 11 in the following definition.

The ***Eulerian numbers*** are nonnegative integers $S_k(i)$, given inductively for integers k, i, with $k \geq 1$, by the rule

$$\begin{aligned} &S_1(0) = 1, \quad S_1(i) = 0, \quad i \neq 0; \\ &S_k(i) = (k-i)S_{k-1}(i-1) + (i+1)S_{k-1}(i), \quad k \geq 2. \end{aligned} \tag{36}$$

[12]With some extra numbers written along diagonal lines indicating, in cryptic fashion (for a clue see (36)), how each horizontal row of Eulerian numbers is obtained from its predecessor.

Notice that we have given an *inductive* rule (with respect to k) for defining $S_k(i)$. The numbers along the diagonals in Figure 11 are the coefficients $(k-i)$, $(i+1)$ in (36). It is easy to show, inductively, that, for a fixed k, $S_k(i)$ is positive inside the range $0 \le i \le k-1$ and zero outside it.

Theorem 3 below asserts some elementary properties of Eulerian numbers (in fact, you may have already noticed some of these properties). Notice that, in the third part of the theorem, we suppress the details of the limits on the summation, since there is no doubt as to exactly what is meant.[13] At this stage, you should regard Theorem 3 as a form of light relief from some rather heavy mathematics. All mathematicians need such relief from time to time.

Theorem 3

(i) $S_k(0) = S_k(k-1) = 1$;

(ii) $S_k(i) = S_k(k-1-i)$;

(iii) $\sum_i S_k(i) = k!$.

Proof Part (i) follows by an obvious induction on k—just look at how each new value on the boundary is obtained.

Part (ii) (*The Symmetry Property*) also follows by induction (it obviously holds for $k = 1$)—just look at how one forms the new entries in positions symmetric with respect to an imaginary vertical line bisecting the array of numbers. More formally, we present the proof by using a PARTICULAR BUT NOT SPECIAL CASE, say $k = 6$ (which is the first missing row in Figure 9). Thus we wish to prove that $S_6(i) = S_6(5-i)$, assuming the symmetry property holds for $k = 5$.[14]

$$\begin{aligned} S_6(i) &= (6-i)S_5(i-1) + (i+1)S_5(i), \quad \text{by (36)}, \\ &= (6-i)S_5(5-i) + (i+1)S_5(4-i), \\ &\qquad\qquad \text{by the inductive hypothesis,} \end{aligned}$$

[13] Here we are using our *principle of licensed sloppiness* (see Chapter 9), but we would stress that we are only free to use this principle if we understand, and convey, exactly what is going on. In this case all should be clear, since $S_k(i)$ is zero outside $0 \le i \le k-1$.

[14] It may even be helpful to you here to look at a particular value of i.

$$= (j + 1)S_5(j) + (6 - j)S_5(j - 1), \quad \text{where } j = 5 - i,$$
$$= (6 - j)S_5(j - 1) + (j + 1)S_5(j)$$
$$= S_6(j), \quad \text{by (36)},$$
$$= S_6(5 - i), \quad \text{by substituting } j = 5 - i.$$

We readily see that if we replace 6 by k we have a general proof.

We also prove part (iii) by induction, since it obviously holds for $k = 1$. Again we look at the PARTICULAR BUT NOT SPECIAL CASE $k = 6$. Thus we suppose the truth of (iii) has been verified for $k = 5$. We observe that

$$\sum_i S_6(i) = \sum_i [(6 - i)S_5(i - 1) + (i + 1)S_5(i)], \quad \text{by (36)},$$
$$= \sum_i S_5(i)[(6 - i - 1) + (i + 1)],$$
REORGANIZING according to $S_5(i)$,
$$= 6 \sum_i S_5(i)$$
$$= 6(5!), \quad \text{because we know that the statement(iii) is true when } k = 5,$$
$$= 6!, \quad \text{so that the theorem holds for } k = 6.$$

Of course, a general proof of (iii) would result by simply replacing 6 by k. Don't feel you *must* look at a particular case first if you're comfortable going straight to the general case.

□

We now turn to the property of the Eulerian numbers that concerns us most (see the caption to Figure 11). Notice that the numbers $S_2(i)$ were, in fact, our initial $S_2(2, i)$.

What we need to show, in order to advance our investigation of Question 2, is that the rows in the Eulerian Triangle give us our initial values. Thus we wish to prove the following result about the Eulerian numbers $S_k(i)$. The published proofs of this theorem (see, e.g., [2]) are far too sophisticated even to appear in an undergraduate course; we will give a very much less sophisticated proof.

Theorem 4 $\sum_i S_k(i)\binom{n+i}{k} = n^k$. (See the caption to Figure 11.)

Proof We first observe that the statement of Theorem 4 is certainly true if $k = 1$. For (see (36)) it asserts that $\binom{n}{1} = n^1$. We are going to argue by induction on k. We proceed by taking a PARTICULAR BUT NOT SPECIAL CASE, namely, the particular case $k = 5$. We assume that we know that $\sum_i S_4(i)\binom{n+i}{4} = n^4$, and then we ask, how do we see that this implies that $\sum_i S_5(i)\binom{n+i}{5} = n^5$? We recall that the Eulerian Identity (36) will give us $S_5(i)$ in terms of $S_4(i)$, so we OPTIMISTICALLY use the Eulerian Identity (what else could we possibly do?) to obtain

$$\sum_i S_5(i)\binom{n+i}{5}$$

$$= \sum_i [(5-i)S_4(i-1) + (i+1)S_4(i)]\binom{n+i}{5}.$$

REORGANIZING according to $S_4(i)$ (look at our proof of Theorem 2) we have

$$\sum_i S_5(i)\binom{n+i}{5}$$

$$= \sum_i S_4(i)\left[(4-i)\binom{n+i+1}{5} + (i+1)\binom{n+i}{5}\right].$$

Now be OPTIMISTIC! Since $\sum_i S_4(i)\binom{n+i}{4} = n^4$ and we are trying to prove that $\sum_i S_5(i)\binom{n+i}{5} = n^5$, we *hope* that the quantity in the square brackets above will equal $n\binom{n+i}{4}$, so we factor $\binom{n+i}{4}$ from both terms inside those brackets, obtaining

$$\binom{n+i}{4}\left[(4-i)\frac{n+i+1}{5} + (i+1)\frac{n+i-4}{5}\right].$$

[14]Steps such as this may not be obvious to you—in fact, they may be seen as trickery if you were inclined to simply *slog it out*. Of course, the more pedantic approach requires less thought, but carrying out all the required steps blindly would require a great deal of energy and consummate accuracy so that even the most industrious student would eventually get bogged down. Remember that good mathematics saves work, and that we should always be on the lookout for shortcuts. If you study devices like these, try to use them, and in-

So far, so good! Now we *hope* that what is inside the new square brackets will turn out to be n. We REORGANIZE[15] inside the square brackets, to obtain

$$\binom{n+i}{4}\left[(4-i)\frac{n+i+1}{5}+(i+1)\frac{n+i+1-5}{5}\right]$$

from which it follows, by bringing $(n+i+1)$ on the top into prominence, that we have

$$\begin{aligned}\binom{n+i}{4}&\frac{(n+i+1)(4-i+i+1)-5(i+1)}{5}\\ &=\binom{n+i}{4}\big((n+i+1)-(i+1)\big)\\ &=\binom{n+i}{4}n.\end{aligned}$$

Thus, as hoped,

$$\begin{aligned}\sum_i S_5(i)\binom{n+i}{5}&=\sum_i S_4(i)\binom{n+i}{4}n\\ &=n\sum_i S_4(i)\binom{n+i}{4}, \quad \text{since } n \text{ is fixed,}\\ &=n(n^4)=n^5,\end{aligned}$$

because we assumed the theorem was true when $n = 4$. Replacing 5 by k results in the *general* proof. □

Theorem 4 gives us our start to a full and positive answer to Question 2. We have already seen, in the special case $k = 2$, that we may define numbers $S_2(r, i)$, $r \geq 2$, by (34), with $S_2(2, 0) = S_2(0) = 1$, $S_2(2, 1) = S_2(1) = 1$, to satisfy (35). We will show that a similar statement is true for any k. That is, we define numbers $S_k(r, i)$,

---

vent others for yourselves, then they will lose their mysterious nature and become your own useful devices. The optimistic strategies we describe above are discussed clearly and systematically in [13].

$r \geq k$, by $S_k(k, i) = S_k(i)$, and the Transition Identity

$$S_k(r, i) = S_k(r-1, i-1) - S_k(r-1, i), \quad r \geq k+1;$$

and our theorem will read

Theorem 5 $\sum_i S_k(r,i)\binom{n+i}{r} = n^k$, *for all* $k \geq 1$.

The proof simply proceeds by first observing that we already know that Theorem 5 holds in the special case $r = k$, since this is just Theorem 4. Then the *proof* of Theorem 2 may be taken over, word for word, simply replacing S_2 by S_k, to yield a proof of Theorem 5 by induction on r. Notice that the induction on r starts with $r = k$.

We call the numbers $S_k(r, i)$ ***pseudo-Eulerian coefficients***. Let us discuss their numerical values in the case $k = 3$.

The weights $S_3(r, i)$ (those numbers in parentheses in Figure 12) are obtained in just the same way as the weights for $S_2(r, i)$ were obtained in Figure 9. That is, we start with the weights in the row $r = 3$, which are obtained from the row $k = 3$ of the Eulerian Triangle, and then use the Transition Identity to obtain the weights in successive rows. It is for this reason that we call these weights the ***pseudo-Eulerian coefficients***, so that the array of coefficients when $k = 2, 3$, as shown in Figures 9, 13, respectively, serve as examples of pseudo-Eulerian coefficients. The explicit formula shown in Figure 13 for $S_3(r, i)$ will be proved in Section 6. Of course, we may construct an array of pseudo-Eulerian coefficients for any $k \geq 1$. (What happens when $k = 1$?)

• • • **BREAK**

In the spirit of trying to bring this material within the reach of as many readers as possible, we outline here a "proof without words" of the facts embodied in the Transition Identity and Theorem 5.

In Figure 14(i) the capital letters represent entries that obey the Pascal Identity. First we assume (and this, in fact, is guaranteed by Theorem 4) that there exist *initial* coefficients a, b, c (shown in Figure 14(i) as lower case letters in parentheses),

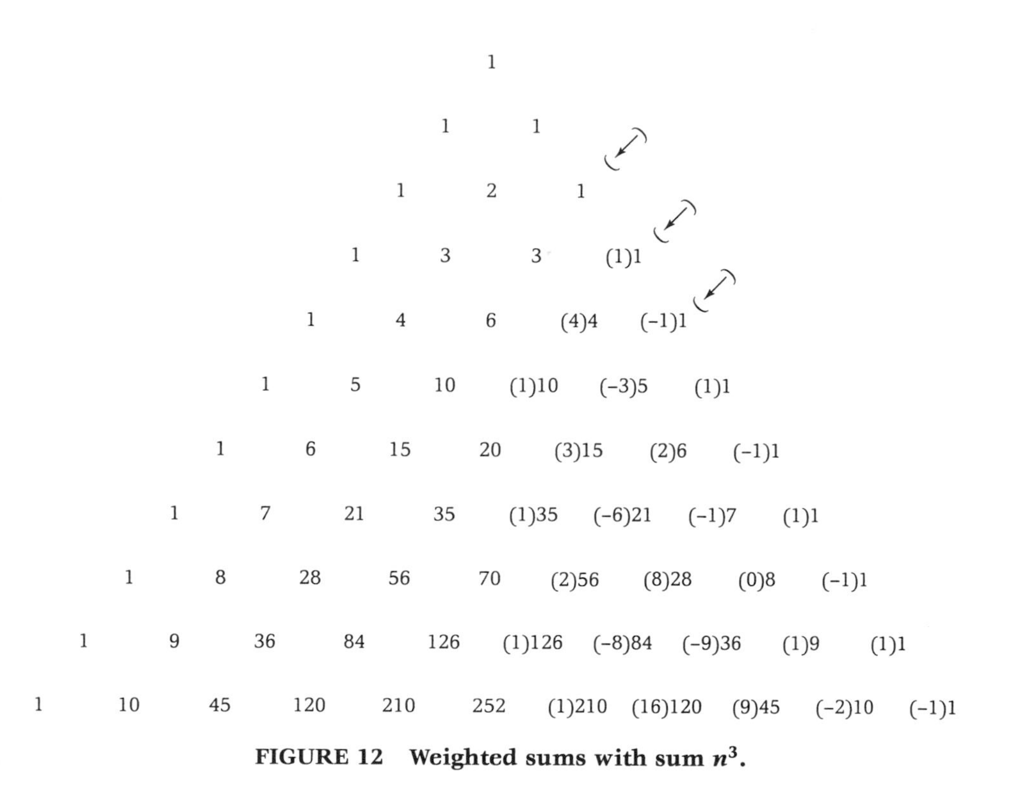

FIGURE 12 Weighted sums with sum n^3.

along a row where $r = 3$ such that

$$\begin{aligned} aA + bB + cC &= n^3 \\ aB + bC + cD &= (n+1)^3 \\ aC + bD + cE &= (n+2)^3 \\ \text{etc.} \end{aligned} \tag{37}$$

(we suppose we are in the case $k = 3$, that is why we start with *three* initial coefficients; the generalization is clear). Next we assume that the "top entry in the next row" (X in our case) is given the coefficient[16] $(-a)$ and that the remaining coefficients in this row are obtained by using the Transition Identity shown in Figure 14(ii).

[16]Our readers will notice that this is consistent with the Transition Identity if we imagine, as we should, that Figure 13 is embedded in a sea of zeros.

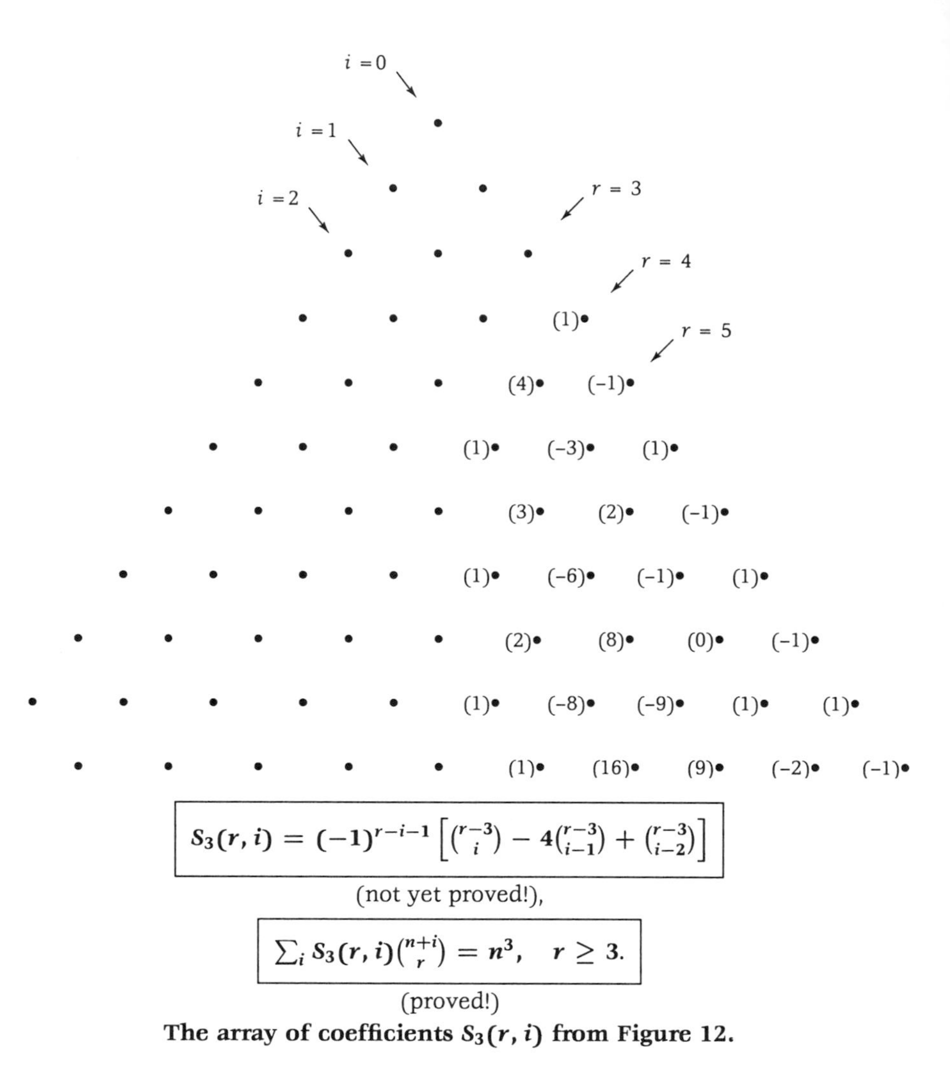

$$S_3(r,i) = (-1)^{r-i-1}\left[\binom{r-3}{i} - 4\binom{r-3}{i-1} + \binom{r-3}{i-2}\right]$$

(not yet proved!),

$$\sum_i S_3(r,i)\binom{n+i}{r} = n^3, \quad r \geq 3.$$

(proved!)

The array of coefficients $S_3(r,i)$ from Figure 12.

FIGURE 13

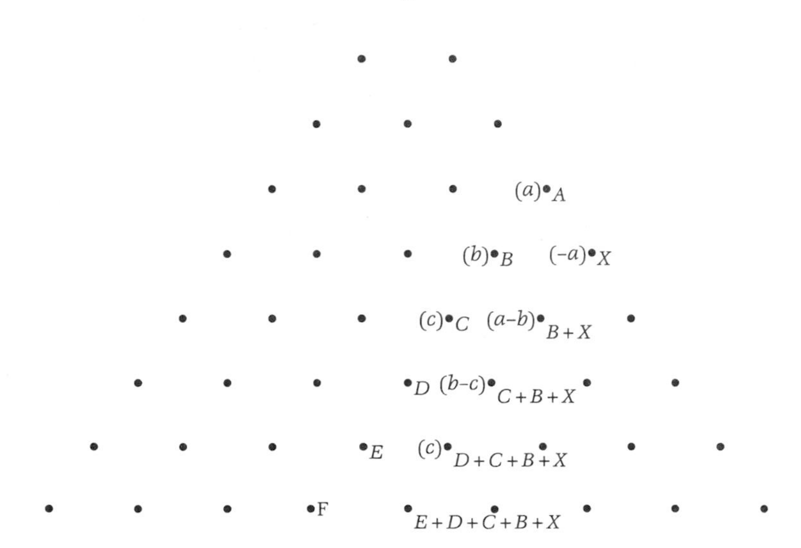

The relations among entries (upper case) and among coefficients (lower case) in the Pascal Triangle.

(i)

(t) T

(u) U (v) V

(w) W

$U + V = W$ **[or** $\binom{n-1}{r-1} + \binom{n-1}{r} = \binom{n}{r}$, $r \geq 1$**], Pascal Identity,**

$w = t - u$ **[or** $S_k(r, i) = S_k(r-1, i-1) - S_k(r-1, i)$, $r \geq k+1$**],**

Transition Identity.

The identities.

(ii)

FIGURE 14

Now, to demonstrate Theorem 5, all we need to show is that we may deduce from (37) the relations

$$
\begin{aligned}
&(-a)X + (a-b)(B+X) + (b-c)(C+B+X) \\
&\qquad\qquad + c(D+C+B+X) = (n+1)^3 \\
&(-a)(B+X) + (a-b)(C+B+X) + (b-c)(D+C+B+X) \\
&\qquad\qquad + c(E+D+C+B+X) = (n+2)^3 \\
&\text{etc.} \qquad\qquad (38)
\end{aligned}
$$

We suggest you carry out these easy calculations!

6.6 EVEN DEEPER MYSTERIES

To sum up the previous section, we showed there that, if we started with the Eulerian numbers $S_k(i)$, and computed certain new numbers, called ***pseudo-Eulerian coefficients***, $S_k(r, i)$, $r \geq k$, by means of the Transition Identity, with initial values $S_k(k, i) = S_k(i)$, thus

$$
\left.
\begin{aligned}
&S_k(k,i) = S_k(i), \\
&S_k(r,i) = 0, \qquad i < 0 \quad \text{or} \quad i \geq r, \\
&S_k(r,i) = S_k(r-1, i-1) - S_k(r-1, i), \qquad r \geq k+1,
\end{aligned}
\right\} \qquad (39)
$$

then Theorem 5 holds, that is, repeating its statement,

Theorem 5 $\sum_i S_k(r,i)\binom{n+i}{r} = n^k$, *for any* $k \geq 1$, *where* $r \geq k$.

We now proceed to enunciate other facts about these pseudo-Eulerian coefficients, but we will change gear from the previous section and trust that our readers have now acquired the mental agility to follow the arguments, or, for those teachers studying our text, that they can now provide for themselves the pedagogical strategies required to make the mathematical arguments palatable to their own students. We start with some very elementary properties; we will omit some proofs, simply remarking that an easy induction on r is required.[17]

[17]Note again that $S_k(r, i)$ is only defined for $r \geq k$, so an inductive argument on the variable r starts with $r = k$. Readers may debate whether it would have been better to denote $S_k(r, i)$

Theorem 6

(i) $S_k(r,0) = (-1)^{r-k}$, $S_k(r,r-1) = 1$;

(ii) $S_k(r,i) = (-1)^{r-k} S_k(r,r-i-1)$.

We now prove a deeper result, giving an expression for $S_k(r, i)$ simply in terms of binomial coefficients. (We will later give another formula for $S_k(r, i)$ which also involves Eulerian numbers.)

Theorem 7

$S_k(r,i) = (-1)^{r-k} \sum_{j=0}^{i} (-1)^j \binom{r+1}{j}(i+1-j)^k$, $i \geq 0$.

Proof The great advantage of having quantities (e.g., $S_k(r, i)$) defined inductively (e.g., by the Transition Identity (39)) is that the natural way—probably the only way—to prove general facts about such quantities will be by induction. Thus Theorem 7 will be proved by induction on r, so we must first establish the initial case $r = k$, that is, we must first prove the following fact about Eulerian numbers:

$$S_k(i) = \sum_{j=0}^{i} (-1)^j \binom{k+1}{j} (i+1-j)^k, \quad i \geq 0. \tag{40}$$

However, we can make one simplification at this stage; if we look at Theorem 6(i) we immediately see that Theorem 7 holds if $i = 0$, so we may assume $i \geq 1$. Now the Eulerian numbers $S_k(i)$ are themselves defined inductively, with respect to k, by (36), so that, to prove (40) in the case $i \geq 1$, we must first establish the case $k = 1$, that is, we must show that

$$S_1(i) = \sum_{j=0}^{i} (-1)^j \binom{2}{j} (i+1-j), \quad i \geq 1. \tag{41}$$

But $S_1(i) = 0$, $i \geq 1$, so it is a matter of proving that

$$\sum_{j=0}^{i} (-1)^j \binom{2}{j} (i+1-j) = 0, \quad i \geq 1. \tag{42}$$

---

as, say, $s_k(r-k, i)$, so that we would have been talking of quantities $s_k(m, i)$, $m \geq 0$, with $s_k(0, i) = S_k(i)$.

Now the scene is set! To prove (42) we first consider the case $i = 1$. Then the LHS is $\binom{2}{0}2 - \binom{2}{1}1$, which is certainly zero. We now consider the general case[18] $i \geq 2$. Remembering that $\binom{2}{j} = 0$ if $j \geq 3$, we find that the LHS is

$$\binom{2}{0}(i+1) - \binom{2}{1}i + \binom{2}{2}(i-1),$$

which is again certainly zero. Thus (42), or (41), is established and we are ready for an inductive argument to establish (40). The argument proceeds as follows; recall that $i \geq 1$, so that the first sum on the second line below makes sense,

$$\begin{aligned} S_k(i) &= (k-i)S_{k-1}(i-1) + (i+1)S_{k-1}(i), \quad \text{by (36)}, \\ &= (k-i)\sum_{j=0}^{i-1}(-1)^j\binom{k}{j}(i-j)^{k-1} \\ &\quad + (i+1)\sum_{j=0}^{i}(-1)^j\binom{k}{j}(i+1-j)^{k-1}, \\ &\qquad\qquad \text{by the inductive hypothesis.} \end{aligned}$$

We now prepare to reorganize! We have, writing $(j-1)$ instead of j on the RHS below,

$$\begin{aligned} &(k-i)\sum_{j=0}^{i-1}(-1)^j\binom{k}{j}(i-j)^{k-1} \\ &\qquad = (k-i)\sum_{j=1}^{i}(-1)^{j-1}\binom{k}{j-1}(i+1-j)^{k-1} \\ &\qquad = (i-k)\sum_{j=1}^{i}(-1)^j\binom{k}{j-1}(i+1-j)^{k-1}; \end{aligned}$$

[18]The reader should understand why the case $i = 1$ had to be considered separately.

while

$$(i+1)\sum_{j=0}^{i}(-1)^j\binom{k}{j}(i+1-j)^{k-1}$$
$$= (i+1)^k + (i+1)\sum_{j=1}^{i}(-1)^j\binom{k}{j}(i+1-j)^{k-1}.$$

Thus, organizing by $(i+1-j)^{k-1}$, we have

$$S_k(i) = (i+1)^k + \sum_{j=1}^{i}(-1)^j\left\{(i-k)\binom{k}{j-1} + (i+1)\binom{k}{j}\right\}(i+1-j)^{k-1}.$$

Now

$$(i-k)\binom{k}{j-1} + (i+1)\binom{k}{j} = (i+1)\left\{\binom{k}{j-1} + \binom{k}{j}\right\}$$
$$- (k+1)\binom{k}{j-1}$$
$$= (i+1)\binom{k+1}{j} - j\binom{k+1}{j},$$

using the Pascal Identity,

$$= (i+1-j)\binom{k+1}{j}.$$

We conclude that

$$S_k(i) = (i+1)^k + \sum_{j=1}^{i}(-1)^j\binom{k+1}{j}(i+1-j)^k$$
$$= \sum_{j=0}^{i}(-1)^j\binom{k+1}{j}(i+1-j)^k.$$

completing the inductive step and hence establishing (40).

Now we return to the statement of Theorem 7, having established the initial step $r = k$, and now use induction on r. Again with $i \geq 1$ (notice again the advantage of having disposed of the case $i = 0$), we have

$$S_k(r,i) = S_k(r-1,i-1) - S_k(r-1,i),$$

by the Transition Identity,

$$= (-1)^{r-k-1}\sum_{j=0}^{i-1}(-1)^j\binom{r}{j}(i-j)^k$$

$$-(-1)^{r-k-1}\sum_{j=0}^{i}(-1)^j\binom{r}{j}(i+1-j)^k,$$

by the inductive hypothesis.

Now, by a similar trick to that used in establishing (40), replacing j by $(j-1)$,

$$\sum_{j=0}^{i-1}(-1)^j\binom{r}{j}(i-j)^k = \sum_{j=1}^{i}(-1)^{j-1}\binom{r}{j-1}(i+1-j)^k,$$

while

$$\sum_{j=0}^{i}(-1)^j\binom{r}{j}(i+1-j)^k = (i+1)^k + \sum_{j=1}^{i}(-1)^j\binom{r}{j}(i+1-j)^k.$$

Thus

$$S_k(r,i) = (-1)^{r-k}\sum_{j=1}^{i}(-1)^j\left\{\binom{r}{j-1}+\binom{r}{j}\right\}(i+1-j)^k$$

$$+(-1)^{r-k}(i+1)^k$$

$$= (-1)^{r-k}\sum_{j=1}^{i}(-1)^j\binom{r+1}{j}(i+1-j)^k$$

$$+(-1)^{r-k}(i+1)^k, \text{ by the Pascal Identity,}$$

$$= (-1)^{r-k}\sum_{j=0}^{i}(-1)^j\binom{r+1}{j}(i+1-j)^k,$$

finally establishing the theorem. □

You have been very patient with us, allowing us to inflict on you several rather long proofs, especially this last one. Indeed, it may appear to you at first sight that we involved ourselves in a great deal of work to prove what is at best a rather technical theorem. However, there is a substantial pay-off. In the first place,

Theorem 7 has been proved for *all* nonnegative i and we know that $S_k(r, r-1) = 1$, $S_k(r, i) = 0$, $i \geq r$. Thus we conclude

Corollary 8 $\sum_{j=0}^{r-1} (-1)^j \binom{r+1}{j} (r-j)^k = (-1)^{r-k}$, *provided* $r \geq k$.

Corollary 9 $\sum_{j=0}^{i} (-1)^j \binom{r+1}{j} (i+1-j)^k = 0$, $i \geq r$, *provided* $r \geq k$.

A beautiful feature of these conclusions is that they make no mention of pseudo-Eulerian coefficients! Thus we have achieved some remarkable identities connecting the familiar binomial coefficients by reasoning with new concepts which have then entirely disappeared from the conclusion—and we ourselves know of no other way of obtaining these results. This feature of mathematical method is, we claim, both unique to mathematics and typical of much of the best of it.

Example 8* We give an example of Corollary 8. We fix $r = 3$, so our result states

$$\sum_{j=0}^{2} (-1)^j \binom{4}{j} (3-j)^k = (-1)^{3-k}, \quad \text{for } k = 1, 2, 3.$$

With $k = 1$, this asserts that $3 - (4 \times 2) + (6 \times 1) = 1$.
With $k = 2$, this asserts that $3^2 - (4 \times 2^2) + (6 \times 1^2) = -1$.
With $k = 3$, this asserts that $3^3 - (4 \times 2^3) + (6 \times 1^3) = 1$.

• • • **BREAK**

Repeat Example 8* with $r = 4$.

Before exemplifying Corollary 9 we will slightly restate it. First, we note that we can allow the summation to run to $i + 1$, since the extra term makes no contribution (k is a positive integer). Second, we have $i + 1 \geq r + 1$ and we can reduce the summation to run only to $r + 1$, since $\binom{r+1}{j} = 0$ if $j > r + 1$. Thus, at this stage, we have replaced the LHS by

$$\sum_{j=0}^{r+1} (-1)^j \binom{r+1}{j} (i+1-j)^k.$$

We now write i for $i+1$ and r for $r+1$. The condition $i \geq r$ is unaffected, but the condition $r \geq k$ becomes $r > k$. Thus, finally, Corollary 9 takes on the following more convenient form.

Corollary 9 (Restated) *For fixed i, r with $i \geq r$, we have*

$$\sum_{j=0}^{r}(-1)^j\binom{r}{j}(i-j)^k = 0, \quad \textit{for all } k < r.$$

Notice that we could even simplify the notation, writing $\sum_j$ for $\sum_{j=0}^{r}$.

Example 9* We give an example of Corollary 9 in its new form. We fix $i = 5$, $r = 4$, so our result states

$$\sum_{j=0}^{4}(-1)^j\binom{4}{j}(5-j)^k = 0, \quad \text{for } k = 1, 2, 3.$$

With $k = 1$, this asserts that $5 - (4 \times 4) + (6 \times 3) - (4 \times 2) + 1 = 0$.
With $k = 2$, this asserts that $5^2 - (4 \times 4^2) + (6 \times 3^2) - (4 \times 2^2) + 1^2 = 0$.
With $k = 3$, this asserts that $5^3 - (4 \times 4^3) + (6 \times 3^3) - (4 \times 2^3) + 1^3 = 0$.

Remark Note that Corollary 9 in its *new* form, holds for $k = 0$, since $\sum_{j=0}^{r}(-1)^j\binom{r}{j}$ is the binomial expansion of $(1-1)^r$. Thus we may conclude that

$$\sum_{j=0}^{r}(-1)^j\binom{r}{j}p(i-j) = 0,$$

for any polynomial $p(x)$ of degree $< r$, provided $i \geq r$. Notice, on the other hand, that Corollary 9 in its *old* form would be false if the case $k = 0$ were allowed.

There are many other identities to be obtained. We will only mention one here (see [3] if you want more), since it may be verified from Figures 9 and 13.

Theorem 10 $\sum_i S_k(r,i)\binom{n+i+1}{r+1} = \sum_{j=1}^{n} j^k$.

Proof By the Pascal Identity,

$$\sum_i S_k(r,i)\binom{n+i+1}{r+1} = \sum_i S_k(r,i)\binom{n+i}{r+1} + \sum_i S_k(r,i)\binom{n+i}{r}. \tag{43}$$

Let us call the LHS $F(n)$, so we then want to show that $F(n) = \sum_{j=1}^{n} j^k$. Then, using Theorem 5, we may restate (43) as

$$F(n) = F(n-1) + n^k. \tag{44}$$

Now $F(1) = \sum_i S_k(r,i)\binom{i+2}{r+1}$; but $S_k(r,i) = 0$ if $i \geq r$ and

$$\binom{i+2}{r+1} = 0$$

if $i < r-1$. Thus only $i = r-1$ yields a nonzero term and hence

$$F(1) = S_k(r, r-1)\binom{r+1}{r+1} = 1 = 1^k.$$

Theorem 10 follows immediately from this and (44). □

Remarks (i) As many readers will know, we often need, in studying the calculus, and the theory of polynomials, to evaluate

$$\sum_{j=1}^{n} j^k$$

for various values of k. Thus Theorem 10 may be viewed as providing us with a means of doing this. It is interesting that what we *proved* was that the RHS of Theorem 10 was a means of evaluating the LHS, but we often *use* the theorem in the form that the LHS provides a means of evaluating the RHS. This type of exploitation is not uncommon in mathematics.

(ii) It is striking that the RHS of Theorem 10 does not involve r, so the LHS doesn't depend on r either! If we're going from right to left in applying the theorem it would be natural to choose $r = k$, since the Eulerian numbers are well known and tabulated.

• • • **BREAK**

Use Theorem 10 to verify that

$$\sum_{j=1}^{n} j^3 = \left[\frac{n(n+1)}{2}\right]^2.$$

We may use Theorem 10 to obtain explicit values for the LHS if $k = 1, 2$, or 3, with any value for $r \geq k$. (This is *not* the same point of view as that taken in the break!) Thus

Corollary 11

(i) $\sum_i S_1(r,i)\binom{n+i+1}{r+1} = \frac{n(n+1)}{2}$.

(ii) $\sum_i S_2(r,i)\binom{n+i+1}{r+1} = \frac{n(n+1)(2n+1)}{6}$.

(iii) $\sum_i S_3(r,i)\binom{n+i+1}{r+1} = \left[\frac{n(n+1)}{2}\right]^2$.

Finally, we give the promised explicit formula for $S_2(r, i)$ and $S_3(r, i)$ featured in Figures 9 and 13, respectively. In fact, we will give a general formula for $S_k(r, i)$ in terms of the Eulerian numbers $S_k(i)$ since this is just as easy as tackling the special cases $k = 2, 3$. The technique of proof is one of fairly general application. Notice that the formula we will obtain, which is (47) below, is *different in kind* from that of Theorem 7. The latter gives $S_k(r, i)$ simply in terms of binomial coefficients; our new formula, as you will see, also involves Eulerian numbers. Each formula has its particular advantages.

Given an integer k, we form, for each $r \geq k$, the polynomial[19]

$$f_r(x) = \sum_{i \geq 0} S_k(r, i)x^i.$$

Then the Transition Identity asserts effectively that

$$f_r(x) = (x-1)f_{r-1}(x), \quad r \geq k+1. \tag{45}$$

(Do you see this? Just compare coefficients of x^i on both sides of (45).) Iterating (45), we conclude that

$$f_r(x) = (x-1)^{r-k} f_k(x)$$

[19]This is called the *generating function* for the pseudo-Eulerian coefficients $S_k(r, i)$, regarded as functions of i.

or

$$\sum_{i\geq 0} S_k(r,i)x^i = (-1)^{r-k}(1-x)^{r-k}\sum_{i\geq 0} S_k(i)x^i. \tag{46}$$

We now compare coefficients of x^i on each side of the relation (46), remembering that, by the binomial theorem,

$$(1-x)^{r-k} = \sum_{j\geq 0}\binom{r-k}{j}(-1)^j x^j.$$

We obtain the result

$$\boxed{\mathbf{S_k(r,i) = (-1)^{r-k}\sum_{j=0}^{i}(-1)^j\binom{r-k}{j}S_k(i-j),\ \ r\geq k.}} \tag{47}$$

Let us find explicit expressions in the cases $k = 1, 2, 3$. Since $S_1(0) = 1$, $S_1(i) = 0$, $i \neq 0$, we have, from (47),

$$S_1(r,i) = (-1)^{r-i-1}\binom{r-1}{i}, \quad r \geq 1; \tag{48}$$

and, since $S_2(0) = S_2(1) = 1$, $S_2(i) = 0$, $i \neq 0, 1$, we have, from (47),

$$S_2(r,i) = (-1)^{r-i}\left[\binom{r-2}{i} - \binom{r-2}{i-1}\right], \quad r \geq 2.$$

Finally, since $S_3(0) = 1$, $S_3(1) = 4$, $S_3(2) = 1$, $S_3(i) = 0$, $i \neq 0, 1, 2$, we have, from (47),

$$S_3(r,i) = (-1)^{r-i-1}\left[\binom{r-3}{i} - 4\binom{r-3}{i-1} + \binom{r-3}{i-2}\right], \quad r \geq 3.$$

Notice that we've now proved everything that we left unproved earlier (see Figures 9, 11, and 13). We can rest!

REFERENCES

1. Aldred, Robert, Peter Hilton, Derek Holton, and Jean Pedersen, Baked Beans and Spaghetti, *The Mathematics Educator*, **5**, No. 2, 1994, 35–41.
2. Comtet, L. *Analyse Combinatoire* (Tome Second), Presses Universitaires de France, Paris, 1970.

3. Carroll, Tim, Peter Hilton, and Jean Pedersen, Eulerian numbers, pseudo-Eulerian coefficients, and weighted sums in Pascal's Triangle, *Nieuw Archief voor Wiskunde*, **9**, No. 1, March, 1991, 41–64.
4. Gauss, C.F., *Summatio Quarundam Serierum Singularium*, Werke, Band 2, pp. 16–17.
5. Gould, H.W., A greatest common divisor property of binomial coefficients, *The Fibonacci Quarterly*, **10**, 1972, 579–584, 628.
6. Hoggatt, V.E., Jr. and W. Hansell, The hidden hexagon squares, *The Fibonacci Quarterly*, **9**, 1971, 120, 133.
7. Hilton, Peter, and Jean Pedersen, Looking into Pascal's Triangle: Combinatorics, arithmetic, and geometry, *Mathematics Magazine*, **60**, 1987, 305–316.
8. Hilton, Peter, and Jean Pedersen, Binomial coefficients in the Pascal hexagon, *Kolloquium Mathematik-Didaktik der Universität Bayreuth*, **14**, 1988, 3–24.
9. Hilton, Peter, and Jean Pedersen, Extending the binomial coefficients to preserve symmetry and pattern, *Comput. Math. Appl.*, **17**, No. 1–3, 1989, 89–102. Reprinted in *Symmetry 2: Unifying Human Understanding* (Ed. I. Hargittai), Pergamon Press, Oxford, 1989.
10. Hilton, Peter, and Jean Pedersen, Generalizations of Eulerian numbers and their q-analogoues, *Nieuw Archief voor Wiskunde*, **9**, (Series 4), No. 3, Nov. 1991, 271–298.
11. Hilton, Peter, Jean Pedersen, and William Rosenthal, Pascalian Triangles and hexagons, *Quaestiones Mathematicae*, **13**, Parts 3 and 4, 1990, 395–416.
12. Pólya, George, and Gerald L. Alexanderson, Gaussian binomial coefficients, *Elemente der Mathematik*, **26**, No. 5, 1971, 102–108.
13. Solow, Daniel, *How To Read and Do Proofs*, Wiley, New York, 1990.

CHAPTER 7 Hair and Beyond

7.1 A PROBLEM WITH PIGEONS, AND RELATED IDEAS

Are there two women in the world with the same number of hairs on their heads?

We know that's not an earth-shattering question, but it might stimulate the conversation over lunch if there's nothing much of interest going on at the time. However, it's actually not difficult to answer the question with complete certainty and without leaving the safety of your own room, except to verify that there are lots of women in the world.

Now an unintelligent way of approaching this hairy question is to put an ad in all the papers in the world, asking every woman to count (or have counted) all the hairs on her head, and then getting them to send back their answers. When all of these numbers are assembled, you could go through them and see if two were precisely the same.

This is clearly a very simple experiment in principle, but it does seem to have several serious flaws. First, there is no way to make sure that *every* woman in the world will actually bother to respond, even if they do get to read the ad—which many won't, as a lot of people in the world are unable to read, many don't read a newspaper, and a lot of others totally refuse to respond to anything. Second, there is no guarantee that the answers you will get will be accurate. Have you tried counting the hairs on someone's head? Even your father's bald patch is not easy to cover systematically enough to ensure that every single fine filament on his crown has been accurately accounted for. And third, and perhaps the major reason for not doing this experiment, have you thought how much the whole exercise would cost? Suppose the ad only cost $2 to put in your local classifieds. How many papers are there in the whole world? So how much is the project going to cost?

Suppose that this question—whether there are two women with the same number of hairs on their heads—is the most important one to be resolved right now on this planet, can we get a simple answer of yes or no, without laying out a sum equivalent to the national debt? What's more, having said yes or no, can that response be fully and totally justified?

• • • **BREAK**

Try to answer the question about women's hair.

Now let's look at another question of the same general kind. What day of the week was your birthday this year? How many people would you need to have at a party to be sure that, this year, two of you had your birthday on the same day of the week?

The answer to this last question is clearly not four. It would be possible for four people to use up, say, Monday, Tuesday, Wednesday, and Thusday. So no two of *them* would necessarily have to have had their birthday on the same day of the week. On the other hand, you might think that among 500 people there would *have* to be two who had birthdays on Monday, say. But how could you try to convince yourself that this had to be the case?

• • • **BREAK**

What is the smallest number of people at a party that would force two of them to have their birthdays on Monday? What is the smallest number of people at a party that would force two of them to have their birthdays on the same day of the week?

We can produce a similar problem with months. How many people would we need at the party in order that two of them *had* to have their birthday in the same month? Clearly four people is again not enough. They could all have birthdays in different months. And 500 people would probably be OK, but what is the smallest number of people at the party to force two birthdays in the same month?

Well, let's move up from four. Five people can have birthdays in different months without any problem. The same is true for 6, 7, 8, 9, 10, 11, and 12 people. Even with 12 people we can't guarantee that two people would have a birthday in the same month. They might all have been born in *different* months.

So let's try 13. The worst situation for us, is that the first 12 all had their birthdays in different months. If not, we've got what we wanted. But when the 13th person comes along, there's nowhere to go. All the months have been taken up. So the 13th person *has* to have a birthday month in common with one of the previous 12. The smallest party to ensure that two people have a birthday in the same month must have 13 people.

Can you now see that we need to have 8 people around to be sure that two of them are forced to have their birthdays on the same day of the week? Of course, that day need not be Monday.

• • • **BREAK**

How many people do we need to gather together in order to guarantee that two of them have their birthdays on the same *day* this year?

The discussion so far has been a bit labored in order to stress that we're dealing with a single idea. In fact, we've just given three examples of the ***pigeon-hole principle***. This states that if we have n pigeon-holes or boxes, and we distribute $n + 1$ objects among these boxes, then there *has* to be *at least* one pigeon-hole which ends up containing two or more objects.

In the "months of the year" case above, we have 12 pigeon-holes (the months of the year) and thus required $12 + 1 = 13$ objects (people at a party) in order to ensure that two objects must get in one pigeon-hole (i.e., for at least two people to have their birthday in the same month). This pigeon-hole principle notion can be extended, of course.

• • • **BREAK**

If there are 12 pigeon-holes, how many objects would we need, in order to guarantee that some pigeon-hole contains at least three objects?

Suppose we've got 12 pigeon-holes and we want to ensure that one of them (we don't care which) has at least 5 objects in it. Obviously, it's not good enough to get 5 objects. They might be spread around, one each to a pigeon-hole, with 7 pigeon-holes empty. And even 48 objects wouldn't be enough. They might perversely distribute themselves 4 to a hole.

So how about 49? Well, either 48 of them have been distributed so that there are at least 5 in one pigeon-hole or there are exactly 4 to a pigeon-hole. If it's the former, we're done. If it's the latter, then along comes lucky object number 49. This object has to be put in *some* pigeon-hole and so the occupancy rate of that pigeon-hole has to increase to 5.

Naturally, being mathematicians, we can formalize this into a ***generalized pigeon-hole principle***. Given $mn + 1$ objects and n pigeon-holes, there *has* to be one pigeon-hole with at least $m + 1$ objects in it, and $mn + 1$ is the smallest number of objects with this property.

• • • **BREAK**

Check the generalized pigeon-hole principle with $m = 7$ and $n = 10$.

It's beginning to seem as though this pigeon-hole thing may be able to help us with women's hair. Just suppose for the moment that the number of hairs on a woman's head was a label on a pigeon-hole. There would be one labeled pigeon-hole for every possible number of hairs on a woman's head. So it's possible that there would be pigeon-holes labeled 0, 1, 2, 3, . . ., and so on. It's not clear at the moment what the biggest label would be (or, equivalently, how many pigeon-holes we'd need). Suppose the labels went up to some number N.

We could, metaphorically, think of each woman on earth being assigned to a pigeon-hole (or at least the number on the pigeon-hole). If we could show that there were $N + 1$ women on earth, the pigeon-hole principle would say that two of them had the same number of hairs.

We have reduced the problem to two tasks. First, find the maximum number of hairs on a woman's head. Second, find out how many women there are in the world. You can find those things out for yourself, can't you? Of course, to solve the problem, you only need to know that there are more women in the world than the total number of hairs on *any* woman's head. Actually, you may end up being able to show that there are at least 10 women in the world with the same number of hairs on their heads. Who knows?

It might be interesting to show that there are at least two salaried people in the world, who will earn the same amount of money this year. Make this before tax if you like, and round the salary to the nearest dollar.

My guess is that no one in the world will earn more than \$500 million. Just in case they do, put a safety factor of 100. It's unlikely that even a North American basketball player will earn \$5000 million! But there are more than 5000 million people in the world. So if the people are our objects and the salaries are numbers on pigeon-holes, then at least two people have exactly the same salary.

You can see that we didn't even have to have very accurate figures to get the salary result. Do we have to get the exact values for the number of hairs on a woman's head? An upper bound that might be worth trying for, would be to measure the cross-sectional area of a hair, measure the area of a woman's head, divide the first into the second, and the number is a ballpark figure for the number of hairs on a woman's head. Now multiply by a safety factor of 100 and you should still find that you haven't exceeded the adult female population of the world. Try it.

7.2 THE BIGGEST NUMBER

When you get to start talking about people's salaries, the number of hairs on a person's head, and the number of people in the world, you're starting to get into pretty big numbers. One of the first questions that comes to mind is, what is the biggest number possible?

Of course, it might also be worth asking which type of object in the world, or, if you're really ambitious, in the Universe, is the most numerous. Is it liters of blood, grains of sand, hairs, flies, what? And how many objects of the chosen type are there?

Naturally, if you thought you had the *biggest* number, then we would just add one to it and get an even bigger number. But then you could add one to that, and so on. Clearly numbers keep getting bigger, so there's never any number which can claim to be the Guinness Book of Records largest number.

But we can, and do, think of the biggest numbers of a given type. The highest a woman can jump, the heaviest haggis ever, the furthest a man can throw a javelin, and so on. See, too, the world's biggest banger, as repeated below.

World's Biggest Banger

> An East German butcher claimed an entry in the Guinness Book of Records this week after he made the world's biggest sausage, a two-kilometer monster that used up 24 sides of pork.
>
> The one-tonne banger made by butcher Karl-Heinz Siebert will be given a ceremonial send-off in an Erfurt square tomorrow on a 10 m diameter specially built steel grill.
>
> (*Cape Times*, Republic of South Africa, 15 May, 1993)

In the same way, we may think about smallest positive numbers. The fastest that a man can run the 100 meters, the fastest a woman can swim 200 meters using the butterfly stroke, the time the first baby is born on New Year's Day, and so on.

• • • **BREAK**

Which all suggests that it might be fun to try to find the smallest positive rational number.

A little while ago you thought about which objects there are most of in the Universe. Now everything is made up of atoms, so the number of atoms must be near the top of this list. Though, of course, you might squeeze out a few more electrons or quarks, or something, than there are atoms. But let's stick to atoms to give an order of magnitude anyway. Somebody has estimated that there are 10^{70} atoms in the Universe. Was your guess anywhere near that?

And how far did you go toward finding the *smallest* positive rational number? Many of you didn't seem to find one. That's not surprising because it doesn't exist. If you thought you had a smallest rational number, divide it by two. Then divide that by two, and so on. These numbers get closer and closer together, of course. There must be quite a log jam of numbers at zero. Is it any worse there than anywhere else though?

7.3 THE BIG INFINITY

So far in this chapter, we have given two examples of infinite processes. The first was an increasing process. Take any number, add one, then add another one, and so on. This is just the way we count. The counting process goes on forever, to infinity you might say. Because the process of counting never stops (there is always room for one more), we say that there are an ***infinite*** number of whole numbers. We say that the size, or ***cardinality***, of the natural numbers $\mathbb{N} = \{0, 1, 2, 3, \ldots\}$ is infinity, denoted by $\aleph_0$. The letter $\aleph$ is aleph, the first letter of the Hebrew alphabet. So we read $\aleph_0$ as "aleph nought" or "aleph zero." Some people write ∞ whenever infinity comes up, but this is not a good thing to do, as we will shortly show.

• • • BREAK

Can you think of any other sets of numbers which have infinite cardinality? It would be worth finding, say, four.

Let's have a look at the even numbers for a moment. Call them $\mathbb{E} = \{0, 2, 4, 6, 8, \ldots\}$. It is clear that the even numbers go on forever. Since that is the case, then $|\mathbb{E}|$, the cardinality of $\mathbb{E}$, must be infinite.

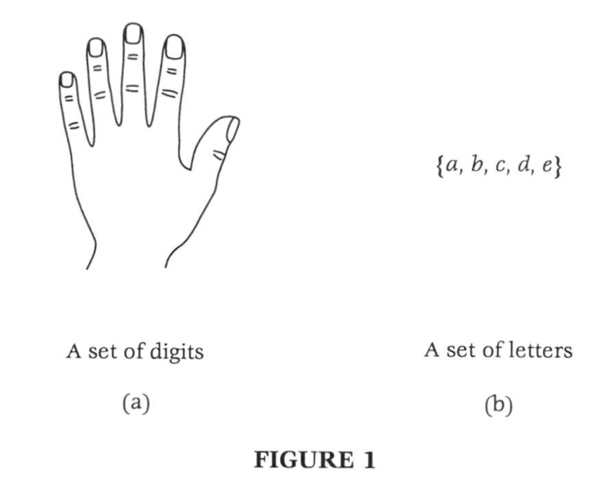

A set of digits

(a)

A set of letters

(b)

FIGURE 1

• • • BREAK

Clearly $\mathbb{E} \subset \mathbb{N}$, $\mathbb{E}$ is a ***proper*** subset of $\mathbb{N}$. Does this mean that $|\mathbb{E}| < |\mathbb{N}|$? If so, there must be more than one size of infinity? Is this possible?

Before we bother with $\mathbb{E}$, $\mathbb{N}$, and their relative sizes, we need to know how to work out when two sets have the same number of elements. For instance, how do we know that the two sets in Figure 1 are equal?

Well, there are probably two ways. The simplest is to count them. In Figure 1(a), there are five digits. In Figure 1(b) there are five letters. Both sets have cardinality 5 and so they are equal in size—they have the same cardinality.

The second approach is to link them directly. We've shown such a linking by the double arrows in Figure 2. To every member of the set of digits of Figure 1(a) we have associated one and only one letter of Figure 1(b). Similarly every letter in Figure 1(b) is linked

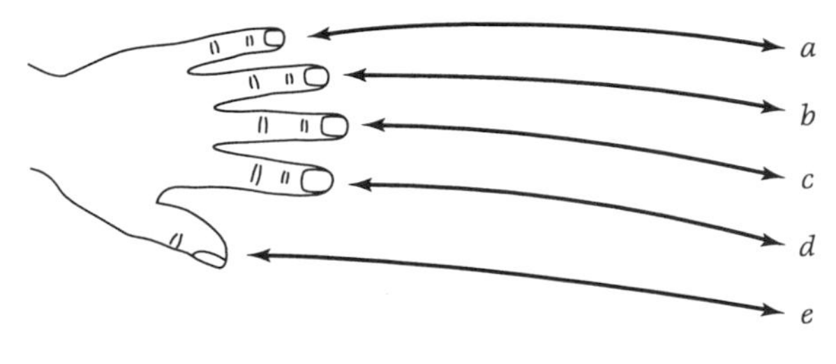

FIGURE 2

to one and only one digit in Figure 1(a). What this does is to show that there is a one-to-one correspondence between the two sets. Hence they have the same number of elements. Indeed, the phrase "the sets A and B have the same cardinality" means precisely that there exists a one-to-one correspondence between them. Sets are called ***equivalent*** if there is a one-to-one correspondence between them. The cardinality of A is then the cardinality of every set in the equivalence class of the set A.

• • • **BREAK**

Show that the relation "A and B have the same cardinality" is indeed an equivalence relation (see Chapter 2).

One-to-one correspondences pair up the elements of two sets. They show that for every member of the first set, there is a corresponding member of the second set, and vice versa. This is shown in Figure 2 by the double-headed arrows, which we have used to emphasize the fact that, with every member of the digit set, we may associate a unique member of the alphabet set, and vice versa. In actual fact, there is no need to use double-headed arrows. After all, the arrows show that there is a map from the whole of one set to the whole of the other, with one element of the first set going to only one element of the other. We could easily, therefore, delete one of the arrowheads—the deleted one can be recreated directly by knowing that exactly one element of the first set goes to any one element of the second set, the map is one-to-one and onto. The ***one-to-one*** property tells us that no members of the first set are mapped to the same member of the second set. This gives the pairing. But if some member of the second set happens to be missed, then we don't have the required one-to-one correspondence. Hence the one-to-one mapping has to be ***onto*** the second set. That is, every member of the second set has to have come from a member of the first set.

So two finite sets have the same number of elements if there is a one-to-one onto mapping between them. Surely it makes sense to apply the same rule to infinite sets too. If we have two sets A and B, and if we can establish a one-to-one mapping (one-to-one correspondence) between them, then we will say that A and B have the same cardinality. If two sets *do* have the same cardinality then there *must* be some way of establishing a one-to-one correspondence. If two sets *don't* have the same cardinality, then, no matter how hard we try to put double arrows between their elements, we won't be able to do it. We'll illustrate this in a minute.

• • • BREAK

Use this one-to-one correspondence idea to show that there are as many corners on a square as there are suits of playing cards. Also show that there are more letters in the Greek alphabet than there are in the Roman alphabet. Let A and B be two *finite* sets. Show that every one-to-one function from A to B, as well as every function from A onto B, is a one-to-one correspondence between A and B.

Let's now use this one-to-one correspondence business to investigate the cardinality of infinite sets. In particular, let's go back to $\mathbb{N}$ and $\mathbb{E}$ and try to establish whether or not $\mathbb{N}$ is bigger. For a start, let's try to find a one-to-one onto mapping between $\mathbb{N}$ and $\mathbb{E}$. If we do find one, many of you will be quite surprised. If we don't find one, it may help us to see how to show that there isn't one.

So let's line up the elements of $\mathbb{N}$ and $\mathbb{E}$ and look for a one-to-one mapping. Things clearly start out OK. We can pair up 0 and 0, 1 and 2, 2 and 4, 3 and 6, 4 and 8, 5 and 10, 6 and 12, 7 and 14, and so on. There is no difficulty putting the elements of $\mathbb{N}$ and $\mathbb{E}$, that we've shown in Figure 3, into a one-to-one correspondence. But, of course, we've only put the first eight members of each set down, so obviously we should have no problem with the correspondence at this stage. The question is, when does this one-to-one correspon-

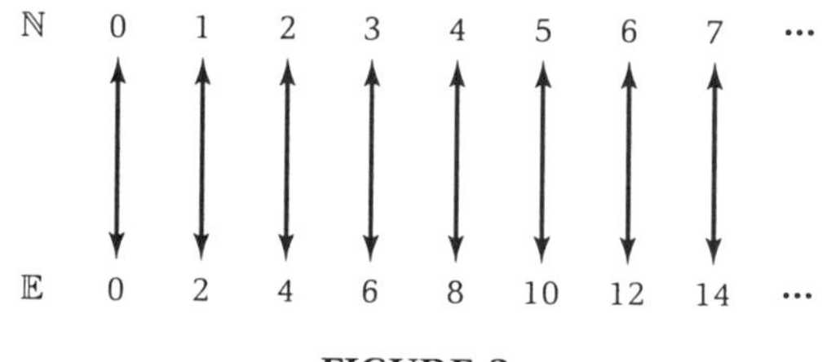

FIGURE 3

dence break down? How about at the number 70? No, because we would pair that up with 140. So, how about 700. No, there's 1400.

• • • **BREAK**

Well, where *does* the one-to-one correspondence between $\mathbb{N}$ and $\mathbb{E}$ break down then?

Hang on. Let's see what we're moving toward. There seems to be a slender chance that $\mathbb{N}$ and $\mathbb{E}$ have the same cardinality. So we're going from $\mathbb{N}$ to $\mathbb{E}$. Now every even number can be written in the form $2n$, where $n \in \mathbb{N}$. The obvious mapping to try (see Figure 4) from $\mathbb{N}$ to $\mathbb{E}$ is therefore the one which takes n to $2n$. Is this mapping one-to-one and onto?

First, is the mapping one-to-one? Can two distinct elements n_1, n_2, of $\mathbb{N}$ map to the same element of $\mathbb{E}$? If they do, then $2n_1$ must equal $2n_2$. Hence $2n_1 = 2n_2$. Clearly this gives $n_1 = n_2$, which contradicts the assumption that they were different. This means that no two *different* elements of $\mathbb{N}$ map to the *same* element of $\mathbb{E}$, so the mapping *is* one-to-one.

Second, there is the problem of onto. Does every even number come from some natural number under the mapping $n \rightarrow 2n$? That's easy. Surely the even number $2m$, whatever that might be, comes from m. Every natural number is therefore covered by the mapping, which therefore *has* to be onto.

Although we instinctively felt that the cardinality of the natural numbers had to be *bigger* than that of the even numbers, this doesn't seem to be the case. No matter what natural number, n, say, we look at, there is a corresponding *even* number $2n$. Between n and $2n$ we can insert the double arrow. This genuine, simple one-to-one correspondence shows that $|\mathbb{E}| = |\mathbb{N}|$!

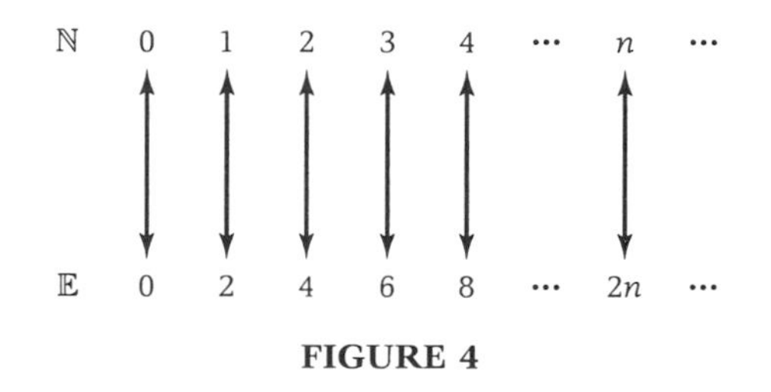

FIGURE 4

While you are getting used to that idea, perhaps it is worth pointing out that there is a difference here between *more* and *bigger*. The set A is *bigger* than the set B if $|A| > |B|$, that is, if it has a larger cardinality. If $A \supset B$, then A contains elements that B doesn't. So we think of A as having *more* elements than B. The way finite sets work is that a finite set with more elements than another is also bigger. But, as we have seen, $\mathbb{N}$ contains *more* elements than $\mathbb{E}$, yet $\mathbb{N}$ is *not* bigger than $\mathbb{E}$. Earlier we defined $|\mathbb{N}|$ to be $\aleph_0$. Now, from the above, we know that $|\mathbb{E}| = \aleph_0$.

• • • BREAK

What other sets have cardinality $\aleph_0$? Do *all* infinite sets have cardinality $\aleph_0$?

Can we tip the scales on $\aleph_0$ by just adding one more thing to it? What we mean by that badly worded statement is, could it just be possible to produce an infinite set that does not have cardinality $\aleph_0$ by taking $\mathbb{N}$ and adding just one element to it? Suppose we let $\mathbb{N}^+ = \mathbb{N} \cup \{-1\}$. Is $|\mathbb{N}^+| = \aleph_0$, or something bigger? The test as always is, does there exist a one-to-one correspondence from $\mathbb{N}$ to $\mathbb{N}^+$?

In Figure 5, we have lined up the two sets one above the other. This suggests that we look at the mapping from $\mathbb{N}$ to $\mathbb{N}^+$ which sends n to $n - 1$. That will certainly send the first five elements of $\mathbb{N}$ onto the first five elements of $\mathbb{N}^+$. But it's not hard to see that this mapping is one-to-one (if n_1 and n_2 are distinct, then so are $n_1 - 1$ and $n_2 - 1$) and onto (any m comes from some $m + 1$). The inescapable conclusion is that $|\mathbb{N}^+| = \aleph_0$.

It doesn't look as if we can "tip the scales" of infinite sets by adding only one new member. Presumably then, we can't change cardinality by adding a finite number of members. After all, adding

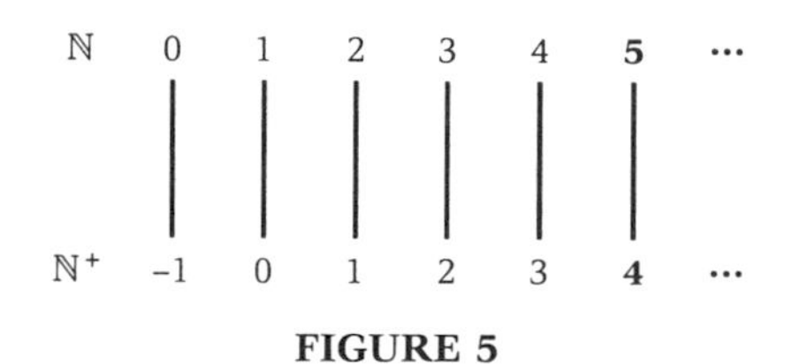

FIGURE 5

a finite number of elements is just the same as adding one element a finite number of times. Can we *prove* that? Can we show that if B is any *finite* set disjoint from $\mathbb{N}$, then $\mathbb{N} \cup B$ *also* has cardinality $\aleph_0$? Can't we just use the old $\mathbb{N}$ to $\mathbb{N}^+$ trick extended a bit? If

$$B = \{a_1, a_2, \ldots, a_r : r \in \mathbb{N}\},$$

then isn't the map we need the one that sends i to a_{i+1}, for $i = 0$ to $r - 1$, and i to $i - r$, otherwise?

• • • **BREAK**

Let A be any set of cardinality $\aleph_0$ and let B be any finite set which is disjoint from A. Show that $A \cup B$ has cardinality $\aleph_0$ too. Is it important to have A and B disjoint?

7.4 OTHER SETS OF CARDINALITY $\aleph_0$

Despite what we had suspected, it looks as if there may only be one size of infinity after all. In that case, what are some obvious candidates to put into one-to-one correspondence with $\mathbb{N}$? The odd numbers, the integers, the rational numbers, and the real numbers all come to mind as things to try. The set of odd numbers $\mathbb{O}$ is $\{1, 3, 5, 7, \ldots\}$, the set of integers $\mathbb{Z}$ is

$$\{\ldots, -3, -2, -1, 0, 1, 2, 3, \ldots\},$$

the set of rationals $\mathbb{Q}$ is $\{\frac{a}{b} : a, b \in \mathbb{Z}, \text{with } b \neq 0\}$, and the set of real numbers $\mathbb{R}$ can be thought of as the set of all decimals, both finite and infinite.

How can we show that there are as many odd numbers as there are natural numbers?

• • • **BREAK**

What one-to-one correspondence can we use to show that $|\mathbb{O}| = |\mathbb{N}|$?

While you're working on the odd numbers, we'll have a go at the integers. Remember, we're trying to line up $\mathbb{Z}$ and $\mathbb{N}$ so that they are in one-to-one correspondence. We want to get a one-to-one mapping between $\mathbb{Z}$ and $\mathbb{N}$. So how can that be done?

This actually looks a little difficult. Lining up 0 and 0, 1 and 1, 2 and 2, 3 and 3, and so on, isn't going to get us anywhere (see Figure 6(a)). If we did that we'd have nowhere to fit the negative integers. We could try mapping the integers 1, 2, 3, ... to the even natural numbers and the negative integers to the odd numbers. But that seems to have left the integer 0 adrift (Figure 6(b)).

Can we modify this idea a little? Why not take 0 to 0, 2 to 1, 4 to 2, 6 to 3, and so on, for the positive integers. That way we send the even integer $2m$ to m. That should leave the odd numbers free to link up with the negative integers. So then we'll map 1 to -1, 3 to -2, 5 to -3, and so on. In general then, if $n = 2m$, we map it to m, while, if $n = 2m + 1$, we map it to $-m - 1$. It looks as if we might have a one-to-one mapping.

The arrow diagram of Figure 7(a) is a real mess. Things are much tidier in Figure 7(b). The algebraic form of the correspondence is given in Figure 7(c), with $m \geq 0$.

We should check that the mapping is one-to-one and onto. Perhaps you could do that. And when you have, you might like to find another one-to-one correspondence between $\mathbb{N}$ and $\mathbb{Z}$.

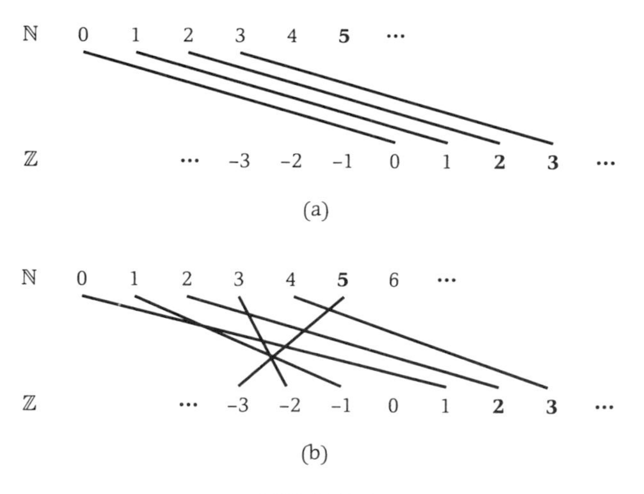

FIGURE 6

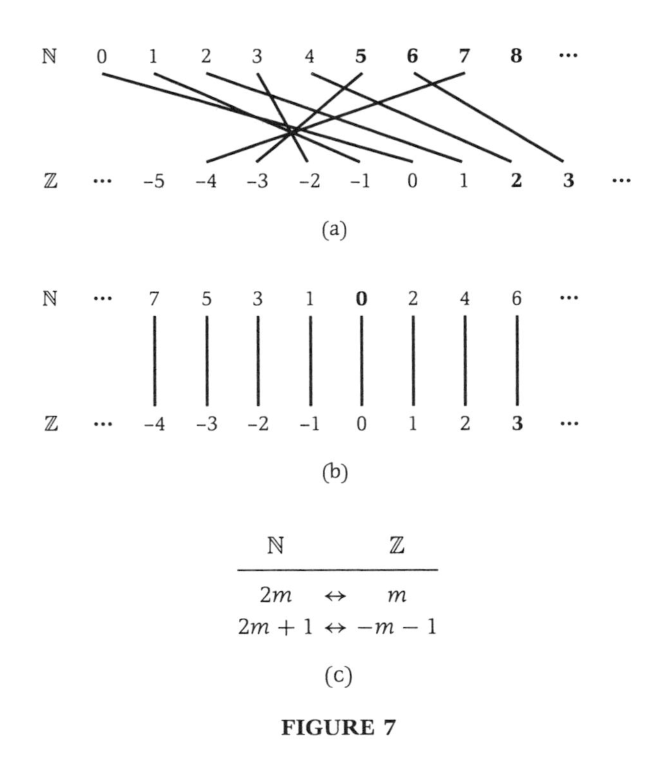

FIGURE 7

• • • **BREAK**

Try to find an infinite number of one-to-one onto mappings between $\mathbb{N}$ and $\mathbb{Z}$.

Going back to $\mathbb{N}$ and $\mathbb{O}$ for a moment, we guess that you probably came up with a mapping which sent a given natural number n to $2n + 1$. This is probably the obvious way to go, given what we've been saying. But there are other one-to-one correspondences that will work just as well, though they are probably more complicated. We show some of these in Figure 8.

We'll leave you to show that each mapping is one-to-one and onto.

• • • **BREAK**

Show that each of the mappings of Figure 8 is one-to-one and onto.

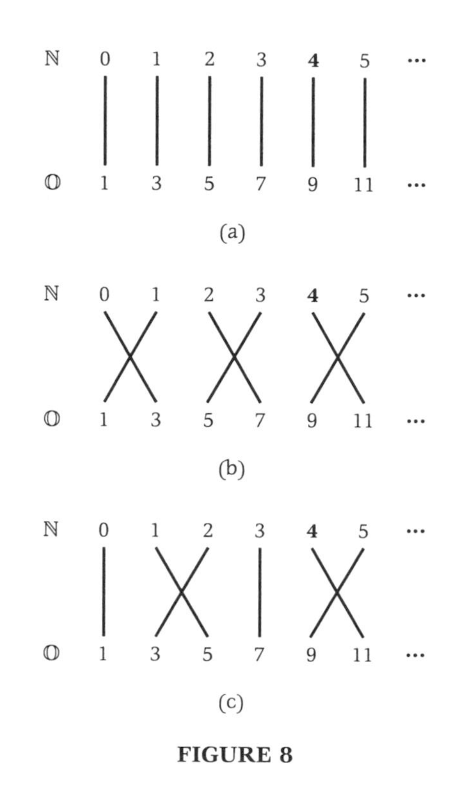

FIGURE 8

And now to the rationals. We won't go into the details here but instead refer you to Figure 9. Can you see how to settle the question of the cardinality of $\mathbb{Q}$ using that diagram? You have to be a little careful and notice that a given rational will occur more than once in the array.

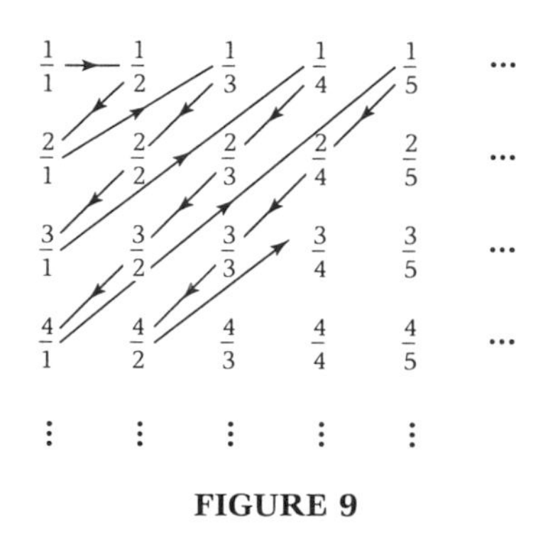

FIGURE 9

Fine, so now we know that $\mathbb{N}$, $\mathbb{E}$, $\mathbb{O}$, $\mathbb{Z}$, and $\mathbb{Q}$ all have cardinality $\aleph_0$. But what about the real numbers $\mathbb{R}$? The first problem with $\mathbb{R}$ is how to describe this set. Since $\mathbb{R}$ is the set of all real numbers, then, as we said earlier, it is the collection of all infinite decimal numbers. But, having got that far, it's not clear yet how to put the elements of $\mathbb{R}$ down in a respectable orderly list. They don't seem to want to be arranged nicely.

To see what we mean, have a look at Figure 10. There we have started out at zero and moved up, 0.01 at a time. Clearly there are lots of gaps. For a start we have totally ignored the negative real numbers, but let's not worry about that for the moment. Then we have put dots after 0.30 to indicate that the reals keep going after 0.30. But we have obviously omitted quite a few real numbers between 0 and 0.01. How to list these numbers systematically is the main problem exercising our minds right now. Can you help? Can you see a way of writing down all the real numbers in some systematic fashion? When that's done we can try to get a one-to-one correspondence between $\mathbb{N}$ and $\mathbb{R}$.

Of course, maybe we don't need a systematic way of writing down the reals in order to get a one-to-one correspondence between $\mathbb{R}$ and $\mathbb{N}$. Maybe we can dream something up off the top of our heads. (You never know your luck!) But how?

There are clearly lots of one-to-one correspondences between *parts* of $\mathbb{N}$ and $\mathbb{R}$. For instance, $1 \to 0.01$, $2 \to 0.02$, $3 \to 0.03$, and so on. Or there's $1 \to 0.1$, $2 \to 0.2$, $3 \to 0.3$, etc., as well. None of these give us $|\mathbb{R}| = |\mathbb{N}|$, which is what we'd be trying to show, because they are not mappings onto the set of *all* real numbers. So maybe there is a crazy correspondence, more like the ones in Figure 8, than the simple $n \to 2n$ that we found between $\mathbb{N}$ and $\mathbb{E}$. Let's put one down (Figure 11) and hope we hit the target. We'll stick to the positive reals for a start, as we said.

Oh, of course, we need to be a little more precise than this. The problem is that decimals can go on forever. So in our list, where

$\mathbb{R}$	0.00	0.01	0.02	0.03	⋯	0.09
	0.10	0.11	0.12	0.13	⋯	0.19
	0.20	0.21	0.22	0.23	⋯	0.29
	0.30	⋮	⋮	⋮	⋮	⋮

FIGURE 10

$$
\begin{array}{rcl}
1 & \leftrightarrow & 194.3726 \\
2 & \leftrightarrow & 0.0543219 \\
3 & \leftrightarrow & 2604.11112667 \\
4 & \leftrightarrow & 5.24 \\
5 & \leftrightarrow & 76629123456273491 \\
6 & \leftrightarrow & 888.8888 \\
7 & \leftrightarrow & 11.1942\overline{234} \\
8 & \leftrightarrow & 16.0 \\
9 & \leftrightarrow & 121.1110908076565421 1 \\
10 & \leftrightarrow & 15.256783145 \\
11 & \leftrightarrow & 99.9999999999999\overline{9} \\
12 & \leftrightarrow & 4471.236 \\
\vdots & \vdots & \vdots
\end{array}
$$

FIGURE 11

we've stopped without a bar over any of the numbers, we assume that the only numbers from there on in the decimal representation are zero. By 194.3726, then, we mean 194.37260000. . . with the zeros going on forever.

On the other hand, where there is a bar over a number or numbers, this implies repetition. We've put bars over the part to be repeated. This means that $11.1942\overline{234}$ is actually 11.1942234234234234. . . with the "234" repeating forever.

One final little problem with real numbers is that $99.\overline{9}$ (i.e., 99 followed by an infinite number of 9's) is the same as 100. It's clear then, that both $99.\overline{9}$ and 100 shouldn't be on our list. To make our list as simple as possible, we agree to put 100 and not $99.\overline{9}$ and we'll do this everywhere we come across a repeated 9's situation. In this way, every real number has a *unique* representation as an infinite decimal.

OK, then. We've set up a magnificent one-to-one correspondence in Figure 10. Are you convinced that *every* positive real is paired to *every* natural number in the required fashion? Things don't look too bad do they? OK, so you don't see your favorite number 33947.128206437915, say), in Figure 11. But we can assure you that we had that in mind to pair up with 13. You say we can't accommodate *all* your favorite numbers at 13? Alright then, we'll take them as you send them in to us on a "first come, first served" basis. The first favorite number gets 13, the next 14 and so on. Are you feeling happier now about our extended list?

• • • **BREAK**

Is it clear that somewhere on that list we've included every *possible* real number?

Right then. So we've extended Figure 11 so that, on the left-hand side, we've put *all* of the natural numbers down in order. On the right-hand side we think we've got *all* of the positive real numbers, in no particular order but they are *all* there. But are they?

Now let's construct a real number. Looking at the first number in the list, we see a 3 in the first decimal place. Let's choose a 0 for the first decimal place of the number we're constructing. The second number on our list has a 5 in the second decimal place, so let's choose a 0 again for the second decimal place of our number.

Keep on going like this. We'll construct a decimal $0 \,.\, a_1a_2a_3 \ldots$, where the number in the kth decimal place, a_k, is zero if the kth decimal place of the kth number in our list is not zero. On the other hand, take $a_k = 1$, if the kth number has a zero in the kth decimal place.

The first twelve decimal places of this new number we're producing gives a decimal equal to 0.000101010111.

But can you see where this new number of ours is on the list? It can't be the first number, because it is different from the first number in the first decimal place. It can't be the second number, because it is different from the second number in the second decimal place. It can't be the third number, because it differs from the third number in the third decimal place. It can't be the fourth number, because ....

We've got a *new* positive real number that wasn't on our list! This contradicts the assumption that *all* positive real numbers were on our list. Hence they weren't. The list we started so confidently in Figure 10 couldn't have been a complete list after all. Indeed, *any* such list that we cared to write would be incomplete for exactly the same reason. The positive reals, $\mathbb{R}^+$, and the natural numbers do *not* have the same cardinality. Hence $|\mathbb{R}^+| \neq |\mathbb{N}|$! So there is, indeed, more than one size of infinity! With an obvious definition of "greater than" for cardinals, we may plainly conclude that $|\mathbb{R}^+| > |\mathbb{N}|$.

Incidentally, the argument we used to show that the reals have a larger cardinality than the naturals is called a ***diagonal*** argument.

You can probably see why. As we constructed the new number, we moved down the list of Figure 11 on a diagonal.

The diagonal argument was initially formulated by Georg Cantor in 1891. It has been used many times since in other situations. For instance, it can be used to show that *the set of subsets of a set is bigger than the original set*, so that there is no largest infinity. There must then be infinitely many infinities (see [3]).

7.5 SCHRÖDER AND BERNSTEIN

In the last section we set out to look at the relation between $|\mathbb{R}|$ and $|\mathbb{N}|$ but, in fact, only succeeded in showing that $|\mathbb{R}^+| > |\mathbb{N}|$. However, since $\mathbb{R}^+ \subset \mathbb{R}$, then $|\mathbb{R}| \geq |\mathbb{R}^+| > |\mathbb{N}|$. So $|\mathbb{R}|$ is indeed greater than $|\mathbb{N}|$. But we still have the problem of deciding whether $|\mathbb{R}|$ is equal to $|\mathbb{R}^+|$ or not. In Question 3 at the end of this chapter, we see how to do this.

Since $\mathbb{N} \subset \mathbb{R}^+ \subset \mathbb{R}$, and $|\mathbb{N}| < |\mathbb{R}^+| \leq |\mathbb{R}|$, it seems that we should look at sets between $\mathbb{N}$ and $\mathbb{R}^+$ to see whether or not there are any whose cardinality is the same as $\mathbb{R}^+$. For example, what about a set like $T = [-1, 1] = \{x: -1 \leq x \leq 1\}$? Is there a one-to-one onto mapping from $\mathbb{R}^+$ to T?

In turns out that we don't have to go that far. Schröder and Bernstein have, in a sense, taken the "onto" out of one-to-one correspondences. They discovered that if you want to show that sets A and B have the same cardinality, then you don't have to find a one-to-one *onto* mapping from one to the other. All you have to do is to find a one-to-one mapping from A to B and a one-to-one mapping from B to A. It turns out that it's usually easier to find one-to-one mappings that aren't onto than those that are, so we gain even if we do have to find two mappings (one from A to B, and one from B to A), rather than one "onto" mapping.

The Schröder–Bernstein Theorem *Let A and B be sets. If there is a one-to-one mapping from A to B and a one-to-one mapping from B to A, then there is a one-to-one correspondence between A and B.*

• • • **BREAK**

What is the intuition behind the Schröder–Bernstein Theorem?

Let's illustrate this by showing that $\mathbb{R}^+$ and $[-1, 1]$ have the same cardinality. Now $f: [-1, 1] \to \mathbb{R}^+$, with $f(x) = x$, is a perfectly good one-to-one mapping from $[-1, 1]$ to $\mathbb{R}^+$. (But it is certainly not an *onto* mapping.) On the other hand, $g: \mathbb{R}^+ \to [-1, 1]$, with $g(x) = \arctan x$, is also one-to-one (but not onto). According to our friends Schröder and Bernstein then, there exists a one-to-one *onto* mapping between $[-1, 1]$ and $\mathbb{R}^+$. These sets then have the same cardinality.

7.6 CARDINAL ARITHMETIC

How do you add together two cardinals, whether they are finite or infinite? Perhaps that's a little too hard straight off the bat. So first let's think about how we add two finite numbers. Adding two finite numbers together is something that we do all the time without any thought. When you think about it though, to add 5 and 7, say, what you really do is to take a set with five elements and a set with seven elements, take their union and count how many elements there are in the combined set. Well, actually you have to be a bit more careful than that. You have to make sure that the smaller sets are *disjoint* or else you'll undercount the sum. So, to add two finite numbers a and b together, we take a set A, of size a, a disjoint set B of size b. Then $a + b$ is just the size of the set $A \cup B$.

There's no reason not to extend this idea to include infinite sets. To add together the cardinals a and b, we'll simply find two disjoint sets A and B and take their union. The cardinality of $A \cup B$ is $a + b$.

Given that definition of the addition of cardinals, let's add $\aleph_0$ to itself. The idea is to find two disjoint sets whose cardinalities are $\aleph_0$. Next we take the union of these two sets and work out the cardinality of the union. This cardinality will be $\aleph_0 + \aleph_0$. What we have to do now, is to find two disjoint sets whose cardinality is $\aleph_0$. That's not too difficult. After all, $|\mathbb{O}| = \aleph_0$ and $|\mathbb{E}| = \aleph_0$ too. The nice thing about these two sets is that their union is $\mathbb{N}$. Now we know that $|\mathbb{N}| = \aleph_0$, so we have reached the inescapable conclusion that $\aleph_0 + \aleph_0 = \aleph_0$. That never happens with finite numbers. Well, not non-zero ones, anyway.

• • • BREAK

1. So $\aleph_0 + \aleph_0 = \aleph_0$. What does $\aleph_0 + n$ equal, where n is any finite cardinal?
2. Verify that our rule for adding cardinals makes sense; that is, the cardinality of $A \cup B$ is unchanged if A, B are replaced by disjoint sets A', B' such that A, A' and B, B' have the same cardinality.

We've actually already shown that $\aleph_0 + 1 = \aleph_0$. Look back to see the cardinality of $\mathbb{N} \cup \{-1\}$. Then $\aleph_0 + n$ shouldn't be too much more difficult.

The interesting thing now is that we can find the cardinality of $\mathbb{Q} \cup \{\pi\}$, without any effort. We know that $\mathbb{Q}$ has cardinality $\aleph_0$, and we know that $\{\pi\}$ has cardinality 1. Since $\mathbb{Q}$ and $\{\pi\}$ are disjoint, the cardinality of $\mathbb{Q} \cup \pi$ is just $\aleph_0 + 1$, which we have just shown to be $\aleph_0$. Bingo! $|\mathbb{Q} \cup \{\pi\}| = \aleph_0$.

7.7 EVEN MORE INFINITIES?

It's bad enough having infinity in this world without having to have *two* infinities. Maybe a worse thought is that there are other infinities as well. Possibly there are an *infinity* of infinities. If there are—and Cantor's argument shows that there are—is that infinity of infinities the infinity of $\mathbb{N}$ or the infinity of $\mathbb{R}$, or some other infinity altogether?

But cardinality gets curiouser and curiouser. It can be proved that there are as many points in the Cartesian plane as there are points on the x-axis. That's the same as saying that there are as many points in the plane as there are real numbers. Having got that far, it will come as no great revelation that there are as many points in three-dimensional space as there are real numbers. In fact, why stop at *three* dimensions?

This business about $\mathbb{R}$ and the plane having equal cardinalities has a strange (no, totally weird and unexpected is more the way to describe it) consequence. If there are as many points on the real line (or x-axis) as there are points in the plane, then it should be possible to *cover* all the points in the plane by a single line (which, of course, could not be straight)!

That's fairly mind-blowing. Does that mean that you can draw a single line through all the points in the plane? Yes, it does. Such lines are called ***space-filling curves***. Peano came up with the first one of these in 1890.

You'll notice we said lines, plural. There are actually an infinite number of space-filling curves. We've tried to show, in Figure 12, one that was devised by Hilbert.

We should add that we are actually only filling the unit square with a line here. However, since there are as many reals in the interval from 0 to 1 as there are overall, this is not a problem.

This space-filling curve is defined iteratively. We have only given the first three steps. You might like to see if you can perform the next iteration. The final space-filling curve is the limit of this iterative process. This all sounds a bit like Koch curves and fractals (see Chapter 8).

• • • **BREAK**

Is the above space-filling curve a fractal? If not, are some fractals space-filling curves?

But what about sets with cardinality greater than that of $\mathbb{R}$? As we have suggested earlier, one way to find a set with larger cardinality than a given set, is to form the set of all its subsets. Thus, if X is a

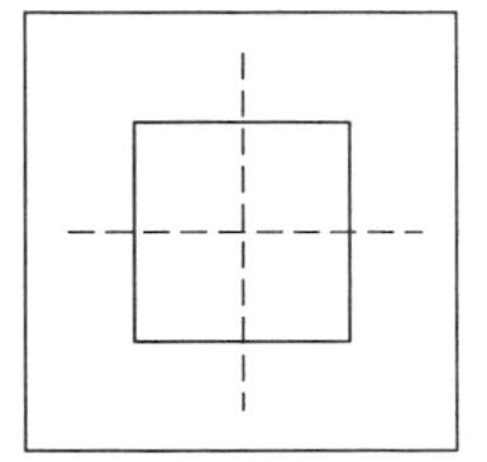
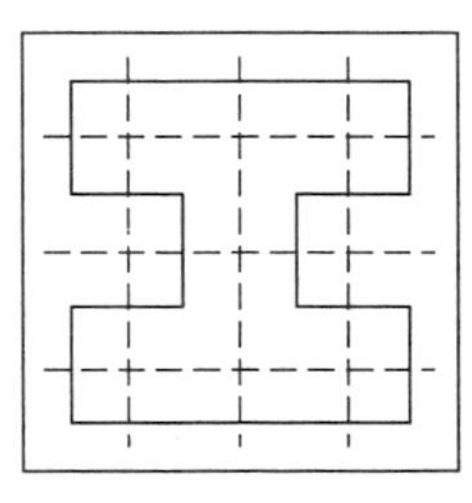
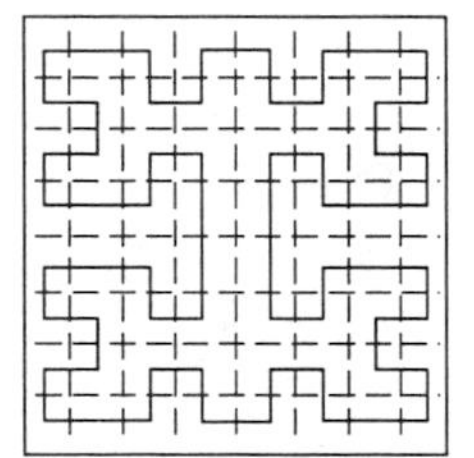

FIGURE 12

given infinite set, then the set of subsets of X has bigger cardinality than X. Using this approach then, it should be clear that there is never any biggest infinite set. Just when you thought you had got such a set, you can get a bigger one by taking the set of all its subsets.

• • • **BREAK**

How many subsets does a set with n elements have?

Now because there is an infinity of infinities, different infinite cardinal numbers have been invented. We have already talked about $\aleph_0$, the smallest, infinite cardinal. From there we have the next cardinal numbers $\aleph_1$, $\aleph_2$, and so on. But does it make any sense to say the "next" cardinal number? Is there always another cardinal number between two given cardinals?

One of the interesting things here is that $|\mathbb{R}|$ is known as c. Now c stands for the ***continuum***, so-called because $\mathbb{R}$ (as compared to $\mathbb{N}$) appears to have no gaps in it. The big question here is, is $|\mathbb{R}| = \aleph_1$? It's all a matter of order, in the following sense. First of all, we know that $|\mathbb{N}| = \aleph_0$, $|\mathbb{R}| = c$, and $|\mathbb{N}| < |\mathbb{R}|$. But we don't know if there exists a set X such that $|\mathbb{N}| < |X| < |\mathbb{R}|$. It turns out, quite remarkably, that the existence of such a set actually depends on the axioms of set theory that we choose.

What is an axiom? Simply, it's an unproven assumption. In mathematics we need to start somewhere. We need to assume certain things exist and have certain properties. Then we can start proving theorems and generating useful algorithms.

In plane Euclidean geometry, for instance, we postulate the existence of points and lines and we axiomatize them by saying things like:

Axiom 1 *There is a unique line through any two distinct points.*

Axiom 2 *Any two nonparallel lines meet at a unique point.*

One of Euclid's controversial axioms concerned parallel lines. This was his fifth axiom.

Axiom 5 *Through any point not on a given line, there exists a unique line which is parallel to the given line.*

Now it turned out that this was indeed an axiom for Euclidean geometry, despite the fact that for centuries mathematicians had tried to show that it was a consequence of Euclid's previous four axioms. In the early nineteenth century, Bolyai and Lobachevsky, independently, came up with new geometries which were based on the first four axioms plus another axiom of their own (see [3]) which contradicted Axiom 5.

So it is important what axioms you use in geometry. In the same way it's important which axioms you use in set theory. With some axiom systems $|\mathbb{R}|$ is indeed equal to $\aleph_1$. In others, this is not the case. Mathematicians have to decide which axioms are most appropriate for what they are doing. Which axioms enable them to model which part of the real world the best? Having chosen their axioms though, certain consequences follow.

The most common axiom system for set theory is that of Zermelo and Frankel. In this axiom system, the so-called ***continuum hypothesis*** ($|\mathbb{R}| = c = \aleph_1$) is actually independent of the other axioms. This follows from work of Cohen and Gödel (see [1] and [2], respectively). They were able to show that the continuum hypothesis does not rely on normal set-theoretical axioms. It is therefore possible that it could play the same role in set theory that Euclid's Axiom 5 plays in geometry, so that different set theories from the one we are used to working with could arise. Only time will tell how important these other sets theories are.

• • • FINAL BREAK

Here are a few problems for you to try out your new skills on.

1. Given 37 natural numbers, is it true that 7 of them can always be chosen so that their sum is divisible by 7? (*Hint*: Use ideas from Chapter 2).

2. A chess master who has 11 weeks to prepare for a tournament, decides to play at least one game every day but, in order not to tire herself, she decides not to play more than 12 games in any 7-day period. Show that there exists a succession of days in which she plays exactly 21 games.

(*Hint*: Try to find $154 = 2 \times 77$ numbers that have to fit into 153 pigeon-holes.)

3. Show that $|\mathbb{R}| = |\mathbb{R}^+|$, where $\mathbb{R}^+ = \{r \in \mathbb{R} \text{ and } r > 0\}$.
4. Prove that $|\mathbb{R}| = |\{x: 0 \leq x \leq 1\}|$.
5. Prove, or find a counterexample to, the following statement:

   The cardinality of $\mathbb{R}^- \cup \mathbb{N}$ is equal to the cardinality of $\mathbb{R}$, where $\mathbb{R}^- = \{r \in \mathbb{R} \text{ and } r < 0\}$.
6. Complete the following addition table:

	n	$\aleph_0$	c
m	$m+n$	$\cdots$	$\cdots$
$\aleph_0$	$\cdots$	$\cdots$	$\cdots$
c	$\cdots$	$\cdots$	$\cdots$

7. Let $\mathbb{N} \times \mathbb{N} = \{(a, b): a, b \in \mathbb{N}\}$. Determine $|\mathbb{N} \times \mathbb{N}|$.

REFERENCES

1. Cohen, P., The independence of the continuum hypothesis I, II, *Proc. Nat. Acad. Sci.*, **50**, 1963, 1143–1148; **51**, 1963, 105–110, 1963.
2. Gödel, Kurt, The consistency of the axiom of choice and the generalized continuum hypothesis, *Proc. Nat. Acad. Sci.*, **25**, 1938, 220–224.
3. Stillwell, John, *Mathematics and Its History*, Springer-Verlag, New York, 1989.

ANSWERS FOR FINAL BREAK

1. Now any number has remainder 0, 1, 2, 3, 4, 5, or 6, when divided by 7. So we can reduce our arbitrary 37 numbers to these residue classes modulo 7. If every residue class is present among the 37 numbers, then we choose a representative of each class. Hence we have numbers whose residues sum to $0 + 1 + 2 + 3 + 4 + 5 + 6 = 21$ and this is congruent to zero modulo 7. So the seven numbers we chose had a sum divisible by 7.

   Hence we can suppose that one residue class is not represented among our 37 numbers. Since there are six residue classes left, by the pigeon-hole principle, one of these classes must occur at least seven times. If we take seven of this residue class and add them, the sum is divisible by seven.

2. The hardest part of this problem is setting it up. One approach is shown below.

   Let a_i be the number of games played up to and including the ith day. Then $a_1 < a_2 < a_3 < \cdots < a_{77}$. Now consider a_1, $a_2, a_3, \ldots, a_{77}, a_1 + 21, a_2 + 21, \ldots, a_{77} + 21$. This is a total of 154 numbers, the largest of which is $a_{77} + 21$.

   Now in any 77 days, the chess master plays at most $12 \times 11 = 132$ games. Hence $a_{77} \leq 132$ and so $a_{77} + 21 \leq 153$. By the pigeon-hole principle, two of the 154 numbers listed above, which are all between 1 and 153 must be the same. Hence for some i and j, $a_i = a_j + 21$. So $a_i - a_j = 21$. There must therefore be a string of days (from day $j + 1$ to day i) when 21 games are played.

3. Use $f\colon \mathbb{R} \to \mathbb{R}^+$, where $f(r) = e^r$.

4. Let $f\colon \mathbb{R}^+ \to \{x\colon 0 < x < 1\}$, where $f(r) = \frac{r}{r+1}$, and let

$$g : \{x\colon 0 \leq x \leq 1\} \to \{x\colon 0 < x < 1\},$$

   where $g(0) = \frac{1}{3}$, $g\left(\frac{1}{3^n}\right) = \frac{1}{3^{n+1}}$, $g(1) = \frac{1}{2}$, $g\left(\frac{1}{2^n}\right) = \frac{1}{2^{n+1}}$, $n \geq 1$ and $g(x) = x$, for other real values. Then use the Schröder–Bernstein Theorem.

   This can also be done by exploiting the equality between the cardinalities of $\mathbb{R}$, $\mathbb{R}^+$, $[-1, 1]$, and $[0, 1]$.

5. Send $\{x: -1 \leq x \leq 0\}$ to $\{x: -1 < x \leq 0\} \cup \mathbb{N}$ using the map which takes n to $-p_n$, the nth prime, and $-p_n^m$ to $-p_n^{m+1}$ and keeps everything else fixed.
6.

	n	$\aleph_0$	c
m	$m+n$	$\aleph_0$	c
$\aleph_0$	$\aleph_0$	$\aleph_0$	c
c	c	c	c

First note from the definition of the sum of two numbers, $a + b = b + a$. Since the commutative law holds, there is no need to worry about the entries under the diagonal.

Clearly adding m and n gives $m + n$.

Now take the disjoint sets $\mathbb{N}$ and $\{-1, -2, \ldots, -m\}$. There union has the same cardinality as $\mathbb{N}$. The mapping $f(r) = r - m$, from $\mathbb{N}$ to $\mathbb{N} \cup \{-1, -2, \ldots, -n\}$, is a one-to-one onto mapping. So $m + \aleph_0 = \aleph_0$.

Take $A = [0, 1]$ and $B = \{2, 3, \ldots, m\}$. We will compare the union of these two sets with $[0, 1]$ itself. Clearly $f: A \to A \cup B$, where $f(x) = x$, is a one-to-one map. If we can get a similar mapping from $A \cup B$ to A, then we can apply the Schröder–Bernstein Theorem. Try sending i to $\frac{1}{i-1}$, for $i \in B$; $\frac{1}{r}$ to $\frac{1}{m-r+1}$, for $r \in N$; and everything else just maps to itself. (Is that map onto as well?) So $m + c = c$.

We have already seen that $\aleph_0 + \aleph_0 = \aleph_0$. (Use the fact that $\mathbb{O}$ and $\mathbb{E}$ are disjoint and their union is $\mathbb{N}$.)

Look at $[0, 1] \cup \{2, 4, 6, \ldots\}$. The mapping to $[0, 1]$ which takes $2n$ to $\frac{1}{2n}$, $\frac{1}{n}$ to $\frac{1}{2n+1}$, and maps all other elements to themselves, is then a one-to-one mapping of $[0, 1] \cup \{2, 4, 6, \ldots\}$ into $[0, 1]$. Since $[0, 1] \subset [0, 1] \cup \{2, 4, 6, \ldots\}$, this shows that $c + \aleph_0 = c$.

For $c + c$ just take $[0, 1] \cup (1, 2]$.

7. Work out a mapping based on $(1, 1) \to 1, (1, 2) \to 2, (2, 1) \to 3, (3, 1) \to 4, (2, 2) \to 5, (1, 3) \to 6, \ldots$. (Compare this with Figure 9.)

CHAPTER 8 An Introduction to the Mathematics of Fractal Geometry

8.1 INTRODUCTION TO THE INTRODUCTION: WHAT'S DIFFERENT ABOUT OUR APPROACH

Among modern topics of mathematics, fractal geometry and the associated theory of chaos have attracted unusual public notice. No doubt the availability of spectacular computer graphics has had the effect of popularizing these theories, in the media and in popular science literature.

We are, of course, very much in favor of good popularization of science and mathematics—and we like pretty pictures, too! However, it has struck us that most of the expositions intended for an audience, or a readership, not averse to mathematics, have also tended to err on the popular, more superficial side. Authors and speakers have stressed applications to the study of coastlines, to the growth of trees and to predator–prey interactions in human and other animal societies, and have shown beautiful, computer-generated colored pictures which have often impressed without conveying real understanding. Thus there has been rather little available[1] in the important area between the specialized publications of research mathematicians active in the field and

[1] Among the best is [7].

popular accounts of the applications in which no real mathematical understanding has been conveyed—or perhaps even intended.

It is our object in this chapter to make a small contribution to filling this gap. It is not our intention to provide a rigorous mathematical treatment—notice the word ***Introduction*** in the title of this chapter. We emphasize that we are not experts in the field (we would particularly like to stress our indebtedness to the texts of Robert L. Devaney [2] and Kenneth Falconer [3]); but we believe that the *mathematics* of fractal geometry should be accessible to the readers of this book, and we are very much concerned to put to rest the dangerous and fallacious view that one can only make a topic interesting by suppressing its mathematical content.[2]

This chapter is structured as follows. In Section 2 we describe how certain familiar, and classical, fractals arise from self-similar transformations, and how one might attempt to attach to such fractals a *(self-similarity) dimension* which would seem better to reflect their density than ordinary topological dimension. In Section 3 we go into details about one particular self-map of the real line $\mathbb{R}^1$, the *tent map*. We explain how the study of the iterations of this map leads naturally to the identification of the *Cantor middle-thirds set* F as the collection of exceptional ("chaotic") points of this map, in the sense that all other points of $\mathbb{R}^1$ are attracted to $-\infty$, while the points of F are endlessly moved around amongst themselves. We also analyze, in Section 3, a general mathematical model, of which that created by the tent map was a special case, so that we have a purely *qualitative* explanation of the phenomena noticed earlier. In particular, we apply our analysis to the *logistic map*, involving the function $\lambda x(1 - x)$, which has many features in common with the tent map and which has long been recognized as an important model in population studies. In this section, we confine our discussion of the logistic function to the situation where $\lambda > 4$, postponing the examination of what happens when $0 < \lambda < 4$ until Section 4.

Section 4 intensifies the mathematics (but we will try to make it as readable as possible)—that is, it explains why the processes

[2] In his record best-seller, *A Brief History of Time*, the author, Stephen Hawking, writes that he was advised against including any equations in his book in order not to reduce the sales, but that he finally restored just one equation. It is not our judgment that this abstemiousness improved the comprehensibility of the book, even if it improved sales.

described in Sections 2 and 3 work as they do and why they fail to work for the logistic map with $0 < \lambda < 4$, and what happens in such a situation.

Thus we begin Section 4 by setting up theories of which the examples of Sections 2 and 3 are particular—and fairly simple—cases. First, we introduce the *Hausdorff dimension*, and show that this is the concept lying behind our ad hoc attempts at framing a self-similarity dimension for the fractals discussed in Section 2. Second, we discuss the fractal associated with a finite set of contractions (for which the Cantor set of Section 2 is an example). Third, we look at the extraordinary dependence of the behavior of iterates of the logistic function $\lambda x(1 - x)$ on the parameter λ where $0 < \lambda < 4$. We close the third part of this section by exhibiting a common context for the (generalized) tent map and the logistic map and detailing some of its remarkable features.

The inquiring reader may well ask at this point, *What precisely is a fractal?* It is remarkable how often this question is avoided in the literature.[3] Some authors seem to use it to describe a set whose Hausdorff dimension is not an integer (though it is very unlikely to be a fraction);[4] others get nearer the mark by using it for sets which are "strange attractors." We would wish to be candid—our view is that the property of being a fractal should not be an *intrinsic* property of the set in question but rather a property of the role it plays in the analysis of the behavior of iterates of a self-map f and of the (finite) family of branches of its inverse f^{-1}. Thus Theorem 2 comes very close, in our view, to providing the framework for an effective working definition. However, since experts have not, usually, adopted this point of view, we have decided not to insist on a precise definition of a fractal in this book. In addition, we do recognize that their "strangeness" is part of the appeal of fractals and therefore, in Section 2, we describe some characteristic

[3]Even Falconer [3] is somewhat devious! His index lists "fractals, definition of xviii–xxii." But these pages contain no definition. He attributes the term to Mandelbrot who used it to refer to objects exhibiting significant geometric irregularity. Devaney [2] is more candid; neither of the terms "definition of a fractal," "fractal, definition of" occurs in the index—and, indeed, he gives no precise definition.

[4]Mandelbrot defines fractal, in [6], as a set for which the Hausdorff dimension is strictly greater than the topological dimension, but he envisages the possibility that the definition may be changed. We will not ourselves adopt Mandelbrot's definition, since we use standard sets to motivate our definition of the self-similarity dimension.

features of some of the fractals which have appeared in popular literature.

We would also like to comment that, while the Hausdorff dimension provides a perfectly secure logical basis from which to describe and explore the fractals F arising from self-similar transformations as in Section 2, we should openly acknowledge that it has the disadvantage of being virtually incalculable except in such elementary examples. Theorem 2 does provide a context in which one can obtain arbitrarily good rational approximations to the Hausdorff dimension—but we do not believe it possible to increase the generality significantly and retain effective computability.

For further reading, the reader may consult [1, 4, 8].

8.2 INTUITIVE NOTION OF SELF-SIMILARITY

We begin with four specific examples of fractals. In each of these examples the fractal F appears as the limit of an iterative process which produces stages F_k, $k \geq 0$, and which has the following three important characteristics:

(a) $\{F_k\}$ is self-similar; that is, if at any stage we look at a small portion of F_k, $k \geq 1$, under suitable magnification, that part will appear identical to a portion of F_{k-1};

(b) $\{F_k\}$ is defined recursively; that is, the rule for obtaining F_{k+1} assumes the construction of F_k and is independent of k;

(c) F has a fine structure; that is, given any open subset U of the ambient Euclidean space, however small, the intersection $U \cap F$ reveals the special structure of F.

You will probably understand (c) better when we discuss the examples below. Typically, a portion of the Euclidean plane $\mathbb{R}^2$ does *not* have fine structure—if we take a small quadrilateral in $\mathbb{R}^2$ and just look at the part of $\mathbb{R}^2$ near to it, we cannot know whether we're on the surface of a sphere or in the plane (remember, there are still many Flat Earthers around!).

As you will see in our examples, despite the regularity of the construction there is something elusive about F; for in every interesting (i.e., nontrivial) case:

> (d) ***The geometry of F is not easily described in classical terms.***

Before we give examples, we recall some important notation. On the real line, the ***open interval*** $a < x < b$ is written (a, b), and the ***closed interval*** $a \leq x \leq b$ is written $[a, b]$. Sometimes we consider ***half-open*** intervals, and use the obvious notation. (There are examples of these ideas and the notation in Chapter 7.)

Our first example is the ***Cantor middle-thirds set***. The construction begins with F_0, the unit interval $[0, 1]$ as shown on the first level of Figure 1. The first iteration deletes the middle-third of the interval $[0, 1]$, that is, the open interval $\left(\frac{1}{3}, \frac{2}{3}\right)$ is removed. This leaves F_1, the union of the two closed intervals shown in the second level of Figure 1. The iteration process continues, deleting the middle third of each of the intervals $\left[0, \frac{1}{3}\right]$ and $\left[\frac{2}{3}, 1\right]$, to obtain the four segments shown in the third level of Figure 1. Continuing in this way we obtain, in the limit, the set known as the Cantor middle-thirds set, or, more simply, the ***Cantor set***.

Observe that the sum of the lengths of the subintervals making up F_k is $\left(\frac{2}{3}\right)^k$; this quantity clearly approaches 0 as $k \to \infty$. Thus it might appear, at first glance, as though there would be no points remaining in the Cantor set. However, we know that we never eliminate any of the subinterval endpoints that occur along the way!

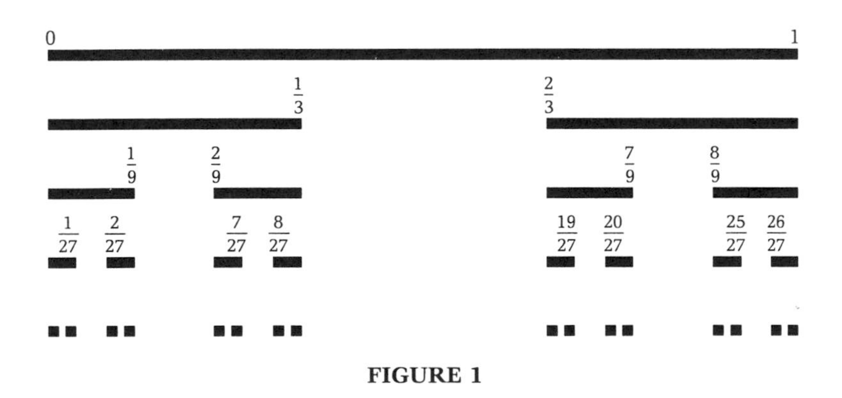

FIGURE 1

All the points in the Cantor set may be written as infinite decimals, in base 3, consisting of a decimal point followed by a sequence[5] of 0's and 2's. To argue the truth of this last statement observe that, for $k \geq 1$, the kth iteration removes all those numbers whose decimal representation (in base 3) has the digit 1 in the kth place after the decimal point. Thus it is also true, conversely, that all infinite decimals consisting only of 0's and 2's are members of the Cantor set.

In fact, the decimal representation (in base 3) of points in the Cantor set gives exact information as to where the point is located. For example, a number which begins

.0 must be to the left of the interval $\left(\frac{1}{3}, \frac{2}{3}\right)$;
.00 must be to the left of the interval $\left(\frac{1}{9}, \frac{2}{9}\right)$;
.000 must be to the left of the interval $\left(\frac{1}{27}, \frac{2}{27}\right)$;
etc.

Similarly, a number which begins

.2 must be to the right of the interval $\left(\frac{1}{3}, \frac{2}{3}\right)$;
.22 must be to the right of the interval $\left(\frac{7}{9}, \frac{8}{9}\right)$;
.222 must be to the right of the interval $\left(\frac{25}{27}, \frac{26}{27}\right)$;
etc.

Now let us consider a slightly more complicated example (it may help to refer to Figure 1). If a number begins

.020. . ., then[6]

– the first digit (0) tells us that the point lies in the ***L***eft-hand subinterval in F_1; call this L_1;
– the second digit (2) tells us that the point lies in the ***R***ight-hand subinterval created from L_1 in forming F_2; call this R_2;

[5]Some points of the Cantor set could be written with 1's. For example, $\frac{1}{3} = .1$, but $\frac{1}{3}$ can also be written as .0222 . . . .

[6]Notice how we use the self-similarity to facilitate our explanation.

- the third digit (0) tells us that the point lies in the ***L***eft-hand subinterval created from R_2 in forming F_3; call this L_3;
- ....

A little reflection on this last example will indicate how to write the base 3 representation of any number in the Cantor set if its location can be determined with regard to the deleted parts at each stage of the iteration. As a special case, consider this example. The point $\frac{1}{3}$ is to the ***L***eft of the first removed middle third, and thereafter it is always to the ***R***ight of the portion removed (in the piece to which it was attached). Thus its successive locations may be abbreviated by the sequence ***LRRR*** .... To get the base 3 decimal representation of $\frac{1}{3}$, systematically replace ***L*** by 0, ***R*** by 2, and affix "." at the beginning. Thus we obtain, as expected, $\frac{1}{3} = .0\overline{2}$. (Remember, the overlined portion is to be repeated indefinitely.)

It is not difficult to see that endpoints from the successive iterative stages must terminate with an infinite sequence of 0's or 2's. But, of course, there must also exist points in the Cantor set located so that they would be represented by an infinite decimal (in base 3) with a *periodic repeat* of the digits 0 and 2 (in which case the point in question represents a *rational* number); and there must exist points located so that the digits in their base 3 representation never repeat (in which case the point in question represents an *irrational* number).

The observation that the points of the Cantor set may be represented as decimals in base 3 (with digits 0 and 2) has a particularly interesting immediate consequence. For we see that, in an obvious way, the points of the Cantor set may be mapped onto the set of *all* points of the original unit interval (*Hint*: Think of the points of the unit interval expressed as decimals in base 2). This shows that, though the Cantor set is so thin that it contains no subintervals, nevertheless, it retains the cardinality of the original unit interval! (This is called the ***cardinality of the continuum***, and written c; see Chapter 7.) Let us just explain characteristic (c), from the list at the start of this section, in this context. The Cantor set (which cannot be drawn) has the property that each of its points lies in an arbitrarily small interval whose endpoints are *not* in the Cantor set. Plainly this extraordinary property, in a set with the cardinality of the continuum, is a *local* property describing the fine structure of the Cantor set.

The convergent sequence $\{F_k\}$ described, together with its limiting Cantor set F, possesses the characteristics (a) (with magnification factor 3), (b), (c), and (d).

Now comes an interesting question. We know that the unit interval we started with was one-dimensional in the usual topological sense and, in fact, so was each of the stages F_k. It is also easy to see that, in the limit, the sum of the lengths of the segments which were removed from the original unit interval was 1; yet, having removed these segments, there appear to be just as many points remaining as there were in the original unit interval! Nevertheless, the topological dimension has dropped suddenly from 1 to 0 as we pass from F_k to its limit F. Thus perhaps the topological dimension is inadequate for describing F, and the question as to what would be a reasonable definition for a dimension of the *limit* (Cantor) set is an intriguing one. Our reflective readers may feel that there are grounds for believing that such a dimension should exist and that its value should lie somewhere between 0 and 1—but precisely where between 0 and 1?

Notice that the passage from F_k to F_{k+1} can also be viewed from the following alternative point of view. Each line segment in F_k is replaced by 2 line segments in F_{k+1} whose lengths are reduced by a factor of 3. We will write r for the reduction factor, and m for the number of reduced copies replacing a single copy at the previous stage. So, for the Cantor set, $r = 3$ and $m = 2$.

When we return to the problem of assigning a new sort of dimension to F the quantities r and m will play a key role.

Our second example is the ***Koch Snowflake***. The construction begins with an equilateral triangle (boundary only) with side of length 1, as shown in Figure 2(a). The first iteration involves removing the center third of each side of the original triangle, replacing it with two line segments of length $\frac{1}{3}$ as shown in Figure 2(b). Notice that whereas the triangle in Figure 2(a) has a perimeter of length 3, the polygon in Figure 2(b) has a perimeter of length $(3)(4)(\frac{1}{3})$. The iteration process continues by removing the center third of each of the sides of the (nonconvex) 12-gon

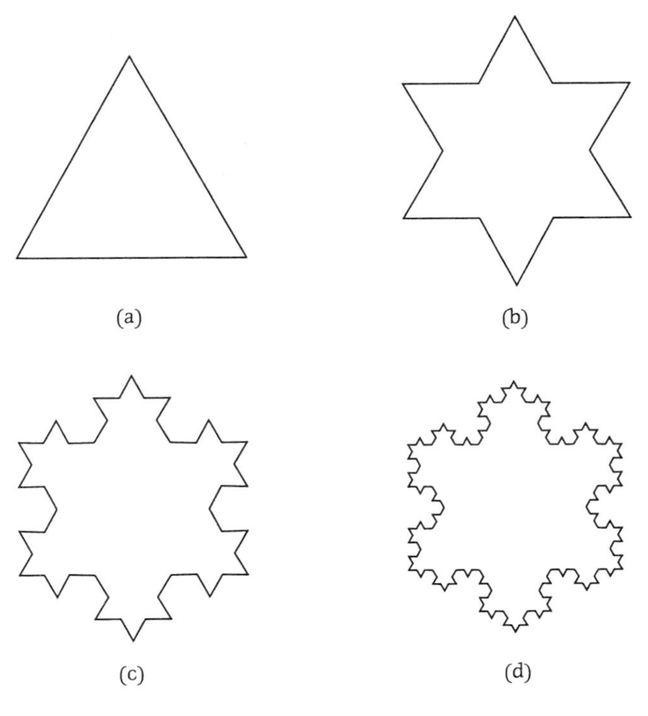

FIGURE 2

and replacing that segment with two line segments of length $\left(\frac{1}{3}\right)^2$ as shown in Figure 2(c). Once again, the perimeter has been multiplied by $\frac{4}{3}$, so Figure 2(c) has a perimeter of length $\left(\frac{4}{3}\right)^2 3$. Continuing in this way we obtain Figure 2(d),. . ., and, in the limit, the curve known as the ***Koch Snowflake***.

Observe that the length of the perimeter at the stage F_k of constructing the Koch Snowflake is given by $\left(\frac{4}{3}\right)^k 3$, and this quantity obviously approaches ∞ as $k \to \infty$. On the other hand, we can see that the area bounded by the curve is finite—encompassing, in fact, no more than the area of the circumcircle of the original triangle.

Again, we have a construction $\{F_k\}$, with $F_k \to F$, which clearly possesses the characteristics (a) (with reduction factor 3), (b), (c), and (d). Adopting the alternative point of view (see the Cantor set example) we see that, for the Koch Snowflake, $r = 3$ and

$m = 4$, since, at each stage, each line segment is replaced by 4 line segments whose lengths have been reduced by a factor of 3.

In this construction, F_k is plainly one-dimensional, yet because its length approaches infinity in the limit, it may seem reasonable to expect that a type of dimension could be defined for the Koch Snowflake which would be a number greater than 1 (which remains the topological dimension of F). Again, we will discuss precisely how this can be done at the end of the section.

In our third example, the ***Sierpinski Triangle*** (or ***Sierpinski Gasket***), we step up the topological dimension. The construction begins with a black equilateral triangle, taken with its interior, of a given area, say A, as shown in Figure 3(a). The first iteration involves removing a central triangle, whose vertices lie at the midpoints of the sides of the original triangle, as shown in Figure 3(b). Notice that the total area of the three remaining black triangles in Figure 3(b) is $\left(\frac{3}{4}\right)A$. The iteration proceeds by removing, as before, the central triangle from each of the three black triangles to obtain the configuration shown in Figure 3(c). The total area of the 9 black triangles in Figure 3(c) is $\left(\frac{3}{4}\right)^2 A$. Continuing in this way we obtain Figure 3(d), ..., and, in the limit, the set of points in the plane

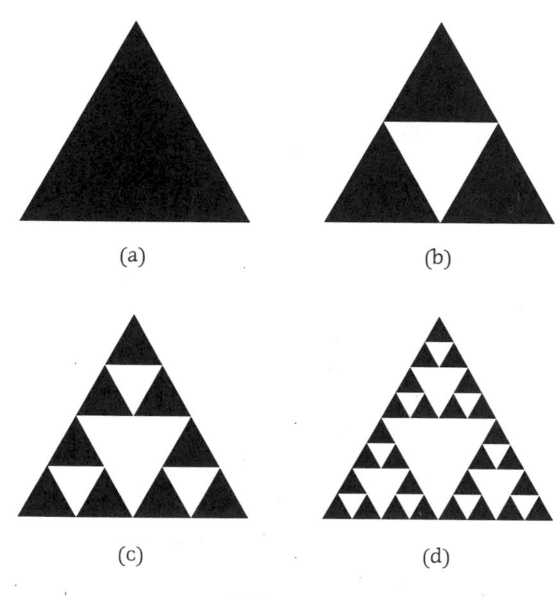

FIGURE 3

known as the Sierpinski Triangle. Observe that the total area of F_k, which consists of the black portions surviving at the kth stage from the original triangle, is $\left(\frac{3}{4}\right)^k A$, and this quantity approaches 0 as $k \to \infty$. Observe also that the one-dimensional line segments on the boundary at every stage must remain in the limit set.

Once again, we have a sequence $\{F_k\}$ with limiting set F which clearly possess the characteristics (a) (with reduction factor 2), (b), (c), and (d). Here $r = 2$ because the *linear* reduction,[7] that is, the reduction in the length of a side of a triangle is 2; and $m = 3$.

Now, on the one hand, we know that the Sierpinski Triangle contains one-dimensional lines, and, as shown above, in the limit the total area of the black triangles approaches zero. On the other hand, we know that, at each stage of the iterative process, the individual triangles are, in fact, two dimensional. Clearly, we may reasonably hope that we can attach to the Sierpinski Triangle a dimension between 1 and 2. Should it be closer to 1 or to 2?

Figure 4 shows two iterations of a ***box-fractal***, in which the iterative process involves, in an obvious way, the removal of $\frac{4}{9}$ of the existing black square(s) at stage k to produce the $(k + 1)$st stage. Except for its shape (and the fact that the reduction factor is now 3), it is entirely analogous to the Sierpinski Triangle and, consequently, we may reasonably expect that it also should have a dimension between 1 and 2. Should the dimension of the box-fractal be larger or smaller than the dimension of the Sierpinski Triangle—or perhaps even equal to it? Notice that $r = 3$, $m = 5$ for the box fractal.

It is common ground that sets F obtained in the way we have described *are* fractals. We now turn to the main mathematical question raised at the end of each of these four examples. That is, how might we reasonably try to define a dimension for a fractal F which

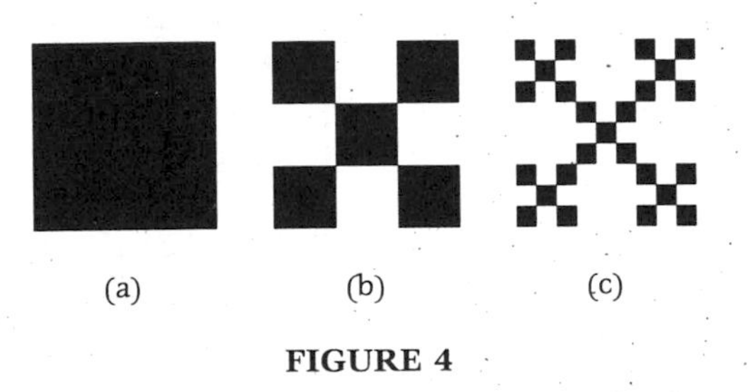

FIGURE 4

[7]Scaling factors are always linear.

satisfies the conditions (a) through (d) given at the beginning of this section. Of course, we would not want the dimension of F to *depend* on the process $\{F_k\}$ by which it was obtained. Nevertheless, it would be natural to use the process in the definition.

We proceed by analogy. In fact, we look at what may be described as the least pathological examples we can find, namely, parts of lines, planes, and space in ordinary Euclidean geometry. We are already familiar with the fact that a line segment has topological dimension 1, a square has topological dimension 2, and a cube has topological dimension 3. Fortunately, just as in the case of our fractals, each of these objects can be broken up into self-similar objects. For the purposes of discussion let us begin with the unit interval I, which we subdivide into r equal subintervals, whose lengths are $\frac{1}{r}$, that is, one rth of the length of the original interval. We call r the ***reduction factor*** or, according to our point of view, the ***magnification factor***. We now subdivide (for any n), the unit hypercube I^n by subdividing I as above along each axis; we will continue to call r the (linear) reduction factor for our subdivision of I^n. Now, using a particular but not special example (with $r = 4$, as shown in Figure 5), we see that the subdivided line segment consists of 4^1 smaller line segments, the subdivided square consists of 4^2 smaller squares, and the subdivided cube consists of 4^3 smaller cubes. Notice that the exponents 1, 2, and 3 are the dimensions of a line segment, a square, and a cube, respectively. This gives us

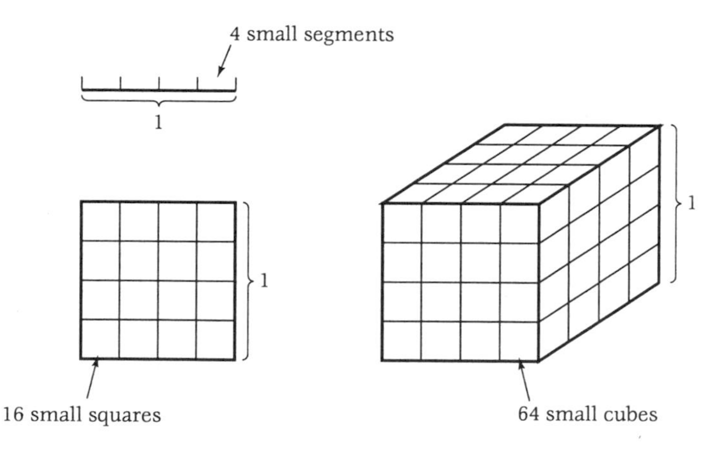

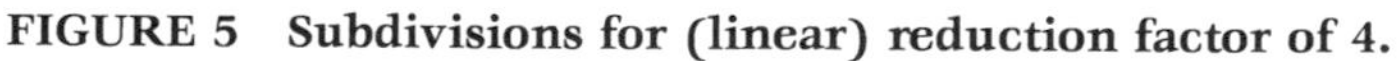
FIGURE 5 Subdivisions for (linear) reduction factor of 4.

the clue. In general, let a fractal F be constructed as described at the start of this section,[8] that is, as the limit of the sequence $\{F_k\}$. Let

$r =$ the reduction factor (or magnification factor);

$m =$ the number of cells formed at stage $(k + 1)$ from a cell at stage k;

and

$d =$ the self-similarity dimension (to be defined!).

We then observe that, for the example in Figure 5, where d must be the usual dimension,

line segment	square	cube
$4^1 = 4$	$4^2 = 16$	$4^3 = 64$

so that, for these examples,

$$r^d = m. \quad (1)$$

Solving this equation for d we see that (without needing to specify the base of the logarithm!)

$$\boxed{d = \frac{\log m}{\log r}} \quad (2)$$

This suggests (2) as an acceptable definition for the ***self-similarity dimension***. Let us see if this definition lives up to our expectations for the four examples given. Table 1 contains the pertinent information for each of the examples. The reader should check that, indeed, the value for d, calculated from (2), does seem plausible in accordance with our discussion.

• • • **BREAK**

1. Check Table 1. Make up an example of your own and calculate r, m, d for your example.

[8]The process described in this paragraph may be thought of as initiating the construction of a fractal in which each F_{k+1} is simply obtained by subdividing F_k, so that $F_1 = F_2 = F_3 = \cdots = F$. We may think of this as a *constant fractal*. Of course, it is *not* a fractal in the sense of Mandelbrot's definition—but we see here a good reason, from the mathematical point of view, for not adopting that definition.

TABLE 1

Fractal	m	r	$d = \frac{\log m}{\log r}$
Cantor middle-thirds	2	3	0.6309 . . .
Koch Snowflake	4	3	1.2619 . . .
Sierpinski Triangle	3	2	1.5849 . . .
Box-fractal	5	3	1.4649 . . .

2. Explain why r can be regarded *either* as a reduction factor *or* as a magnification factor.
3. Give a convincing argument why the box-fractal must have smaller dimension than the Sierpinski Triangle.
4. What is the relation between the self-similarity dimensions of the Cantor set and the Sierpinski Triangle? Between the self-similarity dimensions of the Cantor set and the Koch Snowflake?

8.3 THE TENT MAP AND THE LOGISTIC MAP

We begin with a description of a typical process, involving a continuous real-valued function f, which gives rise in a natural and important way to a fractal as discussed in Section 2. In fact, the function we choose to discuss gives rise to the Cantor set. It is an example of a type of function discussed in greater depth in Section 4.

We consider a particular function f, from $\mathbb{R}^1$ to $\mathbb{R}^1$, that is, the function given by[9]

$$f(x) = \begin{cases} 3x, & x \leq \frac{1}{2}, \\ 3(1-x), & x \geq \frac{1}{2}. \end{cases} \tag{3}$$

This function, called the ***tent function*** for obvious reasons, is shown in Figure 6. As the figure suggests, the most crucial part of the domain of f will be the unit interval $[0, 1]$. First we show how repeated iteration of this function leads us to the Cantor set.

[9]Notice that the rules for $x \leq \frac{1}{2}$ and $x \geq \frac{1}{2}$ agree at $x = \frac{1}{2}$.

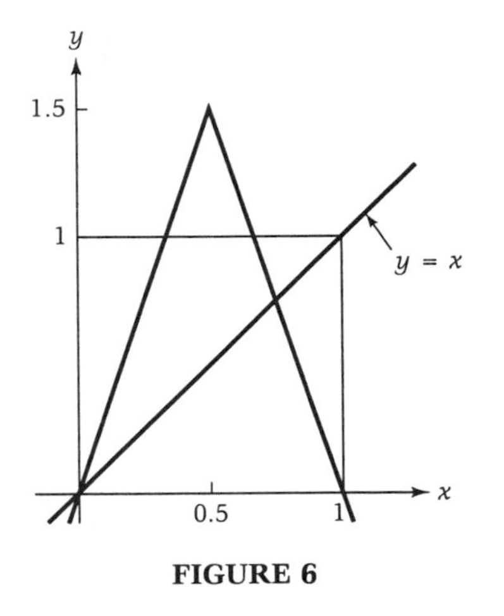

FIGURE 6

Observe (see Figure 7) that an easy geometric way to locate the values of the iterates f^k of a function f at the point x_0 is:

| 1 | move vertically from $(x_0, 0)$ to hit the curve $y = f(x)$ at $(x_0, f(x_0))$;

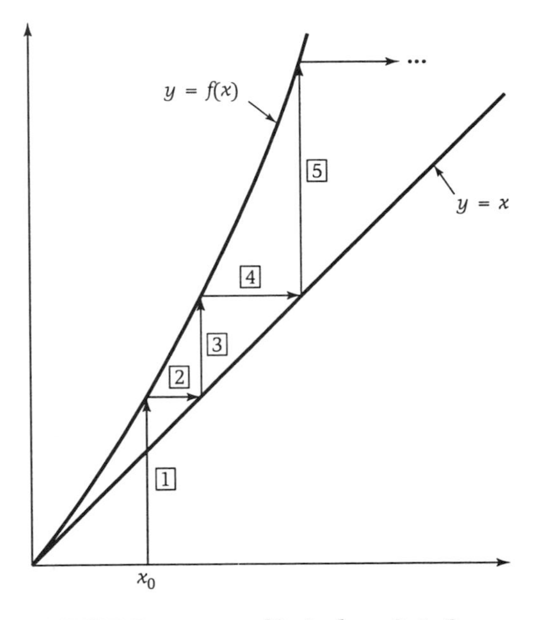

FIGURE 7 $x_{k+1} = f(x_k)$, $k = 0, 1, 2, \ldots$.

2	move horizontally from this point to hit the line $y = x$ at $(f(x_0), f(x_0))$;
3	move vertically from this point to hit the curve $y = f(x)$ at $(f(x_0), f^2(x_0))$;

where $f^2(x) = f(f(x))$; and so on. For example, if $f(x) = x^2$ and $x_0 = 3$, then $f(x_0) = 9$, $f^2(x_0) = 81$, $f^3(x_0) = 6561, \ldots$. It is natural (see Figure 7) to set $x_1 = f(x_0)$, $x_2 = f(x_1), \ldots$, so that, in this example, $x_1 = 9$, $x_2 = 81$, $x_3 = 6561, \ldots$.

Once the iteration process is understood, one may locate the values of f^k, for successive values of k, by simply proceeding geometrically along a zigzag path starting with x_0 and making the appropriate right-angle turns, left or right, on hitting the curves $y = f(x)$ and $y = x$, as illustrated in Figure 7.

• • • **BREAK**

Draw the (approximate) graph of a function $f(x)$, and choose a starting value x_0, so that the iteration process involves turns to the east and to the west, and turns to the north and to the south.

Returning to the tent function, you should now be able to provide a proof that if $x_0 < 0$, or if $x_0 > 1$, then the iterates of f are attracted to $-\infty$, in the sense that $\lim_{k\to\infty} f^k(x_0) = -\infty$. (*Hint*: Discuss $x_0 < 0$ first.) If $x_0 = 0$ or 1 the values of the iterates of f are seen, from (3), to be 0, a fixed point for the function f in the sense that $f(0) = 0$. Thus $f^k(0) = 0$, and $f^k(1) = 0$ if $k \geq 1$.

What happens to the iterates in the interval $0 < x < 1$? It is easy to see from Figure 8, using the geometric technique displayed in Figure 7, that if $\frac{1}{3} < x_0 < \frac{2}{3}$, then the first iteration produces a value > 1 and, consequently, those points in the middle-third of the interval $[0, 1]$ will all be attracted to $-\infty$. When $x_0 = \frac{1}{3}$ or $\frac{2}{3}$ one application of f gives 1, and the second produces 0, the fixed point. This leaves the interval $0 < x < \frac{1}{3}$ and $\frac{2}{3} < x < 1$ to be discussed. But after one application of f each of these intervals is enlarged to cover the interval $0 < x < 1$, and we already know (and can see clearly from Figure 8) that the middle third of that interval is at-

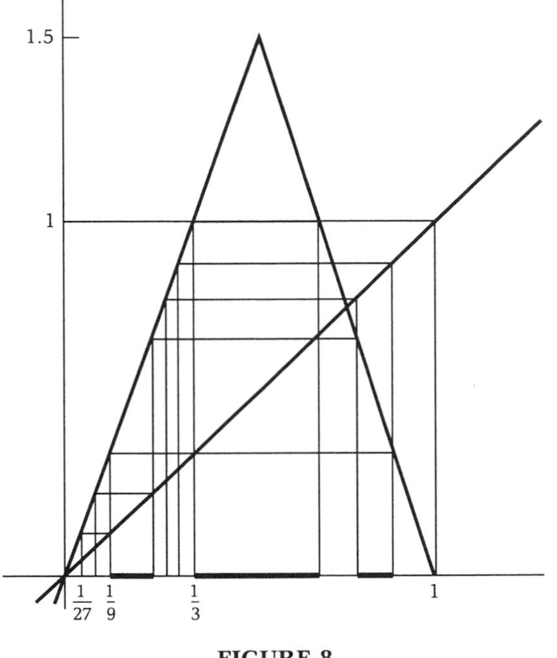

FIGURE 8

tracted to $-\infty$! This means that the middle thirds of $0 < x < \frac{1}{3}$ and $\frac{2}{3} < x < 1$ are attracted to $-\infty$, leaving 4 subintervals $0 < x < \frac{1}{9}$, $\frac{2}{9} < x < \frac{1}{3}$, $\frac{2}{3} < x < \frac{7}{9}$, $\frac{8}{9} < x < 1$, to be discussed. The Cantor set is emerging!!

We now make more precise how the Cantor set is associated with the tent map f as what we might call its ***invariant fractal set***; this will provide us with a prototype for the more general discussion in Section 4.

Consider the functions $S_1, S_2: E \to E$, where[10] E is the closed unit interval $[0, 1]$, given by

$$S_1 x = \frac{1}{3}x, \quad S_2 x = 1 - \frac{1}{3}x. \tag{4}$$

Then S_1 is a homeomorphism of E onto $\left[0, \frac{1}{3}\right]$ which reduces by a factor of 3; and S_2 is a homeomophism of E onto $\left[\frac{2}{3}, 1\right]$ which also

[10]We call the unit interval E here, because E will play a key role in the theory developed in Section 4(ii).

reduces by a factor of 3 (and flips). Their relation to the first stage of the construction of the Cantor set F is evident (see Figures 1 and 8)—what is also of great importance here is that S_1 and S_2 are the two branches of f^{-1}, the inverse of the tent function, on E; that is,

$$f(x) = y \in E \quad \text{if and only if} \quad x = S_1 y \text{ or } S_2 y.$$

Write SX for the union $S_1 X \cup S_2 X$, for any $X \subseteq E$, and write

$$S^k E = S(S^{k-1}E), \quad k \geq 1, \quad \text{with } S^0 E = E.$$

Then the Cantor set F is $\bigcap_k S^k E$, and $S^k E$ is the kth stage F_k of the self-similar process, as described earlier. Notice that

$$SF = F; \tag{5}$$

however, F has a further striking invariance property, namely,[11]

Theorem 1 $fF = F$.

Proof[12] If $x \in S^k E$, then $x = S_1 y, y \in S^{k-1}E$, or $x = S_2 z, z \in S^{k-1}E$. Thus $fx = y$ or z, so $fx \in S^{k-1}E$. Thus if $x \in S^k E$ for all k, then $fx \in S^k E$ for all k, showing that $fF \subseteq F$.

Now let $x \in F$. Then $S_1 x \in F$ and $fS_1 x = x$, so that $F \subseteq fF$, proving the theorem. □

We noted in the proof that $x \in S^k E \Rightarrow fx \in S^{k-1}E$. But, conversely, if $fx \in S^{k-1}E$ then, as already stated (since $S^{k-1}E \subseteq E$), $x = S_1 fx$ or $S_2 fx$, so $x \in S^k E$. Thus

$$x \in S^k E \quad \Leftrightarrow \quad fx \in S^{k-1}E. \tag{6}$$

But (6) immediately implies

Theorem 2 $x \in S^k E \Leftrightarrow f^k x \in E$.

From this theorem we deduce the behavior of the points of $\mathbb{R}^1$ under iteration of the tent map f.

[11] Henceforth, we often adopt the simplified notations fF, fx for $f(F)$, $f(x)$, where no ambiguity should arise. We have already adopted such notations with S, S_1, and S_2.

[12] We adopt arguments here which lend themselves to generalization, as we will see later. Simpler arguments are available if we are *only* interested in the tent map (see the forthcoming break).

Theorem 3 *All points of $\mathbb{R}^1 - F$ are attracted to $-\infty$ under iteration of the tent map.*

Proof (See Figure 8.) If $x < 0$, then $fx = 3x < x$. Thus

$$f^k x = 3^k x \to -\infty \quad \text{as } k \to \infty.$$

If $x > 1$, then $fx < 0$ so, as above, $f^k x \to -\infty$ as $k \to \infty$.

Now let $x \in E - F$, that is, $x \in E$ but $x \notin F$. Then $x \notin S^n E$ for some n. Thus, by Theorem 2, $f^n x \notin E$. Hence, by what we have proved, $f^k x \to -\infty$ as $k \to \infty$. □

We put Theorems 1 and 3 together to see the picture—the points of F are moved around[13] among themselves by f, while all other points are attracted to $-\infty$.

We remark that the set F is not only determined by the sequence $\{F_k\}$; it is also determined as the unique closed, bounded nonempty set satisfying (5). It is thus uniquely determined by the function f itself. This characterization of F will be of critical importance in Section 4.

Note, finally, that we could modify the tent map and obtain approximately the same analysis, so long as its general appearance resembled that of Figure 6; it would, for example, suffice to have two "legs" $y = \lambda x$, $x \le \frac{1}{2}$, $y = \lambda(1 - x)$, $x \ge \frac{1}{2}$, provided $\lambda > 2$. All that would change would be the explicit description of the associated fractal $F = \bigcap_k S^k E$ as a set of points—it would no longer be the Cantor set. The qualitative argument would be unaffected, and the analogues of Theorems 1 and 3 would hold. Of course, we may generalize the tent map to allow any $\lambda > 0$, but if $\lambda < 2$ the behavior is very different.

• • • **BREAK**

Use the representation of points of F in terms of decimals in base 3 to prove Theorem 1 and property (5). (This argument does *not* generalize!)

[13] Of course, $f \mid F$, the restriction of f to the subset F of its domain, is a two-one function. That is why we do not speak of a permutation.

We have already commented on the (generalized) tent map as an example of the situation described in Theorems 2 and 3. As a further example, we consider the ***logistic map*** $y = f_\lambda(x)$, where

$$f_\lambda(x) = \lambda x(1 - x), \quad \lambda > 0. \tag{7}$$

We write f for f_λ if no ambiguity would result; remember we are not now talking about the tent map! The case $\lambda = 4$ is very special—Figure 9(b) shows you why. It is the unique value of λ for which the curve $y = f_\lambda(x)$ touches the line $y = 1$. The curve rises above this line if and only if $\lambda > 4$, so there is a certain resemblance between the generalized logistic map if $\lambda > 2$ and the tent map if $\lambda > 4$ (compare Figures 8 and 9(a)).

We now suppose $\lambda > 4$. Then $f(x) = 1 \Leftrightarrow x = \frac{1}{2} \pm \sqrt{\frac{1}{4} - \frac{1}{\lambda}}$. Set $a = \frac{1}{2} - \sqrt{\frac{1}{4} - \frac{1}{\lambda}}$, so that $1 - a = \frac{1}{2} + \sqrt{\frac{1}{4} - \frac{1}{\lambda}}$. Then f is increasing on $[0, a]$ and hence bijective from $[0, a]$ to $[0, 1]$; similarly, f is decreasing on $[1 - a, 1]$ and hence bijective from $[1 - a, 1]$ to $[0, 1]$. Thus the two branches of f^{-1} on $[0, 1]$ are given by

$$S_1 x = \frac{1}{2} - \sqrt{\frac{1}{4} - \frac{x}{\lambda}}, \quad S_2 x = \frac{1}{2} + \sqrt{\frac{1}{4} - \frac{x}{\lambda}}. \tag{8}$$

An easy argument, based on the mean value theorem, shows that S_1, S_2 are contractions[14] if $\lambda > 2 + \sqrt{5}$. However, for those of you who have not yet studied differential calculus, here is an argument which you should be able to understand.

It plainly does not matter whether we consider S_1 or S_2 as

$$|S_1 x - S_1 y| = |S_2 x - S_2 y| = \left| \sqrt{\frac{1}{4} - \frac{x}{\lambda}} - \sqrt{\frac{1}{4} - \frac{y}{\lambda}} \right|.$$

Let us write S_* for either! Then

$$|S_* x - S_* y| = \frac{\left| \left(\frac{1}{4} - \frac{x}{\lambda}\right) - \left(\frac{1}{4} - \frac{y}{\lambda}\right) \right|}{\sqrt{\frac{1}{4} - \frac{x}{\lambda}} + \sqrt{\frac{1}{4} - \frac{y}{\lambda}}} = \frac{|x - y|}{\lambda\left(\sqrt{\frac{1}{4} - \frac{x}{\lambda}} + \sqrt{\frac{1}{4} - \frac{y}{\lambda}}\right)}.$$

[14] A precise definition of a contraction is given as item (14) of Section 4(ii); but just think of it as a function which shortens distances between points.

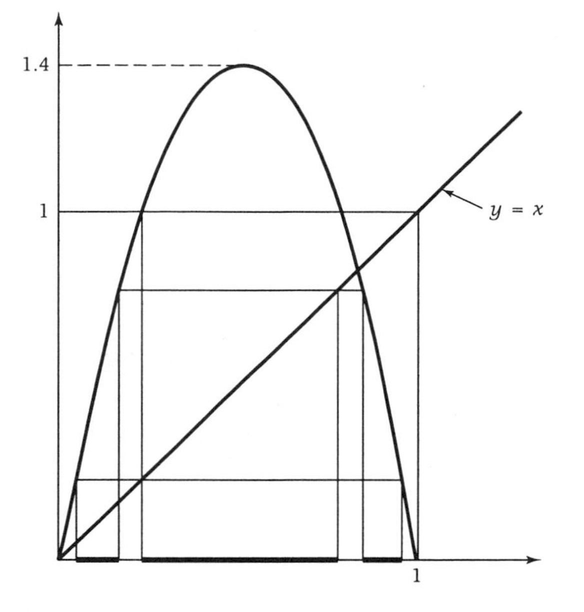

Graph showing $f_\lambda(x) = \lambda x(1 - x)$ for $\lambda = 5.6$ (compare Figure 8).

(a)

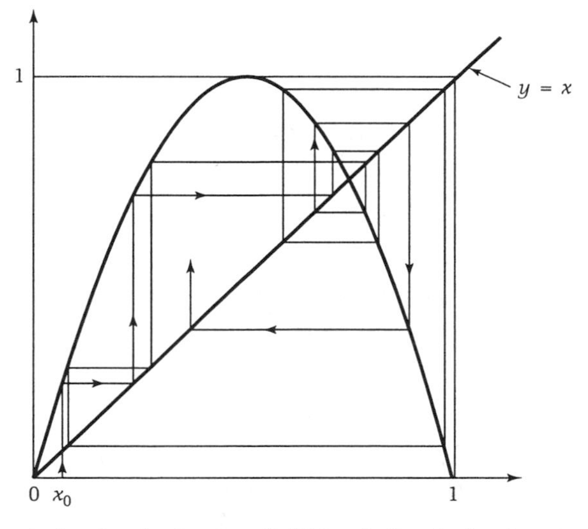

Graph showing the iterates of $f_\lambda(x) = \lambda x(1 - x)$ for $\lambda = 4$.

(b)

FIGURE 9

Now $x, y \in [0, 1]$, so $\sqrt{\frac{1}{4} - \frac{x}{\lambda}}, \sqrt{\frac{1}{4} - \frac{y}{\lambda}} \geq \sqrt{\frac{1}{4} - \frac{1}{\lambda}}$. Thus

$$\frac{1}{\lambda\left(\sqrt{\frac{1}{4} - \frac{x}{\lambda}} + \sqrt{\frac{1}{4} - \frac{y}{\lambda}}\right)} \leq \frac{1}{2\lambda\sqrt{\frac{1}{4} - \frac{1}{\lambda}}} = \frac{1}{\sqrt{\lambda^2 - 4\lambda}}.$$

Thus S_* is a contraction provided $\lambda^2 - 4\lambda > 1$, that is, provided $\lambda > 2 + \sqrt{5}$.

Thus, in the range $\lambda > 2 + \sqrt{5}$, we get an analysis of the logistic map very similar to that of the tent map, leading to a unique fractal F such that $F = S_1F \cup S_2F$ and $fF = F$. If $4 < \lambda \leq 2 + \sqrt{5}$, we no longer have a proof of uniqueness,[15] but we may still define $F = \bigcap_{k=1}^{\infty} S^k[0, 1]$ and regard this as the ***fractal associated with f***. Of course, F is not a fractal of the (relatively!) simple kind discussed in Section 2, that is, a self-similar fractal. However, its relation to the logistic curve is just like that of the Cantor middle-thirds set to the tent map.

It is when we consider values of λ in the range $0 < \lambda < 4$ that we encounter really astonishing behavior. We will discuss those values in the next section.

• • • **BREAK**

Try to imitate our arguments for the tent map, but replacing it by the logistic map $f_\lambda(x)$, with $\lambda > 2 + \sqrt{5}$ (see Figure 9(a)), to produce analogues of Theorems 1, 2, and 3.

⋆ 8.4 SOME MORE SOPHISTICATED MATERIAL

4(i) Hausdorff Dimension

In this section we will show that the self-similarity dimension as defined in (2) is really a special case of a known geometrical invariant (not, of course, a topological invariant). At this stage, we cannot even be sure that the self-similarity dimension depends only on the limiting set (fractal) F and not on the sequence of sets $\{F_k\}$ which gives rise to it.

[15] The general statement, and its proof, are given in Theorem 7 in the next section.

Think for a moment about what you know about length, area, and volume. We want to stress here certain properties of these familiar notions which may not have been in the forefront of your mind.

Property 1 You must ask the right question! By this we mean that if, for example, you ask for the area of something, that something had better be two-dimensional; and here we mean *intrinsically* two-dimensional. It need not lie in a plane, but it should look locally very much like a piece of the plane containing some interior (see Figure 10). If you ask for the area of something whose dimension is too *small* for the question to be sensible ("the area of a straight-line segment," for example) then the only reasonable answer you can expect is "zero"; and if you ask for the area of something whose dimension is too *big* for the question to be sensible ("the area of a solid ball," for example), then the only reasonable answer you can expect is "infinity." Similar remarks apply to crazy questions like the volume of a square with its interior, or the length of a square with its interior; if you ask a question appropriate to a figure of *greater* dimension, you get the answer "zero," and if you ask a question appropriate to a figure of *smaller* dimension, you get the answer "infinity."

Property 2 Suppose that f is a nice one-to-one mapping from the figure E_1 onto the figure E_2 which brings points nearer together by a factor of at least c. By this we mean that, if x, y

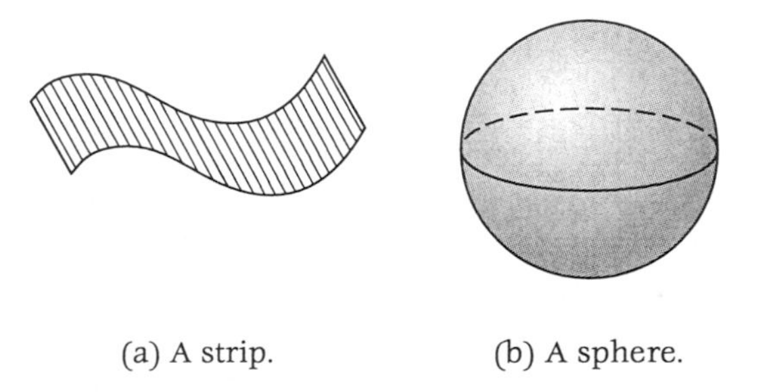

(a) A strip. (b) A sphere.

FIGURE 10 Two two-dimensional configurations.

are any two points of E_1 and if ***dist*** stands for distance, then

$$\text{dist}(fx, fy) \leq c\,\text{dist}(x, y).$$

The conclusion we may draw is that, if E_1, E_2 are s-dimensional, then

$$\mu_s(E_2) \leq c^s \mu_s(E_1). \tag{9}$$

Here μ_s is the (appropriate) s-dimensional measure, so that

$$\mu_1 = \text{length}, \quad \mu_2 = \text{area}, \quad \mu_3 = \text{volume}.$$

Notice, however, two important features of (9). First, it remains true, of course, if we take an inappropriate measure; this looks unimportant but will turn out to be very important for our purposes. Second, (9) remains true even if points aren't brought closer together by f, that is, even if $c > 1$. We will immediately exploit this to prove a consequence of (9) which may, in fact, be more familiar to you than (9) itself.

We say that the mapping f from E_1 onto E_2 is a ***similarity*** (there's that word again) with ***scaling factor c*** if we have the stronger property

$$\text{dist}(fx, fy) = c\,\text{dist}(x, y) \tag{10}$$

for all points x, y of E_1. Clearly a similarity must be one-to-one.

Theorem 4 *If f, from E_1 to E_2, is a similarity with scaling factor c, then*

$$\mu_s(E_2) = c^s \mu_s(E_1).$$

Proof It follows from (9) that $\mu_s(E_2) \leq c^s \mu_s(E_1)$. Now consider the inverse mapping f^{-1}, from E_2 to E_1. Then

$$\text{dist}(f^{-1}a, f^{-1}b) = c^{-1}\,\text{dist}(a, b), \quad a, b \in E_2.$$

Thus, again by (9), $\mu_s(E_1) \leq c^{-s} \mu_s(E_2)$ or $\mu_s(E_2) \geq c^s \mu_s(E_1)$. When this inequality is combined with the first inequality of the proof, the theorem is established. □

Notice that a *scaling* factor is the reciprocal of a *reduction* factor as defined in Section 2.

• • • **BREAK**

Show that the transformation

$$X = -3y,$$

$$Y = 3x,$$

is a similarity with scaling factor 3; and that if E_1 is the square with vertices $(\pm 1, \pm 1)$, then

$$\mu_2(E_2) = 3^2 \mu_2(E_1),$$

verifying Theorem 4.

We come now to the third property, which we are sure you have used yourselves very often.

Property 3 Suppose that F is made up of s-dimensional pieces,

$$F = F_1 \cup F_2 \cup \cdots \cup F_m,$$

such that any two pieces intersect in a piece of smaller dimension[16] than s (if they intersect at all). Then F is s-dimensional and

$$\boxed{\mu_s(F) = \sum_{i=1}^{m} \mu_s(F_i)} \tag{11}$$

Again we remark that (11) remains valid even if we take an inappropriate measure.

In all that we have said so far we have used dimensionality to determine the appropriate measure. Logically, however, we could just as well use the appropriate measure to determine dimensionality; for, as we have seen, if the volume of a figure is zero and its length is infinite, then the figure is two-dimensional. It is this rather strange point of view which we now exploit to introduce a notion of dimension which will prove to be just what we need to understand properly the self-similarity dimension of Section 2.

[16]Such an intersection is said to be *thin*. In any particular case, it is crucial to verify that the sets F_i have the *thin intersection property* in order to apply (11).

In fact, the notion of dimension we are about to describe is due to the German mathematician Felix Hausdorff (1868–1942). He introduced, for any real nonnegative number s, the notion of an ***s-measure***, which we will designate $\mathcal{H}_s$ in honor of its inventor. Then $\mathcal{H}_s$ is equivalent to the measure μ_s which we have been discussing. This measure $\mathcal{H}_s$, like its prototype μ_s, has the property that, for any figure F, if $\mathcal{H}_s(F)$ is positive and finite, then $\mathcal{H}_t(F)$ is zero for $t > s$, and infinite for $t < s$. Moreover, for reasonable figures (certainly including the fractals of Section 2) there *will* be a value of s such that $\mathcal{H}_s(F)$ is positive and finite. Thus, as suggested by our previous discussion, we may define the ***Hausdorff dimension***[17] of F, written $\dim_H(F)$, to be that unique s making $\mathcal{H}_s(F)$ positive and finite. For the figures of everyday life (most fractals don't qualify for *this* description!) the Hausdorff dimension is the ordinary dimension, as you should now see.

Moreover, the Hausdorff s-measure has Properties[18] 2 and 3—and we don't have to apply only the s-measure when we're looking at s-dimensional figures; indeed, we use Properties 2 and 3 to determine the Hausdorff dimensions of the figures we're looking at. Here is the important theorem we will now prove.

Theorem 5 *Let S_i, $i = 1,2,\ldots,m$, be a similarity from a figure E to itself with scaling factor c_i, let F be contained in E, and let*

$$F = \bigcup_{i=1}^{m} S_i F,$$

where any two of S_iF, S_jF has a thin intersection. Then the Hausdorff dimension of F is the solution s of the equation

$$\boxed{\sum_{i=1}^{m} c_i^s = 1} \tag{12}$$

[17]This is also sometimes called the Hausdorff–Besicovich dimension. A.S. Besicovich was another great mathematician—he flourished in Cambridge in the first half of the twentieth century.

[18]The definition of $\mathcal{H}_s$ leads naturally to Property 2, rather than to the statement of Theorem 4.

Remark In the light of Property 3 it will be plain what we mean by a "thin intersection." Remember, two figures of dimension d have a ***thin intersection*** if the dimension of the intersection is less than d.

Proof of Theorem 5 By Property 3, $\mathcal{H}_s(F) = \sum_{i=1}^m \mathcal{H}_s(S_i F)$. But, by Theorem 4, $\mathcal{H}_s(S_i F) = c_i^s \mathcal{H}_s(F)$. Thus

$$\mathcal{H}_s(F) = \left(\sum_{i=1}^m c_i^s\right) \mathcal{H}_s(F).$$

Now if $s = \dim_H(F)$, then $\mathcal{H}_s(F)$ is positive and finite. Thus we may divide it out, getting, as an equation to determine $\dim_H(F)$,

$$\sum_{i=1}^m c_i^s = 1. \qquad \square$$

Notice that, if each similarity S_i involves the *same* scaling factor c, then (12) becomes

$$mc^s = 1,$$

so that

$$\boxed{\dim_H(F) = s = -\frac{\log m}{\log c}} \tag{13}$$

This shows that our self-similarity dimension (see Section 2) is nothing other than the Hausdorff dimension. For our procedure in constructing fractals in Section 2 involved us in a reduction factor r, which, by its definition, was the reciprocal of the scaling factor c, and a number m which represented the numbers of cells (or copies) formed at stage $(k+1)$ from a cell at stage k. Thus

$$F_{k+1} = \bigcup_{i=1}^m S_i(F_k),$$

whence, in the limit,

$$F = \bigcup_{i=1}^m S_i(F).$$

Then (13) tells us that $\dim_H F = -\frac{\log m}{\log c} = \frac{\log m}{\log r}$, as before. Indeed, we may now considerably generalize the construction in Section 2, by allowing the reduction factor to vary with the cell at each stage, by a rule which is independent of the stage. Let us give an example of this generalization.

The Modified Koch Snowflake

Start with F_0 the unit interval, $[0, 1]$, and repeatedly replace the middle of any interval of proportion $\frac{1}{4}$ by the other two sides of an equilateral triangle. Here we take a scaling factor of $\frac{3}{8}$ twice and a scaling factor of $\frac{1}{4}$ twice. Thus, by (12), the Hausdorff dimension of the resulting fractal is the solution s of the equation $\left(\frac{3}{8}\right)^s + \left(\frac{1}{4}\right)^s = \frac{1}{2}$. (See Figure 11.) This gives a value of s around 1.2.

• • • **BREAK**

Try to get a more accurate estimate of the value of s above. As a check on your efforts $\left(\frac{3}{8}\right)^s + \left(\frac{1}{4}\right)^s = 0.500000108$ with $s = 1.195882$.

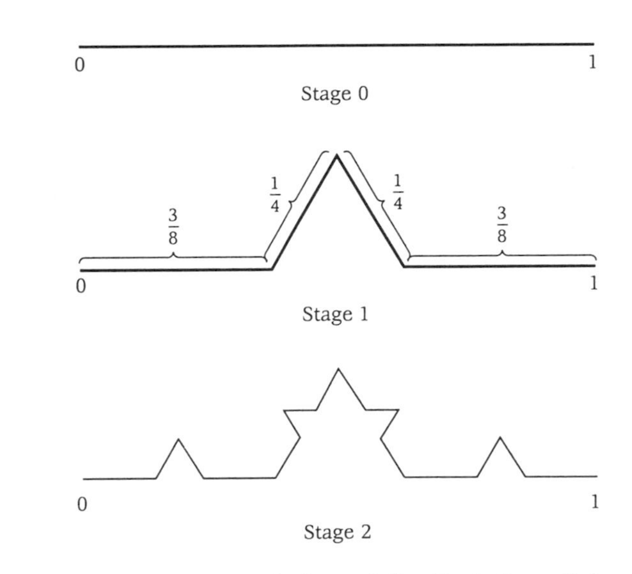

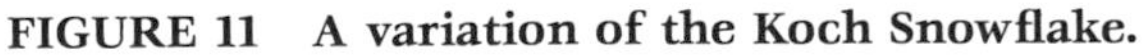
FIGURE 11 A variation of the Koch Snowflake.

Unfortunately, as you can tell from this rather simple example, all this is rather better in theory than in practice—the Hausdorff dimension is terribly difficult to calculate. Of course, a computer program could be developed (it probably has been) to solve (12), but even this is a very special case of the Hausdorff dimension.

If you want to know more about Hausdorff dimension—in particular, the actual definition of Hausdorff measure—you should consult [3] or [5].

(ii) The Fractal Associated with a (Finite) Set of Contractions

We saw in Section 3 that the tent map f had a branched inverse on $[0, 1]$, consisting of the two branches $S_1, S_2\colon [0,1] \to [0,1]$, given by

$$S_1 x = \frac{1}{3}x, \quad S_2 x = 1 - \frac{1}{3}x.$$

Then the Cantor set F was the unique compact nonempty subset of the real line $\mathbb{R}^1$ such that

$$fF = F, \quad F = S_1 F \cup S_2 F.$$

We will now show that this is a very standard situation. We first consider a nonempty closed subset D of[19] $\mathbb{R}^n$ and a finite set $S_1, S_2, \ldots, S_m$ of *contractions* of D. Here the function S_*, from D to D, is a ***contraction*** if there is some positive number $c < 1$ such that

$$\boxed{|S_* x - S_* y| \le c|x - y|, \quad x, y \in D} \tag{14}$$

Of course, the functions S_1 and S_2 discussed above are contractions of $[0, 1]$; we may take $c = \frac{1}{3}$ with either of them. Now let us write

$$SX \quad \text{for} \quad S_1 X \cup S_2 X \cup \cdots \cup S_m X$$

for any subset X of D. Our first result is

Theorem 6 *If $S_1, S_2, \ldots, S_m$ are contractions of D, then there exists a compact nonempty subset E of D such that $SE \subseteq E$.*

[19] By all means take $n = 1, 2$, or 3 if that would make you more comfortable! By the way, a set is *closed* if it contains its limit points, and *open* if its complement is closed.

Remark The force of this theorem is that E is compact (i.e., closed and bounded). D itself may not be compact—for example, D could be the whole of $\mathbb{R}^n$. In fact, in our example of the tent map, we may think of S_1 and S_2 as defined on the whole of $\mathbb{R}^1$ and we may take E to be the unit interval $[0, 1]$.

Proof of Theorem 6 Suppose that $|S_i x - S_i y| \leq c_i |x - y|$, $c_i < 1$, and let $c = \max(c_1, c_2, \ldots, c_m)$. Then $c < 1$. Choose a point $a \in D$ and let $p_i = |S_i a - a|$. Choose a number $r \geq \frac{p_i}{1-c}$ for all i, and set

$$E = D \cap B_r(a),$$

where $B_r(a)$ is the closed ball, center a, radius r, in $\mathbb{R}^n$, that is, the set of points x in $\mathbb{R}^n$ with $|x - a| \leq r$. We claim that, for each i, $S_i E \subseteq E$. For if $x \in E$, then

$$|S_i x - S_i a| \leq c_i |x - a| \leq cr,$$

so that $|S_i x - a| \leq |S_i x - S_i a| + |S_i a - a| \leq cr + p_i \leq r$. Thus $S_i x \in E$, as required. □

Having introduced the notation SX, it is natural to extend it to

$$S^2 X = S(S(X)), \quad S^2 X, \ldots .$$

Then E, SE, $S^2 E, \ldots$ is a descending sequence of nonempty, compact sets,

$$E \supseteq SE \supseteq S^2 E \supseteq \cdots .$$

It is now a standard result, and certainly highly believable, that the intersection F of such a descending sequence is itself a nonempty compact set. Of course, in our example of the tent map, F is precisely the Cantor set. Suppose now that, as in that example, each S_i is one-to-one, and the images of the S_i are disjoint. Then F has the special property

$$F = SF. \tag{15}$$

For obviously, $SF \subseteq F$, so it remains to show that $F \subseteq SF$. It is here (and only here) that we need the supplementary hypotheses on the S_i above. Let $x \in F$. Then for every $k \geq 0$, $x \in S(S^k E)$. Moreover, since the S_i have disjoint images, there exists a unique

i, independent of k, such that $x = S_i y_k$, $y_k \in S^k E$. However, since S_i is one-to-one, y_k is, in fact, independent of k. Thus we may write y for y_k, and we have $x = S_i y$, where $y \in S^k E$, for all $k \geq 0$. Thus $y \in F$, so that $x \in SF$ and hence (15) is proved. Thus we have the very important theorem:

Theorem 7 *Given contractions $S_1, S_2, \ldots, S_m$ of a closed region D in $\mathbb{R}^n$, subject to the supplementary hypotheses that: (a) each S_i is one-to-one; and (b) the images of the S_i are disjoint, then there is a unique compact nonempty subset F of D such that $F = SF$.*

Sketch of Proof of Uniqueness We don't want to burden you with details, but we think you'll find the idea of the proof intriguing.[20] It is based on the notion of the *distance* between two nonempty compact subsets X and Y of $\mathbb{R}^n$. First, we write X_δ for the *δ-shadow* of X, that is, the set of points of $\mathbb{R}^n$ within a distance δ of some point of X. Then, by definition, the distance $d(X, Y)$ between X and Y is given by

$$d(X, Y) = \min\{\delta \mid X \subseteq Y_\delta \quad \text{and} \quad Y \subseteq X_\delta\}, \tag{16}$$

that is, the smallest δ such that each of the sets X, Y is in the δ-shadow of the other (see Figure 12).

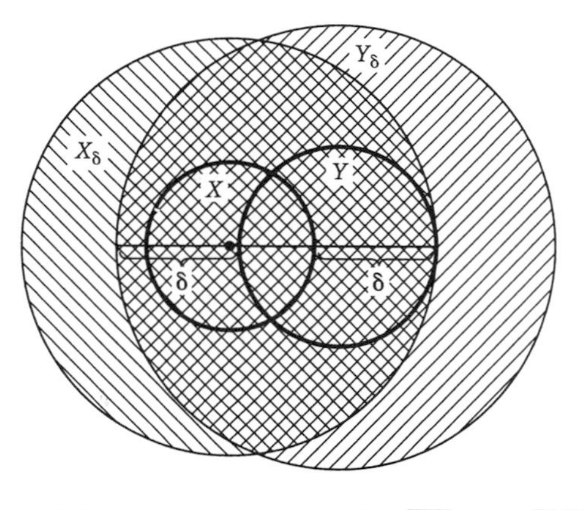

Sets X, Y, and their δ-shadows, X_δ = [diagonal hatching], Y_δ = [diagonal hatching], where $\delta = d(X, Y)$.

FIGURE 12

[20] Assuming, of course, that you've stayed with us this far!

It is important to remark that (16) defines a genuine distance function, satisfying all the usual requirements of a distance (e.g., the triangle inequality $d(X, Y) + d(Y, Z) \geq d(X, Z)$). Of special importance to us is the property

$$d(X, Y) = 0 \quad \Rightarrow \quad X = Y; \tag{17}$$

for this property it is essential that our sets be compact. Further, it is not difficult to see that if $A = \bigcup_i A_i$, $B = \bigcup_i B_i$, then

$$d(A, B) \leq \max_i d(A_i, B_i).$$

Suppose now that F, G are both compact nonempty subsets of D, such that $F = SF$, $G = SG$. Then

$$d(F, G) = d\left(\bigcup_i S_i F, \bigcup_i S_i G\right) \leq \max_i d(S_i F, S_i G).$$

However, since S_i is a contraction such that

$$|S_i x - S_i y| \leq c_i |x - y|,$$

it is easy to deduce that

$$d(S_i F, S_i G) \leq c_i d(F, G). \tag{18}$$

We again set $c = \max(c_1, c_2, \dots, c_m)$ so that $c < 1$. Then we have

$$d(F, G) \leq c d(F, G).$$

But, with $c < 1$, this is only possible if $d(F, G) = 0$. Hence, by (17), $F = G$. □

Remarks **(i)** Some of you may have heard of the Contraction Mapping Theorem (see [9]). We are here using a rather sophisticated version of the ideas behind a proof of that theorem.

(ii) Notice that the *uniqueness* of F does not require the supplementary hypotheses of Theorem 7, only the *existence* of F.

(iii) The supplementary hypotheses are satisfied whenever the contractions $S_1, S_2, \dots, S_m$ are the branches of the inverse of a function $f: E \to E$. To reinforce this connection, we give the generalization below (Theorem 8) of facts which are already clear for the tent map.

(iv) We know now how to associate, in a unique way, an invariant set F with a (finite) set of contractions $S_1, S_2, \ldots, S_m$ of the region D. Those authors who regard the notion of a fractal as describing an *intrinsic* property of the set F might now say that "F is often a fractal." We prefer[21] to take the point of view that the set F, which we have associated with the contractions $S_1, S_2, \ldots, S_m$, is the *fractal associated with this set of contractions*, so that the property of being a fractal is *not* intrinsic but refers to the role F plays in the analysis of this set of contractions. If $S_1, S_2, \ldots, S_m$ are the branches of the inverse of some map f from D to D, restricted to E, then we may say that F is the fractal associated with f. Thus, in a very precise sense, *the Cantor set is the fractal associated with the tent map.*

We noted in Section 3 that if f is the tent map, then $fF = F$; it is also, of course, true that $f^{-1}F = F$. We show now that this is a quite general phenomenon.

Theorem 8 *Let f, from D to D, be a function such that, for some nonempty compact set E in D, $E \subseteq fE$, and $f(D - E) \subseteq D - E$. Let*

$$S_1, S_2, \ldots, S_m, \quad \textit{from } E \textit{ to } E,$$

be the branches of the inverse of f on E. Then, for any $F \subseteq E$,

$$SF = F \quad \Leftrightarrow \quad fF = F \quad \textit{and} \quad f^{-1}F = F.$$

Remark You should verify that the hypotheses of the theorem are satisfied for the tent map f from $\mathbb{R}^1$ to $\mathbb{R}^1$ with $E = [0, 1]$.

Proof of Theorem 8 We know that, for $a, b \in E$,

$$fa = b \quad \Leftrightarrow \quad a = S_1 b \text{ or } S_2 b \text{ or } \cdots \text{ or } S_m b, \quad a, b \in E. \tag{19}$$

Now suppose $SF = F$ and let $x \in F$. Then $x = S_i y$ for some i and some $y \in F$. Thus, by (19), $fx = y$, so $fF \subseteq F$. Again, since

[21] We are by no means insisting on this definition of a fractal.

$E \subseteq fE$, we have $x = fy$, $y \in E$; but then $y = S_i x$ for some i, so $y \in F$ and $x \in fF$. So $F \subseteq fF$, and hence $F = fF$.

Now let $fy \in F$. Since $f(D - E) \subseteq D - E$ and $F \subseteq E$, it follows that $y \in E$. Then our argument above shows that, in fact, $y \in F$. Thus $f^{-1}F \subseteq F$. Since, certainly, $F \subseteq f^{-1}F$, it follows that $f^{-1}F = F$.

Now suppose $fF = F$, $f^{-1}F = F$; and let $a \in F$. Then $b = fa \in F$. Thus, by (19), $a = S_i b$, for some i, so $a \in SF$ and hence $SF \supseteq F$. Finally, let $a \in SF$, say, $a = S_i b$, $b \in F$. Then $a \in F$ and, by (10), $fa = b$. Hence $a \in f^{-1}F = F$, so $SF \subseteq F$, $SF = F$. □

Thus we see that many of the results on the tent map in Section 3 apply much more broadly. They apply to the logistic map $y = \lambda(1 - x)$, provided $\lambda > 2 + \sqrt{5}$. Indeed, Theorem 8 applies provided $\lambda > 4$, if we take E to be $[0, 1]$. We may then take F to be $\bigcap_{k=0}^{\infty} S^k E$, even though the mappings S_1, S_2 which are the branches of f^{-1} on E may fail to be contractions (for $4 < \lambda \leq 2 + \sqrt{5}$). (Of course, the fractal F is now no longer the Cantor set.)

(iii) The Logistic Map $y = \lambda x(1 - x)$ for $0 < \lambda < 4$

What was essential in our qualitative analysis of the tent map and of the logistic map $f(x) = \lambda x(1 - x)$, $\lambda > 4$, was that (with $E = [0, 1]$) we should have a function which rises to a single peak, of height > 1, as x goes from 0 to 1, and then descends again; we also require that $f(\mathbb{R}^1 - E) \subseteq \mathbb{R}^1 - E$. We could, as we have said, modify the tent map and still achieve this; thus we could define a generalized tent map, for any $\lambda > 2$, by the formula

$$f_\lambda(x) = \begin{cases} \lambda x, & x \leq \frac{1}{2}, \\ \lambda(1 - x), & x \geq \frac{1}{2}. \end{cases}$$

However, extraordinary things happen to the iterates of the logistic map when $\lambda < 4$, and to the modified tent map with $\lambda < 2$. Let us describe these bizarre phenomena, in fairly informal terms, in the case of the logistic map.

We plan, then, in this final part of the final section of this chapter, to study iterations of the logistic map

$$f_\lambda(x) = \lambda x(1 - x)$$

when $0 < \lambda < 4$; we will write f for f_λ if no ambiguity would result. We will again write f^k for the kth ***iterate*** of f, so that $f^1 = f$, $f^k(x) = f(f^{k-1}(x))$, $k \geq 2$. The ***orbit*** of the point x is the collection of all points $\{f^k(x)\}$, $k \geq 0$, where (as you would expect) f^0 is the identity map, so that $f^0(x) = x$. Let $m_k(x)$ be the slope of the tangent to the curve $y = f^k(x)$ at the point x.

Now we say that x is a ***period k point of f*** (with $k \geq 1$) if

$$f^k(x) = x, \quad \text{with } k \text{ minimal for this property} \tag{20}$$

(a period 1 point is a ***fixed point*** of f).

Of course, all these definitions make sense for any function f from $\mathbb{R}^1$ to $\mathbb{R}^1$; and the following theorem is a cornerstone of the theory of iterations of differentiable functions.

Theorem 9 *If x is a period k point of f, then x is stable and attracts nearby orbits if $|m_k(x)| < 1$, x is unstable and repels nearby orbits if $|m_k(x)| > 1$.*

Rather than proving this, let us be content to explain what it means in informal terms. It says that if $|m_k(x)| < 1$, then orbits which get close to the orbit of x stay close to the orbit of x—they are *captured* by the orbit of x; but if $|m_k(x)| > 1$, then orbits which get close to the orbit of x move away again—they *escape* from the orbit of x.

So let us look at the logistic map with $0 < \lambda < 1$ first. It may be seen easily that the only fixed point of f on $[0, 1]$ is 0 itself, and that $m_1(0) = \lambda$. Thus, by Theorem 9, 0 is an attractive fixed point. Indeed, it attracts the orbits of all points in $[0, 1]$ (see Figure 13).

Now assume $1 < \lambda < \lambda_1 = 3$. Now 0 is an unstable fixed point (and remains so as λ increases further). But there is now a new fixed point on $[0, 1]$, namely, $1 - \frac{1}{\lambda}$, and $m_1\left(1 - \frac{1}{\lambda}\right) = 2 - \lambda$, so this fixed point is a stable attractor in this range of values of λ. In fact, it attracts all orbits except for the orbit consisting of the pair $1, 0$ (see Figure 14).

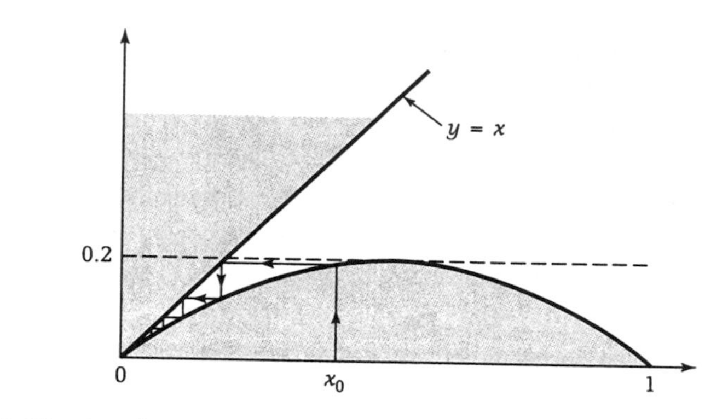

FIGURE 13 **Graph showing the iterates of $f_\lambda(x) = \lambda x(1 - x)$ for $\lambda = 0.8$.**

It turns out (using Theorem 9, in fact) that the next value of λ at which the behavior of the function f_λ undergoes a sudden change is given by $\lambda_2 = 1 + \sqrt{6}$. If $\lambda_1 < \lambda < \lambda_2 = 1 + \sqrt{6}$, then $1 - \frac{1}{\lambda}$ is unstable (and remains so as λ increases further); but a stable orbit of period 2 arises to which all but countably many points of $(0, 1)$ are attracted. We say that the fixed point $1 - \frac{1}{\lambda}$ ***splits*** into a stable orbit of period 2 (see Figure 15).

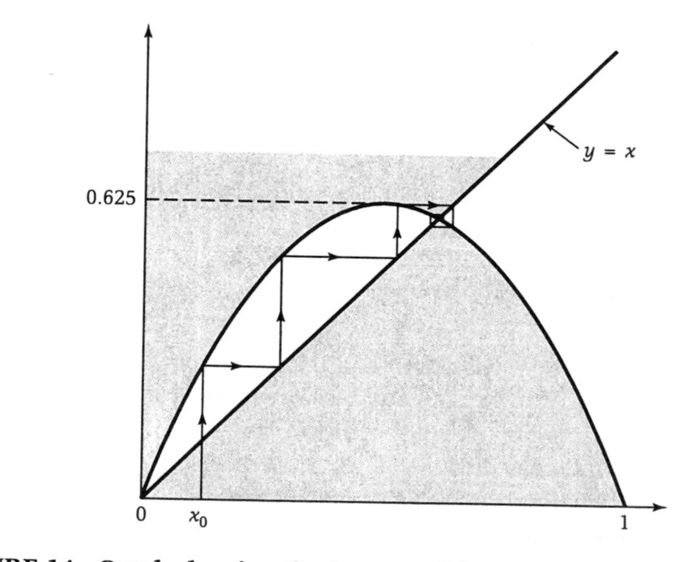

FIGURE 14 **Graph showing the iterates of $f_\lambda(x) = \lambda x(1 - x)$ for $\lambda = 2.5$.**

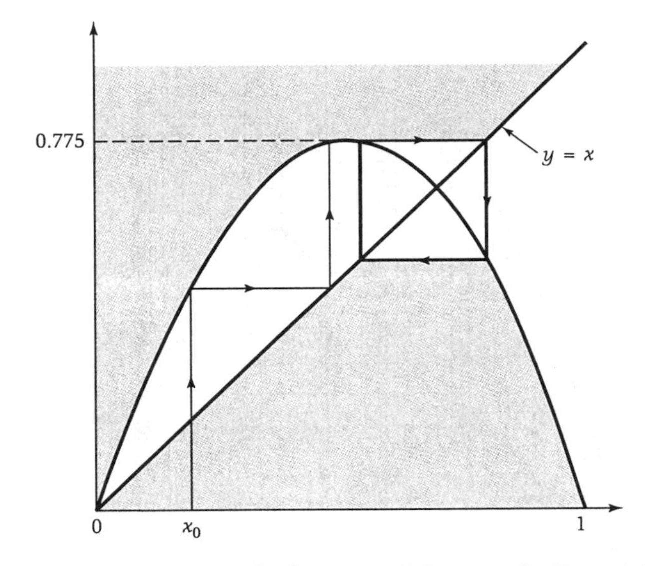

FIGURE 15 Graph showing the iterates of $f_\lambda(x) = \lambda x(1 - x)$ for $\lambda = 3.1$.

The story now continues. If $\lambda_2 < \lambda < \lambda_3$, for a certain critical value λ_3, the stable orbit of period 2 splits into a stable orbit of period 4; and the ***period-doubling*** continues, with increasing frequency, as λ increases. We obtain a sequence of critical values of λ, namely, $\lambda_1, \lambda_2, \ldots$, and $\lambda_q \to \lambda_\infty$ which is approximately 3.570. When $\lambda = \lambda_\infty$ the attractor F is a set of Cantor type, invariant under f_{λ_∞}, with all but a countable number of points of (0, 1) attracted to F. A strange attractor has arisen!

For $\lambda_\infty < \lambda < 4$, the behavior becomes even more extraordinary (see Figure 16). There is a set K of values of λ such that f_λ has a truly chaotic attractor of positive length. However, in the "windows" between the stretches of K, period-doubling again takes place, but now starting with orbits of period 3, 5, 7,. . . .

However, in a qualitative sense, the logistic map, the (generalized) tent map, and, indeed, *any* family of transformations $\lambda f(x)$, where $f(x)$ has a unique maximum, located between $x = 0$ and $x = 1$, all behave similarly. Of course, the actual critical values λ_1, $\lambda_2, \ldots$, depend on f, but the rate at which these values approach

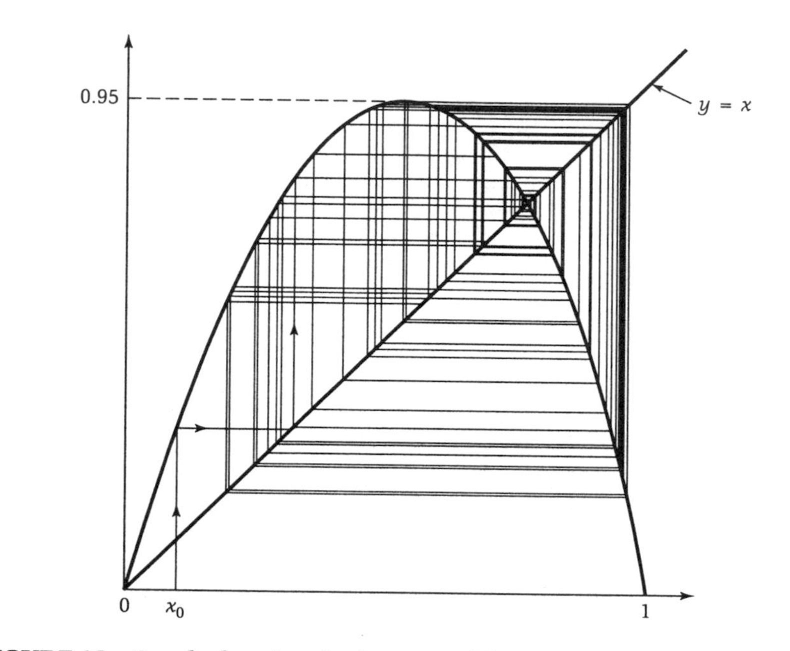

FIGURE 16 Graph showing the iterates of $f_\lambda(x) = \lambda x(1 - x)$ for $\lambda = 3.8$. Chaos has arrived!

λ_∞ does not; we have, *always*,[22]

$$\lambda_\infty - \lambda_q \sim c\delta^{-q},$$

where

$$\delta = 4.6692, \ldots,$$

approximately. We call δ the ***Feigenbaum constant***, after its discoverer. It deserves to take its place alongside π, e, and Euler's γ as a new universal constant. From a mathematician's point of view, its discovery is one of the most exciting features of fractal geometry.

• • • FINAL BREAK

1. Suppose the Koch Snowflake has area A at stage 0 (see Figure 17), show that as $k \to \infty$ the area approaches $\frac{8}{5}A$.

[22] That is, λ_q approaches λ_∞ at a rate similar to that at which δ^{-q} approaches 0.

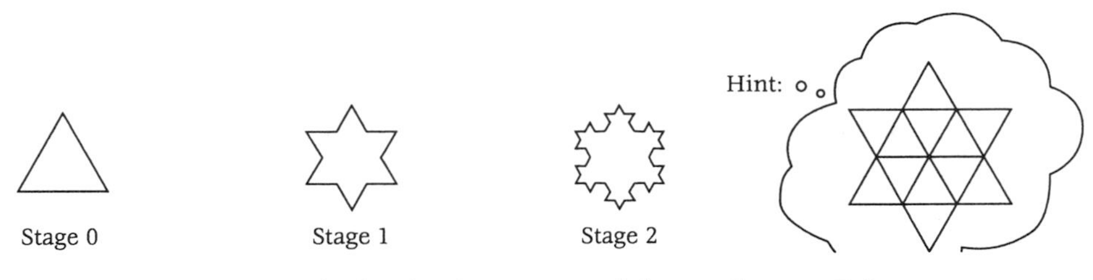

FIGURE 17 The beginning stages of the Koch Snowflake.

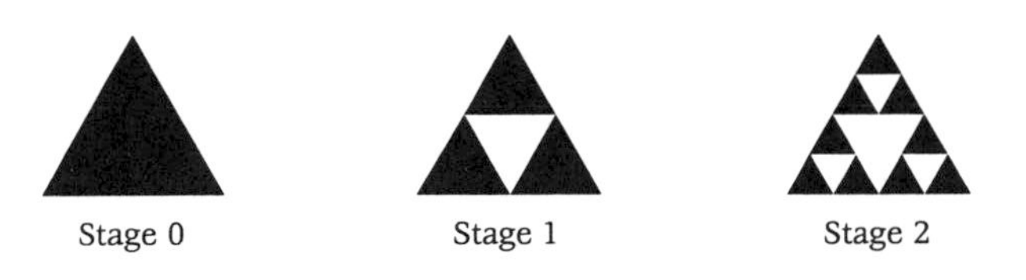

FIGURE 18 The beginning stages of the Sierpinski Triangle.

2. Suppose the Sierpinski Triangle has perimeter P at stage 0 (see Figure 18). What is the ratio of the sum of the perimeters of the black triangle at stage k to the sum at stage $k-1$?
3. Suppose the triangular fractal shown in Figure 19 has area A and side length ℓ at stage 0.[23]
   (a) What is the relationship between A and ℓ?
   (b) Let the process of going from stage k to $k+1$ involve attaching an equilateral triangle, with a side length half that of a triangle constructed at the previous stage, at

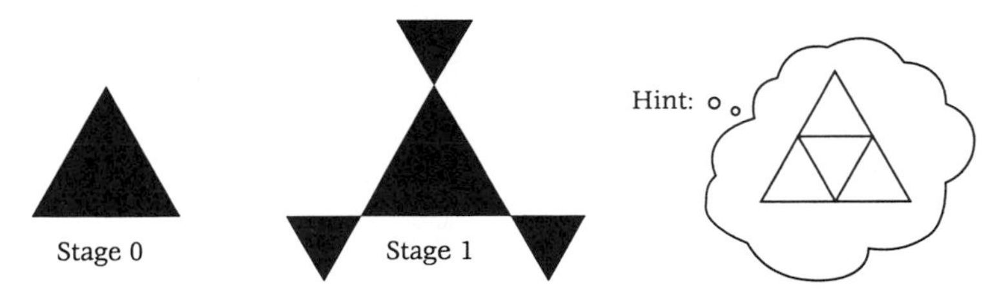

FIGURE 19 The beginning stages of the triangular fractal.

[23]This problem can easily be generalized in two ways. One way is to begin with another regular polygon (as in Problem 4)—and the other way is to vary the size of the congruent polygons which get attached to the vertices. But make sure there is no overlap.

each available vertex (see Figure 19). Calculate what the total area and the perimeter become as $k \to \infty$.

4. Suppose the square fractal of Figure 20 has area A at stage 0; and hence area $2A$ at stage 1. Calculate what the area becomes as $k \to \infty$.

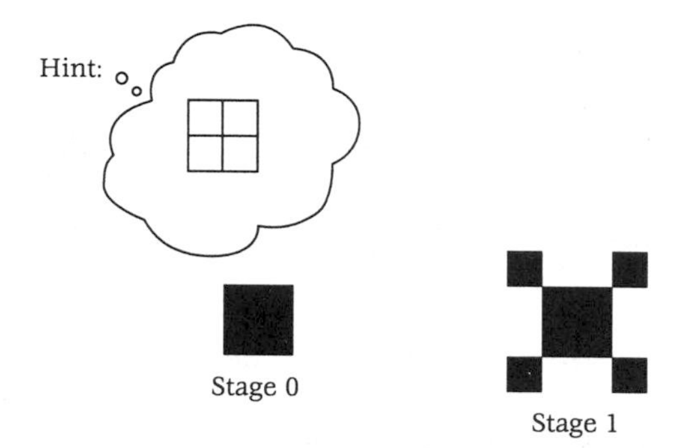

FIGURE 20 The beginning stages of the square fractal.

5. Let S_0 represent a square region with side of length 1. Let the process of going from stage k to stage $k+1$ be described thus: Subdivide each side into three equal parts and erect a square region on the center part (as shown in Figure 21).

   (a) Show that as k tends to ∞ the area approaches 2.

   (b) What shape is the limiting fractal region?

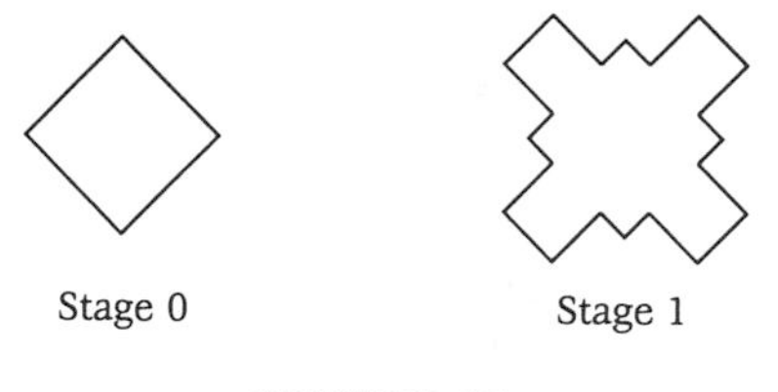

FIGURE 21

6. Let F_0 be a regular tetrahedron T. Let the process of going from stage k to stage $k+1$ be described thus: Subdivide each triangular face into four congruent equilateral triangles and erect a regular tetrahedron on the inner (shaded) triangle as shown in Figure 22.

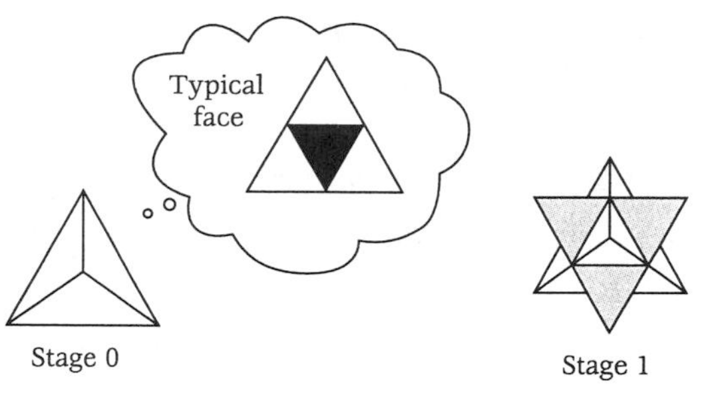

FIGURE 22

(a) Show that as k tends to ∞ the surface area tends to ∞. Now assume that the edge length of T is ℓ.

(b) As k tends to ∞, what happens to the volume? (You'll be particularly interested in the case $\ell = \sqrt{2}$.)

(c) What is the limit of the tetrahedral ***solids***?

(d) What is the dimension of the resulting fractal "surface"?

7. (a) Give your own proof, using base 3 decimals, that the tent map sends the Cantor set F onto itself in 2–1 fashion.

   (b) When we have a function f from a set E to itself, recall that x is a ***periodic point, with period k***, if k is the smallest positive integer such that

$$f^k(x) = x.$$

   Show that the tent map f, restricted to the Cantor set F, has 2 fixed points, 2 points of period 2, and 6 points of period 3. (In fact, there are 2^k points x such that $f^k(x) = x$.)

REFERENCES

1. Barnsley, Michael F., *Fractals Everywhere*, Academic Press, Boston, 1988.

2. Devaney, Robert L., *Chaos, Fractals, and Dynamics: Computer Experiments in Mathematics*, Addison-Wesley, Menlo Park, CA, 1990.
3. Falconer, Kenneth, *Fractal Geometry*, Wiley, New York, 1990.
4. Gleick, James, *Chaos: Making a New Science*, Viking, New York, 1987.
5. Hilton, Peter, and Jean Pedersen, *An Introduction to the Mathematics of Fractal Geometry*, Akademie-Vortäge **14**, Akademie für Lehrenfortbildung, Dillingen, Germany, 1993.
6. Mandelbrot, Benoit B., *Fractals, Form, Chance, and Dimension*, W.H. Freeman, San Francisco, 1977.
7. Peitgen, H.-O., H. Jurgens, and D. Saupe, *Fractals in the Classroom*, Springer-Verlag, New York, 1992.
8. Peitgen, H.-O., and P. Richter, *The Beauty of Fractals*, Springer-Verlag, New York, 1985.
9. Wilde, Carroll O., The contraction mapping principle, *UMAP* Unit 326 COMAP, Inc. (60 Lowell Street, Arlington, MA 02174), 1978.

ANSWERS FOR FINAL BREAK

Several problems in this set involve summing a geometric series. So before giving the solutions we remind you that

$$a + ar + ar^2 + \cdots + ar^{k-1} = a\left(\frac{1 - r^k}{1 - r}\right), \quad r \neq 1, \qquad (*)$$

which tells us that if $|r| < 1$, then as $k \to \infty$ the sum on the LHS gets closer and closer to $\frac{a}{1-r}$. This may be expressed more formally as

$$\lim_{k\to\infty}(a + ar + ar^2 + \cdots + ar^{k-1}) = \frac{a}{1-r}, \quad \text{when}\||r| < 1. \quad (**)$$

Now for the solutions.

1. At stage k the total area enclosed by the Koch Snowflake, denoted $A_{Koch}(k)$, may be expressed as

$$A_{Koch}(k) = A + \underbrace{\frac{1}{3}A + \frac{4}{3^3}A + \frac{4^2}{3^5}A + \frac{4^3}{3^7}A + \cdots + \frac{4^{k-1}}{3^{2k-1}}A.}$$

   Applying (**) to the terms bracketed on the RHS, with $a = \frac{1}{3}$ and $r = \frac{4}{3^2}$, we obtain

$$\lim_{k\to\infty} A_{\text{Koch}}(k) = A + \frac{\frac{1}{3}A}{1 - \frac{4}{3^2}} = \frac{8}{5}A.$$

2. The perimeter of the Sierpinski Triangle at stage k, denoted $P_{ST}(k)$ is $\left(\frac{3}{2}\right)^k P$. Thus

$$\frac{P_{ST}(k)}{P_{ST}(k-1)} = \frac{\left(\frac{3}{2}\right)^k P}{\left(\frac{3}{2}\right)^{k-1} P} = \frac{3}{2}.$$

3. (a) From trigonometry $A = \frac{1}{2}\ell^2 \sin 60° = \frac{\ell^2\sqrt{3}}{4}$. (Of course, there are elementary geometrical methods available, too.)

   (b) At stage k the perimeter, denoted $P(k)$, is seen to be $P(k) = \frac{3\ell}{2}(3k + 2)$. Thus

$$\lim_{k\to\infty} P(k) = \infty.$$

At stage k the total area, denoted $A(k)$, is given by

$$A(k) = A + 3\left(\frac{A}{4}\right) + 6\left(\frac{A}{4^2}\right) + \cdots + 3 \cdot 2^{k-1}\left(\frac{A}{4^k}\right), \qquad k \geq 1,$$

$$= A + \frac{3A}{4}\underbrace{\left(1 + \frac{1}{2} + \frac{1}{2^2} + \cdots + \frac{1}{2^{k-1}}\right)}.$$

Applying (**) to the terms bracketed on the RHS, with $a = 1$ and $r = \frac{1}{2}$, we obtain

$$\lim_{k\to\infty} A(k) = A + \frac{3A}{4}(2) = \frac{5A}{2}.$$

4. At stage k the total area, denoted $A(k)$, is given by

$$A(k) = A + A + 3\left(\frac{A}{4^1}\right) + \cdots + 3^{k-1}\left(\frac{A}{4^{k-1}}\right)$$

$$= A + A\underbrace{\left[1 + \frac{3}{4} + \left(\frac{3}{4}\right)^2 + \cdots + \left(\frac{3}{4}\right)^{k-1}\right]}.$$

Applying (**) to the terms bracketed on the RHS, with $a = 1$ and $r = \frac{3}{4}$, we obtain

$$\lim_{k\to\infty} A(k) = A + A\left[\frac{1}{1 - \frac{3}{4}}\right] = A + 4A = 5A.$$

5. (a) The enclosed area at stage k, denoted $A_S(k)$, is given by

$$A_S(k) = 1 + \underbrace{\frac{4}{9} + \frac{4 \cdot 5}{9^2} + \frac{4 \cdot 5^2}{9^3} + \cdots + \frac{4 \cdot 5^{k-1}}{9^k}}.$$

Applying (**) to the bracketed portion on the RHS, with $a = \frac{4}{9}$, $r = \frac{5}{9}$, we obtain

$$\lim_{k\to\infty} A_S(k) = 1 + \frac{\frac{4}{9}}{1 - \frac{5}{9}} = 2.$$

(b) The limiting fractal is a square with edge length $\sqrt{2}$.

6. (a) Suppose the original surface area is A, then the surface area at stage k, denoted $A(k)$, is given by

$$A(k) = A\left(\frac{3}{2}\right)^k.$$

Clearly, as k tends to ∞, so does $A(k)$.

(b) As k tends to ∞, with $\ell = \sqrt{2}$, the volume approaches 1 (as may be verified using (**)).

(c) You may *see* this by constructing such a model and observing how successive iterations fill up more and more of the cube circumscribing the original tetrahedron—without ever protruding beyond the faces of that cube. Figure 23 illustrates the placement of the regular tetrahedron inside the cube that circumscribes it.

7. (a) We adopt the notation that
$\tilde{a} = 2$ if $a = 0$,
$\tilde{a} = 0$ if $a = 2$.
If $x \in F$, $x < \frac{1}{2}$, then $x = .0a_2a_3a_4\cdots$
and $f(x) = .a_2a_3a_4\cdots$.
If $x \in F$, $x > \frac{1}{2}$, then $x = .2a_2a_3a_4\cdots$
and $f(x) = .\tilde{a}_2\tilde{a}_3\tilde{a}_4\cdots$.

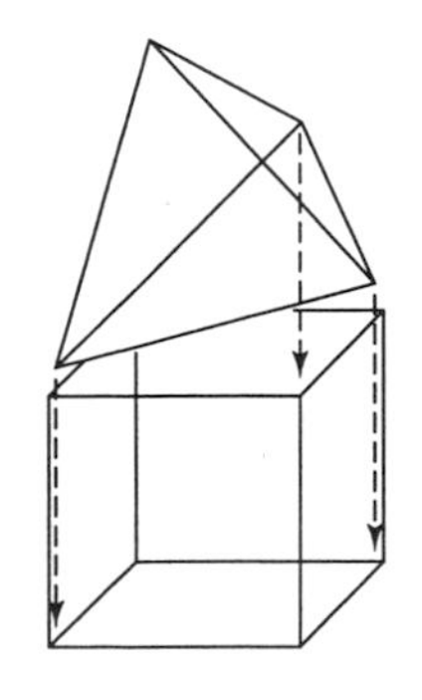

FIGURE 23

Thus f maps F into F.
If $x \in F$, then $x = .a_1a_2a_3 \cdots$
and $x = f(y_1)$ where $y_1 = .0a_1a_2a_3 \cdots$,
and $x = f(y_2)$ where $y_2 = .2\tilde{a}_1\tilde{a}_2\tilde{a}_3 \ldots$.

(b) Recall the notation that a bar over the top of a sequence of digits in a decimal indicates that the sequence is repeated indefinitely.
$f(0) = 0$, $f(.\overline{20}) = .\overline{20}$,
$f^2(.\overline{0220}) = .\overline{0220}$, $f^2(.\overline{2200}) = .\overline{2200}$,
$f^3(.\overline{00220}) = .\overline{00220}$;
similarly, $.\overline{022200}$, $.\overline{222000}$, $.\overline{200}$, $.\overline{020}$, $.\overline{220}$ are periodic points of period 3. You may obtain these periodic points in various ways; one way (which was not ours) is to obtain them by traditional algebra and then convert to base 3 decimals.

9 CHAPTER Some of Our Own Reflections

How should mathematics be done? Of course, the answer to this question cannot take the form of a set of precise prescriptions which guarantee, if they are followed, that we will be successful every time we undertake a piece of mathematics—no such guarantee is ever available, even for the most expert and talented performers on the mathematical stage. Nevertheless, we believe it may be helpful to our readers to formulate certain principles which can be discussed by them, and which may prove of practical use in improving their success rate when actually doing mathematics—and we emphasize that the readers of this book should be expecting to get a lot of satisfaction from doing mathematics (one of our principles is that mathematics is something we *do*, not just something we read and try to learn).

So we will provide a set of principles here which may be useful to you. In fact, we provide *two* sets. The first set of principles is of a general nature, and refers to the overall approach we should take to doing mathematics. The second set of principles is more specific and, we hope, will be found useful when you are actually involved in some particular piece of mathematics. For those of you who have a special interest in teaching mathematics we include a short section on principles of mathematical pedagogy.

9.1 GENERAL PRINCIPLES

1 Mathematics is only done effectively if the experience is enjoyable

Certainly, mathematics is a "serious" subject, in the sense that it is very useful and important. But that does not mean that a mathematical problem is to be tackled in a spirit of grim earnestness. Let us be frank—mathematicians are very happy that mathematics is important—but, for very few mathematicians, is this its prime attraction. We might say it is the reason why they are paid to do mathematics, but not really the reason why they do it. They do it because it brings them joy, fulfillment, and excitement; they find it fun even when it is deep, difficult, and demanding—indeed, this is a large part of its appeal. Our readers may like to read [1] on this theme.

2 Mathematics usually evolves out of communication between like-minded people

Mathematics is not, by its nature, a solitary activity; it is only our obsession with traditional tests and the determination of (some of) us to ensure the honesty of our students which have led us to insist so much on mathematics always being done by students working on their own. Like any other exciting and engrossing activity, mathematics is something to be talked about among friends, to be discussed informally, so that insights and ideas can be shared and developed, and so that many can enjoy the sense of achievement, even triumph, that the successful solution of a challenging problem brings. Much of the most important work that mathematicians do is done at conferences and in the staffroom over coffee.[1]

This crucial kind of social mathematical activity requires that we feel free to chat about mathematical ideas without the restraining necessity to be absolutely precise—we will have more to say about this later. It is one of the (many) advantages that human beings have over machines that they can communicate informally without having to be pedantic. Indeed, a stronger statement is broadly true—***human beings can only communicate interesting***

[1] A great contemporary mathematician, Paul Erdös, has defined a mathematician as a device for turning coffee into mathematics.

ideas informally,[2] while ***machines can only communicate*** (with each other or with human beings) ***pedantically***.

Finally, let us admit that there is a stage, in the solution of a mathematical problem or the successful completion of a piece of mathematical research, when solitude and silence are essential. The working out of detail can only be accomplished under really tranquil conditions. But mathematics is more than the working out of detail.

3 Never be pedantic; sometimes, but by no means always, be precise

When a new idea is introduced—and especially when a new idea is presented to students or colleagues—it is necessary to make that idea precise. If, for example, we want to discuss whether an equation has a solution, we must specify what kind of numbers we allow. If we want to know the quotient in a division problem, we need to know if we are referring to the *division algorithm* (involving both a quotient and a remainder) or *exact division* (involving only a quotient)—or even some other version of division.

However, once the precise idea has been conveyed, it is no longer necessary, or desirable, to insist on precision in subsequent discussion. We benefit, in all our conversations, from informality—without it there can be no ease of communication, no ready exchange of ideas. If you are talking to your friends about a dog called Jack, you need, at the outset, to specify to which "Jack" you are referring; once you've made that clear, subsequent references can—and should be—simply to "Jack." We refer to this refinement of the original principle as the

Principle of Licensed Sloppiness.

Our readers may find many examples of this principle in the earlier chapters of this book.

They may also be interested in the following logical consequence of the statement of the principle in its original form.

Corollary *Precision and pedantry are different things.*

[2]Unfortunately, when we communicate our results in research journals we are compelled to write very formally.

4 Elementary arithmetic goes from question to answer; but genuine mathematics also, and importantly, goes from answer to question[3]

What we mean to imply by this slogan is that a false picture of the true nature of mathematics in action is conveyed by our earliest contact with elementary arithmetic, when we are given addition, subtraction, multiplication, or division problems to do, and have to provide exactly correct answers to score maximum points. The questions are uninteresting (and completely standard); and they are not *our* questions. We provide the equally uninteresting answers by carrying out an uninteresting algorithm.

In genuine mathematical activity answers suggest new questions, so that, in an important sense, mathematical work is never complete. It is, moreover, to be thought of as investigation and inquiry rather than the mere execution of mechanical processes. Here again we see a vital difference between human beings and machines—and we see that elementary arithmetic is fit for machines and not for human beings! Our readers should find many examples of this principle in action in the earlier chapters of this book.

5 Algorithms are first resorts for machines, but last resorts for human beings

Consider the problem 31×29. A calculating machine tackles this problem by recognizing it as a multiplication problem involving the product of two positive integers and applies its programmed algorithm for solving such problems. The intelligent human being may reason as follows:

$$31 \times 29 = (30 + 1)(30 - 1) = 30^2 - 1 = 900 - 1 = 899.$$

(But he, or she, probably does these steps mentally.) For the human being this is a great gain in simplicity, and hence a significant saving of time and effort, compared with the standard hand-algorithm for multiplication; for the machine it would be absurd to look for a short cut, when it can do the routine calculation in a flash.

[3]The perceptive reader may object that there had to be a question in the first place to generate the answer. We would respond that "In the beginning there was elementary arithmetic."

Thus the natural procedures of machines and human beings when faced with an arithmetical or algebraic problem are diametrically opposed. The machine identifies the problem as belonging to a certain class, and then applies an algorithm suitable for solving any problem in that class. Intelligent human beings look for special features of the problem which make it possible to avoid the use of a universal algorithm, that is, to employ a short cut.

Students need to understand why the traditional algorithms work, but it is absurd to drill them so that they can use them ever more accurately and faster. For the machine will always be much more accurate and much faster.

6 Use particular but not special cases

This principle has many applications, both in teaching mathematics and in doing mathematics. When we want to think about a mathematical situation, it is usually a good idea to think about examples of this situation which are particular but typical. For example, if we are asked for the sum of the coefficients in the binomial expansion of $(1 + x)^n$, we might look at the cases $n = 3$, $n = 4$. Then

$$(1 + x)^3 = 1 + 3x + 3x^2 + x^3, \qquad 1 + 3 + 3 + 1 = 8,$$

$$(1 + x)^4 = 1 + 4x + 6x^2 + 4x^3 + x^4, \qquad 1 + 4 + 6 + 4 + 1 = 16.$$

We might well be led to the conjecture that, in general, the sum of the coefficients is 2^n; and our experiments with $(1 + x)^3$, $(1 + x)^4$ should suggest to us how to prove this conjecture. Notice that we do *not* experiment with $(1 + x)^0$; the case $n = 0$ is too special.

The technique is, as we have said, a very good way of detecting patterns, and hence of generating conjectures. It is also a useful tool in understanding mathematical statements and arguments, and thus in explaining them to others. Of course, in testing conjectures already formulated, special cases may be used—but particular, nonspecial cases are usually more reliable indicators.

7 Geometry plays a special role in mathematics

We really have in mind here geometry at the secondary and undergraduate levels. We claim that it is a serious mistake to regard geometry as just one more topic in mathematics, like algebra, trigonometry, differential calculus, and so on. In fact, geometry and algebra, the two most important aspects of mathematics at these levels, play essentially complementary roles. Geometry is a source of questions, algebra is a source of answers. Geometry provides ideas, inspiration, insight; algebra provides clarification and systematic solution.

Thus it is particularly absurd to teach geometry and algebra in separate watertight compartments, as is so often done in the United States. Geometry without algebra leaves the student with questions without answers, and hence creates frustration; algebra without geometry provides the student with answers to questions nobody would ask, and hence creates boredom and disillusion. Together, however, they form the basis of a very rich curriculum, involving both discrete and continuous mathematics.

Some may argue that there are methods in geometry (while not disputing that method is the characteristic of algebra). By "in geometry" they mean "in synthetic geometry" and refer to the method of proof by exploiting symmetry, similar triangles, properties of the circle, etc. Of course, it is true that one does exploit these key ideas in any form of geometrical reasoning, but, practically always, one needs a clever trick—for example, an ingenious construction—to complete the argument by "Euclidean" means. Roughly speaking, each geometrical problem, if solved by purely geometric methods, requires its own special idea. And none of us is bright enough to function, in any aspect of our lives, with such an enormous idea-to-problem ratio; we have to make a good idea go a long way. Fortunately, in mathematics, it does! So remember

All Good Ideas in Mathematics Show Up
in a
Variety of Mathematical and Real-World Contexts

8 Symmetry is a pervasive idea in mathematics

It is not only in geometry that we should look for opportunities to exploit symmetry—though the importance of the idea of symmetry in understanding geometrical situations and solving geometrical problems cannot be overemphasized. Symmetry also plays a very important role in algebra—consider, for example, the problem of determining the coefficient of a^3b^7 in the binomial expansion of $(a+b)^{10}$; whatever the answer is, considerations of symmetry show that the coefficient of a^7b^3 must be the same. You will find symmetry much exploited in Chapters 5 and 6. You will also find it playing a very significant role in Chapter 4, both in the geometrical and in the number-theoretical topics of that chapter. We doubt, indeed, if it is absent from *any* chapter. Our recommendation is—***always look for symmetry***.

There are, of course, many other characteristic properties of mathematics, but we will not go into detail about them here. We might mention two such properties, however. First, we often know something can be done without knowing how to do it. There's a wonderful example of this in Chapter 4 where we quote Gauss's discovery of which regular convex polygons can be constructed with straightedge and compass; but his argument gives no rule for carrying out the constructions. There is another example in Chapter 2, where we show that every residue modulo m, which is prime to m, has an order—but our proof provides no means of calculating the order. A second property characteristic of mathematics is that new concepts are introduced to help us to obtain results about already familiar concepts, but play no part in the statement of those results. There are lovely examples of this in Chapter 3, where we introduce certain irrational numbers α and β in order to establish identities connecting Fibonacci and Lucas numbers, which are, of course, integers; and in Chapter 6 where we use *pseudo-Eulerian coefficients*, just introduced, to prove identities relating binomial coefficients.

However, rather than listing these characteristic properties of mathematics systematically, we prefer to turn to certain more specific principles. Of course, the distinction between these and the principles above, which we have called "general", is not absolute. The reader should think of the principles above as relating more to the general *strategy* of doing mathematics, while those that follow relate more to the *tactics* to be used in trying to solve a particular problem.

9.2 SPECIFIC PRINCIPLES

1 Use appropriate notation, and make it as simple as possible

It is obviously important to use good and clear definitions in thinking about a mathematical problem. But it is remarkable the extent to which we can simplify our thinking by using the appropriate notation. Consider, for example, the summation notation where

$$1 + 2 + 3 + \cdots + n$$

is replaced by the more concise expression

$$\sum_{i=1}^{n} i,$$

which is read "the sum from $i = 1$ to n of i." This notation obviously simplifies algebraic expressions; and, where we may even regard the range of the summation as understood and omit it, we achieve further valuable simplification.

Good notation also often results in reducing the strain on our overburdened memories. For example, in considering polynomials it is usually much better to use the notation

$$a_0 + a_1x + a_2x^2 + \cdots + a_mx^m$$

rather than a notation which begins

$$a + bx + cx^2 + \cdots.$$

(What is then the coefficient of x^n?) To take this point further, in discussing the product of two polynomials (compare the discussion of the third principle, below), it is very nice to write the general rule in the form

$$\left(\sum a_nx^n\right)\left(\sum b_nx^n\right) = \sum c_nx^n,$$

where $c_n = \sum_{r+s=n} a_rb_s$.

Let us see what we have achieved. First, by adopting the summation notation $\sum$, we have been free to write only one typical term to denote a general polynomial. Second, by adopting the convention that $a_n = 0$ if the term x^n doesn't occur, we have not had to worry about the degree of the polynomial; and, since we have a coefficient for each n, we also don't have to stipulate the limits of the summation. Third, the rule for calculating c_n becomes completely general and does not have to take account of the degrees of the polynomials being multiplied. Just compare the resulting simplicity with how we would have to state the rule for multiplying a cubic polynomial by a quadratic polynomial[4] in the "bad old notation." Our polynomials would be

$$a_0 + a_1x + a_2x^2, \quad b_0 + b_1x + b_2x^2 + b_3x^3.$$

(This notation is not at all as bad as that found in many algebra texts!) and the rule would be

$$(a_0 + a_1x + a_2x^2)(b_0 + b_1x + b_2x^2 + b_3x^3)$$
$$= c_0 + c_1x + c_2x^2 + c_3x^3 + c_4x^4 + c_5x^5,$$

where

$$\begin{aligned} c_0 &= a_0b_0 \\ c_1 &= a_0b_1 + a_1b_0 \\ c_2 &= a_0b_2 + a_1b_1 + a_2b_0 \\ c_3 &= a_0b_3 + a_1b_2 + a_2b_1 \qquad \text{(remember } a_3 = 0\text{)} \\ c_4 &= a_1b_3 + a_2b_2 \\ c_5 &= a_2b_3. \end{aligned}$$

And even then we would only have a rule for multiplying a cubic by a quadratic, whereas our statement $c_n = \sum_{r+s=n} a_rb_s$ is completely general and even applies to the multiplication of power series.

As another example of good notation, consider the study of quadratic equations. Such an equation is usually presented as $ax^2 + bx + c = 0$ with solutions

$$x = \frac{-b \pm \sqrt{b^2 - 4ac}}{2a}.$$

How much better to take the general equation as

$$ax^2 + 2bx + c = 0$$

with solutions

$$x = \frac{-b \pm \sqrt{b^2 - ac}}{a}.$$

[4]You may view this argument as an application of our General Principle 6.

Compare the two formulae, for example, when you want to solve the equation

$$x^2 - 10x + 16 = 0.$$

Those of you who have already become familiar with the quadratic formula in its first form may decide the effort to change is not worthwhile. We would not argue with that—but you should look out for trouble when you come to study quadratic forms later.

Similarly, the general equation of a circle should be given as

$$x^2 + y^2 + 2gx + 2fy + c = 0$$

so that the center is $(-g, -f)$ and the radius $\sqrt{g^2 + f^2 - c}$; yet many texts in the United States give the equation of a circle as $x^2 + y^2 + ax + by + c = 0$, so that the center is $\left(-\frac{a}{2}, -\frac{b}{2}\right)$, and the radius $\frac{1}{2}\sqrt{a^2 + b^2 - 4c}$. ***UGH***!

There is one small price to pay for adopting good notation. Over the course of an entire book, we must often use the same symbol to refer to quite different mathematical objects. We cannot reserve the symbol F_k for the kth Fibonacci number if we want also to talk about the kth Fermat number and about a sequence $\{F_k\}$ of geometric figures converging to the fractal F. We depend—as we do in everyday life—on the context to make clear which meaning the symbol has. Thus, in ordinary conversation, if somebody says "Can Jack join us on our picnic?," it is presumably clear to which Jack she refers. There are not enough names for us to be able to reserve a unique name for each human being—and there are not enough letters, even if we vary the alphabet and the font, to reserve a unique symbol for each mathematical concept. Thus the fear of repetition should never deter us from adopting the best notation.

2 Be optimistic!

This may strike the reader as banal—surely one should always be optimistic in facing any of life's problems. But we have a special application of this principle to mathematics in mind. When you are trying to prove something in algebra, assume that you are making progress, and keep in mind what you are trying to prove. Suppose, for example, that we wish to prove that

$$\sum_{r=1}^{n} r^3 = \left(\frac{n(n+1)}{2}\right)^2.$$

We argue by induction (the formula is clearly correct for $n = 1$) and find ourselves needing to show that

$$\left(\frac{n(n+1)}{2}\right)^2 + (n+1)^3 = \left(\frac{(n+1)(n+2)}{2}\right)^2. \qquad (1)$$

At this stage we should not just "slog out" the LHS of (1). We should note that we are hoping it will equal the RHS, so we take out the factor $\left(\frac{n+1}{2}\right)^2$, present in all three terms of (1), getting, for the LHS,

$$\left(\frac{n+1}{2}\right)^2 (n^2 + 4n + 4); \qquad (2)$$

and this plainly equals the RHS.

You will find that when you have reached the stage of applying this principle automatically you will be much more confident about your ability to do mathematics successfully—such confidence is crucial.

There is, however, one other sense (at least) in which optimism is a valuable principle when one is actually making mathematics, that is, making conjectures and trying to prove them; and we feel we should mention it explicitly. One should be as ambitious as possible! This means that one should test the validity of strong rather than weak statements, and one should formulate conjectures in their most general (but reasonable) form. There are many examples of this aspect of optimism in Chapter 4. Thus we hope to find a means of constructing arbitrarily good approximation to *any* regular convex polygon, or even to *any* regular star polygon. Then, when we try to determine which convex polygons can be folded by the 2-period folding procedures, we quickly decide to look at rational numbers $\frac{t^{m+n}-1}{t^n-1}$, for *any* integer $t \geq 2$, instead of merely the numbers $\frac{2^{m+n}-1}{2^n-1}$ emerging from our paper-folding activities. The resulting gains, in both cases, are immense—not only do we get better results, but we find ourselves inventing new and fruitful concepts.

3 Employ reorganization as an algebraic technique

A simple example of such reorganization may be seen in the strategy of factorizing the polynomial $x^4 + 4$. Polynomials are usually organized, as in this case, by powers of the indeterminate x. However, factorization is often achieved by exhibiting the expression as a difference of two squares. We recognize $x^4 + 4$ as consisting of two "parts" of the expression for $(x^2 + 2)^2$. Thus we reorganize the expression by writing

$$x^4 + 4 = (x^4 + 4x^2 + 4) - 4x^2$$

and hence achieve the factorization

$$x^4 + 4 = (x^2 + 2x + 2)(x^2 - 2x + 2).$$

(Let us insert here a small anecdote. At the meeting of the New Zealand Association of Mathematics Teachers in Christchurch, New Zealand, in September, 1993, which all three authors attended, the question was going around—Can you show that $n^4 + 4$ (where n is a positive integer) is only prime if $n = 1$? It is clear that this fact follows from the factorization of $x^4 + 4$ above. For this factorization shows that $n^4 + 4$ is not prime unless the smaller factor of $n^4 + 4$, namely, $n^2 - 2n + 2$, is equal to 1. But the equation $n^2 - 2n + 2 = 1$, with n a positive integer, has the single solution $n = 1$.)

4 Look for conceptual proofs

Conceptual proofs are not only almost always simpler than algebraic proofs; by their very nature, they also almost always give us better insight into the reason why a statement is true. An algebraic proof may compel belief in the truth of the statement being proved; but, so often, it does not convey the genuine understanding which makes the student confident in using the result. Let us give a couple of examples.

We claim that if you multiply together r consecutive positive integers, the result is divisible by $r!$. Here's a proof. Let the integers be $n, n-1, n-2, \ldots, n-r+1$, with $n \geq r$ (this is perfectly general). Now

$$\frac{n(n-1)\cdots(n-r+1)}{r!} = \binom{n}{r},$$

the binomial coefficient. Since $\binom{n}{r}$ may be viewed as the number of ways of selecting r objects from n objects, it is, by its nature, an integer. This completes the proof. A purely arithmetical proof could have been given; but such a proof could not have been described, as this proof may fairly be described, as an ***explanation.***

As a second example, also involving binomial coefficients, consider the Pascal Identity

$$\binom{n}{r} + \binom{n}{r-1} = \binom{n+1}{r}, \quad r \geq 1. \tag{3}$$

This could be proved by using the formula in the previous paragraph for $\binom{n}{r}$ and "slogging it out"; but you would not then be any the wiser as to why (3) is true. Instead, you could consider $(n+1)$ objects of which one is marked. A selection of r of these objects might, or might not, contain the marked object. We claim that $\binom{n}{r-1}$ of the selections contain the marked object and $\binom{n}{r}$ do not. (Do you see why?) This observation achieves a conceptual proof. Certainly there are also available arithmetical and algebraic proofs; all three types of proof are given in Chapter 6. In this case, the arithmetical and algebraic proofs have the advantage that they are easily extended to the case where n is *any real number* (while r remains a positive integer). However, while the arithmetical proof continues to provide no insight, the algebraic proof shows how the Pascal Identity may be viewed conceptually in an entirely different light, namely, as a special case of the rule for multiplying power series. Thus the conceptual viewpoint once again triumphs!

You should not think that conceptual proofs are always combinatorial. There are other kinds of conceptual proof, which some might label *abstract*, but which we prefer to characterize as *noncomputational.* Chapter 2, though it is all about numbers, is full of such conceptual proofs; among the advantages we get from using them one should especially mention the aesthetic satisfaction they bring.

Conceptual proofs also serve in the solution of geometrical problems, though, frequently, such nonanalytic proofs are hard to find. But the geometrical viewpoint may often provide the crucial insight into the strategy for proving a mathematical assertion which has no obvious geometrical content.

9.3 APPENDIX: PRINCIPLES OF MATHEMATICAL PEDAGOGY

We have included this Appendix because we invisage that our text may be read by actual and prospective teachers of mathematics. However, we very much hope that all our readers will find it useful.

It is clear that any principle for doing mathematics effectively will imply a principle of sound mathematical pedagogy; and it cannot be necessary for us, writing for readers of this book, to be explicit about how each principle we have enunciated *translates* into a pedagogical principle. It is surely sufficient to enunciate a very general but rather controversial

> ***Basic Principle of Mathematical Instruction***
> *Mathematics should be taught so that students have a chance of comprehending how and why mathematics is done by those who do it successfully.*

However, we believe that there are certain pedagogical principles which do *not* follow directly from principles about doing mathematics—or, at least, from those principles which we have explicitly identified. With the understanding that we offer the following list tentatively as a basis for discussion, that we do *not* claim that it is complete, and that, emphatically, we do *not* claim any special insight as teachers, we append here a brief list of such principles.

P_1 ***Inculcate a dynamic approach.*** Mathematics is something to be *done*, not merely something to be learnt, and certainly not something simply to be committed to memory.

P_2 ***Often adopt a historical approach.*** Make it plain that, over the centuries, mathematics has been something which intelligent adults have chosen to do. Moreover, the mathematical syllabus was not engraved on the tablets Moses brought down from Mount Sinai. No piece of mathematics has always existed. Each piece has been invented in response to some stimulus, some need; and the best pieces have continued to be used.

P_3 ***Recognize the utility of mathematics, but do not underestimate the power of mathematics itself to attract students.*** Thus, applications should be used, both as justification and as inspiration for mathematical ideas; but one should not always insist on dealing extensively with applications in presenting a mathematical topic.[5] We should remember that mathematics has its own natural internal dynamic which should guide its development and sequencing, so that it is often philosophically correct, as well as pedagogically sound, to "stay with the mathematics."

P_4 ***Insist on the proper, and only the proper, use and design of tests and other evaluation instruments.*** Tests are unacceptable unless they contribute to the learning process. Students must never be asked to do mathematics under conditions which no mathematician would tolerate; it follows, of course, that their ability to do mathematics must not be assessed by subjecting them to artificial conditions and restraints. As a minimal requirement, tests must be designed so as not to endanger the crucial relationship between teacher and student.

For further thoughts on tests, the reader may like to consult [2].

P_5 ***Where there are at least two different ways of looking at a problem, discuss at least two.*** Different students look at problems and ideas in different ways. What is clear to one may be far less clear to another, without this being a reflection of their overall mathematical ability. In particular, some students (and mathematicians) visualize discretely, others continuously. By giving attention to more than one approach, the teacher gives more students the chance to benefit, and enhances the prospect of new connections being made in the students' understanding.

[5] We cannot support the concept, prevalent in the United States, of a 'problem-driven' curriculum, that is, a curriculum in which mathematical items are introduced as and when they are needed to solve problems coming from outside mathematics.

In the context of this principle it is particularly important to insist that one must never cut short an explanation or exposition in order to complete an unrealistically inflated syllabus. Alas, how often have we heard a colleague say words to the effect, "I did not really expect the students to understand, but they will need the technique in their physics course next term!" There is no merit in the teacher completing the syllabus unless the students complete it too!

REFERENCES

1. Hilton, Peter, The joy of mathematics, *Coll. Math. J.*, **23**, 4, 1992, 274–281.
2. Hilton, Peter, The tyranny of tests, *Amer. Math. Monthly*, **100**, 4, 1993, 365–369.

Index

Undergraduate Texts in Mathematics

(continued from page ii)

Lang: Undergraduate Algebra. Second edition.
Lang: Undergraduate Analysis.
Lax/Burstein/Lax: Calculus with Applications and Computing. Volume 1.
LeCuyer: College Mathematics with APL.
Lidl/Pilz: Applied Abstract Algebra.
Macki-Strauss: Introduction to Optimal Control Theory.
Malitz: Introduction to Mathematical Logic.
Marsden/Weinstein: Calculus I, II, III. Second edition.
Martin: The Foundations of Geometry and the Non-Euclidean Plane.
Martin: Transformation Geometry: An Introduction to Symmetry.
Millman/Parker: Geometry: A Metric Approach with Models. Second edition.
Moschovakis: Notes on Set Theory.
Owen: A First Course in the Mathematical Foundations of Thermodynamics.
Palka: An Introduction to Complex Function Theory.
Pedrick: A First Course in Analysis.
Peressini/Sullivan/Uhl: The Mathematics of Nonlinear Programming.
Prenowitz/Jantosciak: Join Geometries.
Priestley: Calculus: An Historical Approach.
Protter/Morrey: A First Course in Real Analysis. Second edition.
Protter/Morrey: Intermediate Calculus. Second edition.
Roman: An Introduction to Coding and Information Theory.
Ross: Elementary Analysis: The Theory of Calculus.
Samuel: Projective Geometry. *Readings in Mathematics.*
Scharlau/Opolka: From Fermat to Minkowski.
Sethuraman: Rings, Fields, and Vector Spaces: An Approach to Geometric Constructability.
Sigler: Algebra.
Silverman/Tate: Rational Points on Elliptic Curves.
Simmonds: A Brief on Tensor Analysis. Second edition.
Singer/Thorpe: Lecture Notes on Elementary Topology and Geometry.
Smith: Linear Algebra. Second edition.
Smith: Primer of Modern Analysis. Second edition.
Stanton/White: Constructive Combinatorics.
Stillwell: Elements of Algebra: Geometry, Numbers, Equations.
Stillwell: Mathematics and Its History.
Strayer: Linear Programming and Its Applications.
Thorpe: Elementary Topics in Differential Geometry.
Troutman: Variational Calculus and Optimal Control. Second edition.
Valenza: Linear Algebra: An Introduction to Abstract Mathematics.
Whyburn/Duda: Dynamic Topology.
Wilson: Much Ado About Calculus.